AF696433

EASTERN ASTROLABES

Historic Scientific Instruments of the Adler Planetarium & Astronomy Museum

VOLUME II

Eastern

Bruce Chandler
General Editor

Astrolabes

by David Pingree

Copyright © 2009 Adler Planetarium & Astonomy Museum.
No part of this book may be reproduced in any form without written permission from the publisher.

Library of Congress Cataloging-in-Publication Data
Adler Planetarium & Astronomy Museum.
Eastern astrolabes / by David Pingree ; Bruce Chandler, general editor, Elaheh Kheirandish, editor.
p. cm. — (Historic scientific instruments of the Adler Planetarium & Astronomy Museum ; v. 2)
ISBN 1-891220-02-0
1. Astrolabes. 2. Astrolabes—Catalogs. 3. Adler Planetarium & Astronomy Museum—Scientific apparatus collections—Catalogs. I. Pingree, David Edwin, 1933–2005. II. Chandler, Bruce, 1931– III. Kheirandish, Elaheh. IV. Title. V. Series: Adler Planetarium & Astronomy Museum. Historic scientific instruments of the Adler Planetarium & Astronomy Museum ; v. 2.
QB85.A347 2009
522'.4—dc22
2005028519

All photographs are by Steven Pitkin, Ron Testa, or R.S. Webster, except as noted.
Copy Editing: Andrea Murschel, Clemency Williams, and Elaheh Kheirandish
Design: Pamela Geismar
Composition: Blue Friday Typographics
Printing: Everbest Printing Company

Printed in Hong Kong

Adler Planetarium & Astronomy Museum
1300 South Lake Shore Drive
Chicago, Illinois 60605

We dedicate this volume to our dear friends and patrons,

Roderick and Marjorie Webster

whose earnest study of instruments, devotion to learning,
and support of their colleagues
continue to inspire us in so many ways.

Acknowledgments

by Marvin Bolt, Vice President for Collections

Steve Pitkin's spectacular photography ensures that this volume will receive an enthusiastic reception from specialists and amateurs alike. Bruce Chandler has seen this project evolve from its beginning, and has influenced it considerably; we thank him particularly for his editorial work and seeing it through to the end. We greatly appreciate the insights and observations of Emilie Savage-Smith on the celestial globes. Thanks to Pamela Geismar's careful eye and elegant design, we believe that this volume will grace the shelves of many libraries and private collections. We also thank Elaheh Kheirandish, Sonam Velani, and Clemency Williams for their work on the index, the bibliography, and the reading of the manuscript to ensure its accuracy on short notice after the passing of David Pingree.

To the past and present staff of the Adler History Department — recently transformed into the *Roderick and Marjorie Webster Institute for the History of Astronomy* — we owe our gratitude for bringing this volume to print: Sara Schechner, Liba Taub, Andrea Murschel, Marvin Bolt, Bruce Stephenson, Devon Pyle-Vowles, Jodi Lacy, Charron Brock, Jennifer Brand, Patti Gibbons, Mave Lawler, Misty DeMars, and many others have contributed. We especially thank Paul Knappenberger and Marge Marek for their long-time support of the Adler collections, the *Webster Institute*, and its many programs, including this catalogue series.

Most of all, we thank Madge Webster and the late Rod Webster for their inspiration and encouragement, and for sharing their passion for scientific instruments with us. This publication is a small token of our appreciation for them and for their contagious enthusiasm.

Contents

Foreword

by Paul Knappenberger, President, Adler Planetarium & Astronomy Museum

Eastern Astrolabes is the second in a series of catalogues presenting the wonderful collections at Chicago's Adler Planetarium & Astronomy Museum. It complements the initial volume, *Western Astrolabes.* Astrolabes, which long served as a symbol of astronomy, played an important role in Islamic culture, particularly as a means for determining times to call the faithful to prayer. The Adler is privileged to care for these distinguished instruments, which remind us of the common astronomical heritage of the world's great civilizations.

This volume began in the 1960s, when Adler curators Roderick and Marjorie Webster served on the visiting committee at the University of Chicago's Oriental Institute. There they met David Pingree, who was working with E. S. Kennedy to identify the geographic treatises used by astrolabe makers to produce gazetteers; this information was needed to determine the *qibla*, the sacred direction to Mecca. The Websters indicated that the Adler owned an astrolabe collection, leading to Pingree's first visit to the Adler. In exchange for access to the instruments, Pingree agreed to write this catalogue.

This agreement initiated a fruitful and intriguing routine. On Monday of each week, the Websters delivered an astrolabe from the Adler to Pingree at the Oriental Institute, where he had several days to study it. On Friday the Websters returned the instrument to the Adler. The process repeated weekly until each astrolabe had been scrutinized and described. It is unlikely that such an informal arrangement could occur today, but it worked extremely well then.

We are indebted to Roderick and Marjorie Webster for initiating this catalogue, for their dedicated work, and for their generous support of the Adler's collections.

We especially thank David Pingree for writing such a splendid catalogue. Sadly, Prof. Pingree died before seeing the printed volume, although he had completed his work on it. May it endure as one of many monuments to his scholarship.

Preface

by David Pingree

This second volume of Historic Scientific Instruments of the Adler Planetarium & Astronomy Museum describes the "Eastern" astrolabes and other instruments. The title is convenient, but not completely truthful. The first eight astrolabes and some of the other instruments were made in Spain and western North Africa (the Maghrib), though the vast majority come from West Asia, the Middle East, and South Asia (the Mashriqi). The instruments described in this volume bear inscriptions in Arabic, French, Hebrew, Latin, Persian, Sanskrit, and Turkish. There is, then, no geographic or linguistic term that fully comprehends this material.

I first began examining the Adler's astrolabes some forty years ago; my purpose was simply to obtain some familiarity with medieval astronomical instruments. Soon, however, Rod and Madge Webster persuaded me to undertake the task of preparing a catalogue. Over the years they provided expertise and technical assistance, as well as introducing me to other collections of "oriental" instruments in the Chicago area, such as that at the Time Museum in Rockford, Illinois. Without their help and encouragement this catalogue could not have been written or produced; to them I dedicate it.

I am also indebted to Andrea Murschel and Clemency Williams who miraculously transformed my manuscript into computer print-out and performed many other useful tasks. And finally I thank Bruce Chandler for overseeing the layout and production of this volume.

Providence, Rhode Island
23 June 2005

Introduction

The "Oriental" Astrolabes

by David Pingree

The first non-Greek treatise[1] on the astrolabe that we know of is that composed in Syriac in imitation of Theon's treatise by Severus Sebokht in the middle of the seventh century A.D.;[2] this was clearly based on Greek sources. While Severus was a Christian bishop of Qinnasrīn (south of Aleppo), the tenth century bibliographer of early Arabic literature, Ibn al-Nadīm (died 990), in his *Kitāb al-Fihrist* begins his history of the astrolabe in Islām with the statement:[3] "The instruments were made in the city of Ḥarrān, and from there they were brought to light and appeared, but they increased, and (their) manufacture expanded for makers in the 'Abbāsid rule, from the days of al-Ma'mūn (caliph 813-833)." Ḥarrān, in northeastern Syria, was a pagan city whose populace worshipped the stars, and whose intellectuals, strongly influenced by Greek astronomy, astrology, and philosophy, were naturally interested in the astrolabe. They shared this interest with their fellow Syrians who were Christians of various sects.

Ibn al-Nadīm appends to his brief history a list of astrolabe makers,[4] which includes, besides several who are identified as Ḥarrānians, Bīṭūlus, an apprentice of Khafīf, and six of Bīṭūlus' apprentices. One of them, 'Alī ibn Ṣurad, was a Ḥarrānian and two, Shujā' and the lady astrolabist, al-'Ijlīya, worked for Sayf al-Dawla, Ḥamdānid ruler of Syria from 945 to 967, with his capital at Aleppo. Bīṭūlus, then, lived in Syria (though not necessarily in Ḥarrān) in the early decades of the tenth century. Since Bīṭūlus is not a plausible Arabic name (or Arabic transliteration of a Greek, Coptic, Syriac, or Iranian name) Wiedemann[5] identified him with the Nasṭūlus (or Basṭūlus) mentioned by al-Bīrūnī in two places as an astrolabist. King later discovered that one of al-Bīrūnī's sources, 'Abd al-Jalīl al-Sijzī, gives further particulars about Nasṭūlus' innovations in astrolabe design, and includes the information that his full name was Muḥammad ibn 'Abdallāh known as Nasṭūlus.[6] In 1974 Maddison and Brieux drew attention to an astrolabe made by Basṭūlus (or Nasṭūlus) in A.H. 315 (A.D. 927/8),[7] which is the earliest *dated* Arabic astrolabe, though

1. For the Greek tradition see O. Neugebauer, *A History of Ancient Mathematical Astronomy*, 3 vols., New York 1975, vol. 2, pp. 868-879.

2. F. Nau, "Le traité sur l'astrolabe plan de Sévère Sabokht," Journal Asiatique, Paris, 1899.

3. Ed. R. Tajaddod, 3rd edition (1988), p. 342.

4. *Ibid.*, pp. 342-343.

5. E. Wiedemann, "Ein Instrument, das die Bewegung von Sonne und Mond darstellt, nach al Bīrūnī," *Der Islam* 4, 1913, 5-13, esp 13; reprinted in E. Wiedemann, *Gesammelte Schriften zur arabisch-islamischen Wissenschaftsgeschichte*, vol. 2, Frankfurt-am-Main 1984, pp. 718-726.

6. D. King, "Note on the Astrolabist Nasṭūlus/Basṭūlus," *Archives Internationales d'Histoire des Sciences* 28 (1978) pp. 117-120; reprinted in D. King, *Islamic Astronomical Instruments*, London 1987, as article IV.

7. F. Maddison and A. Brieux, "Basṭūlus or Nasṭūlus? A Note on the Name of an Early Islamic Astrolabist," *Archives Internationales d'Histoire des Sciences* 24 (1974) pp. 157-160.

certainly not the earliest one made. Maddison and Brieux were further able to point to several other astrolabes made by members of the tradition to which Nasṭūlus belonged: Khafīf, Nasṭūlus' master, made two instruments now at Oxford; Aḥmad ibn Khalaf, apprentice along with Khafīf of 'Alī ibn 'Īsā, made one, now at Paris, for the son of the caliph al-Muktafī (902-908); and Ḥāmid ibn 'Alī, apprentice of both Aḥmad ibn Khalaf and 'Alī ibn Aḥmad, who was an apprentice along with Nasṭūlus of Khafīf, made two: one now in Cairo and another in Palermo, of which the second was made in A.H. 343 (954/5). Brieux later identified a second astrolabe, now in Cairo, made by Nasṭūlus. Further, Sezgin found an anonymous treatise preserved in a manuscript in Bombay in which it is said that Muḥammad ibn Muḥammad (*not* 'Abdallāh) known as Nasṭūlus invented the moon-shaped astrolabe in A.H. 280 (A.D. 893/4);[8] this pushes Nasṭūlus' date of birth back to about 860/870. Finally King and Kunitzsch have conjectured that Nasṭūlus is a variant writing of Nasṭūrus, meaning a Nestorian Christian;[9] under this hypothesis the Muslim astrolabist had a Nestorian ancestor.

This establishes firmly the existence of a large group of astrolabists active in Syria in the eighth and ninth centuries whose surviving instruments are among the very earliest known to have been made under Islam. The originator of this tradition, according to the cited passage in Ibn al-Nadīm, was ...ibn Khalaf al-Marwarrūdhī (*i.e.*, from Marw al-Rudh in Khurāsān), who is said by the bibliographer to have made an armillary sphere for the caliph al-Ma'mūn (probably in 830/831)[10] and to have been the master of 'Alī ibn 'Īsā of Ḥarrān, whose *Epistle on Knowing the Science of the Astrolabe*[11] contests for the distinction of being the oldest surviving treatise on the subject in Arabic[12] with al-Khwārizmī's books on the construction and use of the astrolabe.[13] Given that in Ibn al-Nadīm's account al-Marwarrūdhī's first name is missing and that the patronymic ibn Khalaf is assigned to two later astrolabists of the group, Muḥammad and Aḥmad, as well as to al-Marwarrūdhī, it may be suggested that the maker of al-Ma'mūn's armillary sphere and the master of 'Alī ibn 'Īsā was Khālid ibn 'Abdalmalik al-Marwarrūdhī whom (see n. 10) al-Ma'mūn had instructed to make instruments to observe the Sun and the Moon better than those used at Shammāsīya; the instruments were used by Khālid and others in observations made at Mount Qāsīyūn — a hill to the Northwest of Damascus — in 831/832. 'Alī ibn 'Īsā worked under Khālid al-Marwarrūdhī in the observations made at Mt. Qāsīyūn and joined him in observations made at Baghdād in 843/4 to determine the length of spring.[14] Since 'Alī was active at this date if not later, his apprentice, Khafīf, may well have taught Nasṭūlus in *ca.* 875-880.

The foregoing data provides us with three possibilities: the astrolabe was introduced into Arabic by al-Fazārī; by the Central Asian scientists such as al-Khwārizmī, al-Marwarrūdhī, and al-Farghānī; or by Syrian scholars influencing the former two (or just those from Central Asia if al-Fazārī is not really involved) and continuing their older traditions at Ḥarrān and elsewhere. I would choose the last. The claim of al-Fazārī is very weak, and there is no trace of the astrolabe in Sasanian Iran; but

8. F. Sezgin, *Geschichte des arabischen Schrifttums*, vol. 6, Leiden 1978, pp. 178-179 and 288.

9. D. King and P. Kunitzsch, "Nasṭūlus the Astrolabist Once Again," *Archives Internationales d'Histoire des Sciences* 33 (1983) pp. 342-343.

10. This is said to have been the instrument used at al-Ma'mūn's Shammāsīya Observatory in A.H. 213-4 (A.D. 828-30) by A. Sayili, *The Observatory in Islam*, Ankara 1960, p. 72, but it may be that used at Qāsīyūn in 831/32 that al-Ma'mūn instructed Khālid al-Marwarrūdhī to make (*ibid.*, p. 57).

11. C. Schoy, "'Alī ibn 'Īsā, das Astrolab und sein Gebrauch," *Isis* 9 (1927) pp. 239-254.

12. The work attributed to al-Fazārī is dubious (D. Pingree "Fragments of the Works of al-Fazārī," *Journal of Near Eastern Studies* 29, 1970, 103-123), and that ascribed to Māshā'allāh has been shown by Kunitzsch, "On the Authenticity of the Treatise on the Composition and Use of the Astrolabe Ascribed to Messahalla," *Archives Internationales d'Histoire des Sciences* 31 (1981) pp. 42-62, not to be his, but that the Latin form of his name was confused with that of Maslama al-Majrīṭī.

13. J. Frank, "Die Verwendung des Astrolabs nach al-Chwārizmī," *Abhandlungen zur Geschichte der Naturwissenschaften und Medizin* 3, 1922, 1-32; see also D. King, *Al-Khwārizmī and New Trends in Mathematical Astronomy in the Ninth Century*, New York [1983?], pp. 23-25.

14. Sayili, *op. cit.* (n.10), pp. 67-70.

Severus Sebokht did describe the astrolabe in Syriac before 661. Another, apparently Syriac treatise on the astrolabe was written by Abyūn ('Απίων) al-Baṭrīq "a little before or after Islām."[15] Because Apiōn is a name common in Coptic, and because baṭrīq can mean Patriarch as well as Patrician, King suggests that Abyūn was a Coptic Patriarch.[16] However, in the same place King adduces the evidence that al-Bīrūnī states that he had seen the Arabic translation of Abyūn's treatise made by Thābit ibn Qurra. There was no Coptic Patriarch named Apiōn, but there was a man with that name in Syria in the early sixth century: Apion from Amida was the father of Ephraem.[17] Ephraem became *comes Orientis* at Antioch under the Emperor Justin in 522/3[18] and was named Patriarch of Antioch in 527;[19] he served in that office until 545. Whether Apiōn was himself a "patricius" is not known; presumably he had some considerable status since his son attained such a high rank. But his example only serves to show that a Syrian might have been named Apiōn and have borne the title *patricius* in the sixth century, "shortly before Islām" (perhaps even a son of Ephraem), and that he might have written on the astrolabe, in either Greek of Syriac, both of which languages Thābit knew well.

Regardless of the solution of the question of the ancestry of the astrolabe in eastern Islam in the ninth century, it had apparently already traveled to al-Andalus during the reign of 'Abd al-Raḥmān II (821-852).[20] Books on the construction and use of the astrolabe, however, only began to be written there in the second half of the tenth century with the work of Maslama al-Majrīṭī (died *ca.* 1007) on Ptolemy's *Planisphaeirum*; the Arabic original of the earliest text in Latin on the astrolabe, the *Sententie astrolabii*, part of which is borrowed from al-Khwārizmī's treatise;[21] and the works of Maslama's pupils, Aḥmad ibn al-Ṣaffār and Ibn al-Samḥ.[22] The brother of Aḥmad, Muḥammad ibn al-Ṣaffār, made the oldest surviving dated astrolabe in al-Andalus in A.H. 417 (A.D. 1026), now at Edinburgh.[23] An inspection of the plates and descriptions of the astrolabes catalogued in this volume will show the significant difference between astrolabes from the Maghrib and the Mashriq; the former rarely have geographical tables in the front while the latter frequently do, and the backs of the one group differ greatly from those of the other.

The earliest Maghribī astrolabes in the Adler Planetarium are of the thirteenth century and were made in al-Andalus or North Africa. Number 1 is by al-Khamā'irī (Ḥamā'irī) at Seville in 1236/7, number 2 is an apparently unique creation of Muḥammad ibn Yūsuf ibn Ḥātim in 1240/1, and number 3 is a universal astrolabe of the type invented by al-Zarqāllu of Toledo (*ca.* 1029-1087) and named by him al-Shakāziyya.[24] This is the only universal astrolabe among the Oriental instruments at the Adler. Number 5 was made, apparently in 1651, in a French territory in North Africa; the inscriptions are in Arabic, French, and Latin.

The next two examples were made by the prolific Moroccan astrolabist, al-Baṭṭūṭī, in 1720/1 and 1723/4 respectively; the first has a geographical list in the recess of the front derived from the work of Ibn al-Bannā' of Marrākush (died 1321).

15. *op.cit.* (n.3), p. 330.

16. D. King, "The Origin of the Astrolabe According to the Medieval Islamic Sources," *Journal for the History of Arabic Science* 5 (1981) pp. 43-83, esp. 49-50; reprinted in his *Islamic Astronomical Instruments*, London 1987, as article III.

17. A. Harrak, *The Chronicle of Zuqnīn. Parts III and IV*, Toronto 1999, pp. 64 and 113.

18. G. Downey, *A History of Antioch in Syria*, Princeton 1961, p. 519.

19. *Ibid.*, pp. 526-527.

20. J. Samsó, *Las ciencias de los antiquos en al-Andalus*, Madrid 1992, p. 51.

21. P. Kunitzsch, "Al-Khwārizmī as a Source for the *Sententie astrolabii*", in *From Deferent to Equant*, New York 1987, pp. 227-236.

22. *op. cit.* (n.20), pp. 93-98.

23. F. Maddison, "Observatoires portatifs: les instruments arabes à usage pratique," in *Histoire des sciences arabes*, ed. R. Rashed and R. Morelon, vol. 1, Paris 1997, pp. 139-172, esp. 154-155.

24. His treatise on this form of astrolabe is presented by R. Puig, *Al-Sakkāziyya. Ibn al-Naqqāš al-Zarqāllub*, Barcelona 1986.

25. S.R. Sarma, "The Lahore Family of Astrolabists and their Ouvrage," *Studies in History of Medicine and Science*, 13 (1994) pp. 203-224.

26. D. Pingree, "Al-Bīrūnī's Knowledge of Sanskrit Astronomical Texts," *The Scholar and the Saint*, New York 1975, pp. 67-81, and "Brahmagupta, Balabhadra, Pṛthūdaka and al-Bīrūnī," *Journal of the American Oriental Society* 103 (1983) pp. 353-360.

27. S.R. Sarma, "Yantrarāja: the Astrolabe in Sanskrit," *Indian Journal of History of Science* 34 (1999) pp. 145-158.

28. Edited by K.K. Raikva, Mumbayī 1936.

29. In his *Siddhāntadīpikā* in T.S. Kuppana Sastri, *Mahābhāskarīya of Bhāskarācārya*, Madras 1957, pp. 329-331; cf. D. Pingree in *Journal of the American Oriental Society* 87 (1967) pp. 337-339.

The latest instrument in this section is number 9, made by Muḥammad ibn al-Mufaḍḍil ibn Kīrān, the muwaqqit of Wazzān in Morocco, in 1847/8.

The sequence of Mashriqī astrolabes begins with one by Badr ibn ʿAbdallāh, an associate of the well-known Hibatallāh, in 1130/1 for the Saljuq Sulṭān, Mughīth al-Dīn Maḥmūd II, who ruled in Iraq and Western Iran from 1118 till 1131 from his capital in Baghdād. This instrument does not have a geographical list.

After instruments made by Jaʿfar ibn ʿUmar ibn Dawlatshāh of Kirmān in Iran in 1388 (number 11) and by his grandson, Maḥmūd ibn Jalāl ibn Jaʿfar, in 1433/4 (number 12), there is a group of seven Persian astrolabes (numbers 13-20) made in India, the first six by members of the famous family of Lahore astrolabists descended from Allāhdād al-Humāyūnī.[25] The first two were made by his son ʿĪsā in 1601 and 1604/5, and the last four by ʿĪsā's grandson, Ḍiyā' al-Dīn Muḥammad, in 1637/8, 1647/8, 1660/1, and 1662/3. From the early eighteenth century is a group of five instruments (numbers 23, 24, 26, 27, and 28) made, decorated, or both by the prolific Iranian astrolabist, ʿAbd al-A'immah; he made number 26 in 1715. Associated with ʿAbd al-A'immah was ʿAbd al-ʿAlī, the maker of number 29, Khalīl Muḥammad (or Muḥammad Khalīl), the maker of numbers 22 (in 1697/8) and 23, and Muḥammad Amīn ibn Muḥammad Ṭāhir, who may have inspired the forger of number 43. Astrolabe 38 was forged in the name of ʿAbd al-A'immah the Elder.

We have already referred to the Persian astrolabes made by the members of Allāhdād's family at Lahore in the sixteenth and seventeenth centuries. Undoubtedly travelers from the Islamic world brought astrolabes to India long before the sixteenth century. One such would have been al-Bīrūnī, who was in India in the 1020s and 1030s. He claims in his *India* (chapter 13) to be exercising himself in composing a translation (into Sanskrit) of the books of Euclid and of Ptolemy's *Almagest*, and in dictating a treatise on the construction of the astrolabe. It is certain from his own discussions of the contents of Sanskrit mathematical and astronomical works that he would not have been able to produce a work on the astrolabe in Sanskrit by himself,[26] though he may have been able to manage with the assistance of the paṇḍita to whom he was dictating his text (in Persian?). In any case, al-Bīrūnī's treatise, if it was ever completed, had no discernible influence on Indian authors.

The first person whom we know to have promoted the manufacture of astrolabes in India was the Tughluk Emperor, Fīrūz Shāh (1351-1388), who both had astrolabes constructed for his own use and sponsored the writing of the first Sanskrit treatise on the astrolabe,[27] the *Yantrarāja* completed by the Jaina, Mahendra Sūri, in about 1370.[28] The first use of an astrolabe by an Indian astronomer belonging to the Sanskrit tradition was by Parameśvara in a series of observations of solar and lunar eclipses that he made in South India between 1393 and 1432, in which the altitudes of the eclipsed bodies were measured by their co-tangents from the table of the shadows of a seven-foot gnomon on the back of the instrument.[29] Sanskrit astrolabes continued to be made into the first decade of the twentieth century; two of them are included in this catalogue, numbers 48 and 49.

I have made no attempt to deal with the history of Hebrew instruments, which have been dealt with by Goldstein,[30] or of celestial globes, for which see Savage-Smith[31] and Sarma.[32] Savage-Smith's descriptions of the Adler's celestial globes are summarized as Other Instruments Q–W (pp.228–238). Nor has it been possible to write convincingly on the history of Oriental quadrants, sundials, qibla-indicators, etc., in the absence of authoritative investigations of the extant material, though much has been accomplished by King and his colleagues at Frankfurt,[33] where also *A Catalogue of Medieval Astronomical Instruments* is being prepared. Similarly, S.R. Sarma is preparing *A Descriptive Catalogue of Indian Astronomical and Time Measuring Instruments*. These works in progress will clarify many obscurities that remain in the history of the design, manufacture, and use of scientific instruments in the medieval period. It is hoped that this catalogue will have made a small contribution to that history.

30. B.R.Goldstein, "Descriptions of Astronomical Instruments in Hebrew," in *From Deferent to Equant*, New York 1987, pp. 105-141. For the Hebrew astrolabe in the Adler see *idem*, "The Hebrew Astrolabe in the Adler Planetarium," *Journal of Near Eastern Studies* 35 (1976) pp. 251-260.

31. E. Savage-Smith, *Islamicate Celestial Globes*, Washington, D.C., 1985.

32. S.R. Sarma, "From al-Kura to Bhagola. On the Dissemination of the Celestial Globe in India," *Studies in History of Medicine and Science* 13 (1994) pp. 69-85.

33. For the qibla or direction to Mecca see D. King, *World-Maps for Finding the Direction and Distance to Mecca*, London-Leiden 1999, and for time-keeping instruments *idem.*, *In Synchrony with the Heavens*, 2 vols., Leiden 2004–05, and F. Charette, *Mathematical Instrumentation in 14-th Century Egypt and Syria*, Leiden, 2004. King has published lists of early eastern and western Islamic astrolabes in his "Bringing Astronomical Instruments back to Earth-the Geographical Data on Medieval Astrolabes (to *ca.* 1100)," in *Between Demonstration and Imagination*, Leiden 1999, pp. 3-53 (esp. pp. 35-36; cf. also the items listed on pp. 37-40).

The Instrument Makers

1. *'Abd al-'Alī* was the maker of No. 29 and also of another instrument belonging to the Oriental Institute (A4091) that was decorated by 'Abd al-A'immah, a fact which effectively dates 'Abd al-'Alī in the early eighteenth century. It is likely, then, that he is identical with the well-known maker 'Abd al-'Alī ibn Muḥammad Rafī al-Juz'ī, of whose astrolabes at least nine are known [Mayer (1956) pp. 27-28] dated between 1707/8 and 1714/15. His most famous astrolabe is that made for Shāh Sulṭān Ḥusayn al-Ṣafawī (reigned 1694-1722) in September 1712, which was decorated by his brother, Muḥammad Bāqir [Morley (1856) pp. 22-31 and Gunther (1932) 33, pp. 147-148]. Our No. 43 seems to have been in part inspired by Morley's description of that astrolabe. Another fake in the Adler, No. 40, is attributed to 'Abd al-'Alī and is said to have been made for one Shaykh Luṭfallāh in 1638/9. This same instrument also names Mullā Ḥasanī without specifying his relationship to it; the name reminds one of the family of astrolabists which includes Muḥammad Ibrāhīm al -Ḥasanī, the maker of No. 25.

2. *'Abd al-A'immah ibn 'Abd al-Ḥusayn* is one of the most prolific makers and decorators of astrolabes known [Mayer (1956) pp. 23-26 and 86, Suppl. 293 and 295; Gingerich *et al.* (1972)]. It appears that he worked at Iṣfahān for the Ṣafavids from about 1680 till 1722 as part of a group of instrument-makers that included 'Abd al- 'Alī, Khalīl Muḥammad, and Muḥammad Amīn.

The only dated instrument in the Adler is No. 26, which he made in 1715 for the Prince Mahdī Qulī Khān Zanganah [*cf.* Gunther (1932) 40, p. 154 and Gibbs and Saliba (1972) CCA No. 40, pp. 74-77]. He also made Nos. 27 and 28, and he decorated No. 23 for Khalīl Muḥammad. Finally, No. 24 is strikingly similar to No. 23 in decoration, arrangement, and script, so that one must conclude that it was also decorated (if not made as well) by 'Abd al-A'immah.

3. *'Abdallāh* (?) is given as the name of the engraver of the fake astrolabe, No. 41, which is claimed to have been made by Muḥammad Mahdī in 1659/60 but is clearly inspired by Morley's descriptions of a genuine Muḥammad Mahdī instrument and the astrolabe of Shāh Sulṭān Ḥusayn Ṣafawī [Morley (1856)] .

4. *Al-'Abd Amīn of Mashhad,* who made No. 21 in 1669/70 or 1688/9, cannot be the 'Abdallāh ibn Muḥammad Amīn who made an astrolabe in 1715 [Mayer (1956) p. 28], but he may be an older relative. However, other astrolabists bearing the name Amīn are connected with Iranian cities other than Mashhad; thus, Muḥammad Mahdī ibn Muḥammad Amīn Yazdī decorated an instrument [Gunther (1932) 25, pp. 139-140 and Gibbs and Saliba (1984) CCA No 25, pp. 65-68] in 1667/8, and an astrolabe made by Muḥammad Amīn ibn Muḥammad Ṭāhir [Gunther (1932) 31, pp. 145-146 and Gibbs and Saliba CCA No. 31, pp. 68-69] was decorated by 'Abd al-A'immah, presumably at Iṣfahān.

5. *Abū Isḥāq's* name appears on the fake astrolabe No. 44 without any indication of what role he is supposed to have played in its production.

6. *'Alā al-dīn ibn Ḥasan Sanandaj* made a disk, Other Instruments J, for a king named Amān Allāh Khān in 1864/5. A similar instrument made by 'Alā al-dīn in the same year is in the Billmeir Collection in the Museum of the History of Science at Oxford.

7. *'Alī ibn Ṣādiq* was the maker of No. 30, most likely in the eighteenth century. Six of his astrolabes were known to Mayer [Mayer (1956) pp. 43-44].

8. *Badr,* the mawlā (associate? freedman?) of *"the wonder of the age,"* Hibatallāh ibn al-Ḥusayn, a famous astronomer and astrolabist who died in Iṣfahān in 1139/40 [Suter (1900) 278, p. 117], made No. 10 in the winter of 1130/1. A Persian globe signed by Badr ibn 'Abdallāh, the mawlā of *"the wonder of age,"* in 1140/1 and now in Teheran provides his patronymic; see Savage-Smith (1985) No. 59, p. 247. Another astrolabe, formerly in the Paul Klostermann Collection in Munich, is signed: *"Hibatallāh ibn al-Ḥusayn of Baghdād made it in the year 513 [AD. 1119/20] of the Hijra"*; the script on this last instrument is virtually identical to that on No. 10. No. 10 also bears an inscription stating that it belonged to the treasury of the Saljuq Sulṭān Maḥmūd, who ruled from 1118 till 1131; since he died in the year that the instrument was made, it most likely was made for him. The inscription recording Maḥmūd's ownership was imitated by the maker of No. 36, who changed only the name to that of an imaginary Ismā'īl.

9. *Ḍiyā' al-dīn Muḥammad* was a member of a prolific family of astrolabists making instruments in Lahore; their assumption of the title al-Humāyūnī seems to indicate some official relation of an ancestor with the Mughal Emperor Humāyūn (1530-1556). The family tree can be reconstructed as follows:

- Allāhdād
  - 'Īsā
    - Qā'im Muḥammad
      - Ḍiyā' al-dīn Muḥammad
    - Muḥammad Muqīm
      - Ḥāmid Jamāl al-dīn

The Adler has instruments made by both 'Īsā and Ḍiyā' al-dīn Muḥammad.

Ḍiyā' al-dīn Muḥammad made No. 15 in 1637/8; No. 16 in 1647/8; No. 17 in 1660/1; and No. 18 in 1662/3. He is known to have made in all 24 astrolabes and 15 globes between *ca.* 1635 and 1680.

10. *Hṛṣikeśa* made a Sanskrit quadrant (yantracintāmaṇi), Other Instruments C, for the unnamed son of Caṇḍīdatta, perhaps at Multān, in the eighteenth century . The construction follows the prescriptions of, *e.g.*, the *Yantracintāmaṇi* composed by Cakradhara before 1600.

11. *Ḥusayn* painted the wooden artillery level, Other Instruments L, in 1783/4, presumably in North Africa.

12. *'Īsā ibn Allāhdād*, the grandfather of Ḍiyā' al-dīn Muḥammad (q.v. for the family), made No. 13 at Lahore in June 1601 and No. 14 in 1604/5 in Lahore. For another astrolabe made by him see Gunther (1932) 68, p. 187.

13. *Ja'far ibn 'Umar ibn Dawlatshāh of Kirmān* was a member of a well known family of astrolabists; the family's genealogy is given under 15, Maḥmūd ibn Jalāl ibn Ja'far. He made No. 11 in 1388. He is also known to have made four other astrolabes and a globe between 1354 and 1388 [Mayer (1956) pp. 53-54, Gunther (1932) 15, p. 128 and Gibbs and Saliba (1972) CCA No. 15, pp. 64-65].

14. *Khalīl Muḥammad ibn Ḥasan 'Alī* was a member of the group of Iṣfahānī astrolabists of the late Ṣafawī period of whom 'Abd al-A'immah was the most prominent member; in fact, 'Abd al-A'immah decorated several of his astrolabes, including the Adler's No. 23. Khalīl Muḥammad also made the body of No. 22 in 1697/8. More than 23 of his astrolabes were known to Mayer (1956) pp. 54-57 and suppl., p. 294, ranging in date from 1682 till 1707/8.

15. *Maḥmūd ibn Jalāl ibn Ja'far* made No. 12 in Iran in 1433/4 and another instrument belonging to R.S. Webster in 1434/5;[1] he also made another astrolabe, now in Istanbul, in 1484 [Mayer (1956) p. 58]. Mayer has reconstructed from their instruments the following family tree:

Dawlatshāh
|
'Umar
|
Ja'far of Kirmān
(five astrolabes and 2
globes 1354-1388)
|
Muḥammad al-Jalāl
(globe; 1430/1)
|
Maḥmūd

16. *Majd al-dīn Ḥasan of Yazd* made a compass, Other Instruments E, in 1675/6. He seems otherwise to be unknown.

17. *Muḥammad ibn Aḥmad al-Baṭṭūṭī* made astronomical instruments in Morocco; about twenty are known, ranging in date from 1706/7 till 1737/8 (eleven are listed in Mayer (1956) pp. 60-61 and suppl. p. 296); the Adler examples, Nos. 6 and 7, were made in 1720/1 and 1723/4. An astrolabe made by his brother, al-Ḥasan, in 1691/2 is described in Gibbs and Saliba (1972) CCA No. 2568, pp. 160-165.

18. *Muḥammad ibn 'Alī* made a level, Other Instruments N, in Sayl (?) on 22 February 1788. He seems to be otherwise unknown.

19. *Muḥammad ibn al-'Arab* made an artillery level, Other Instruments M, probably in North Africa in the eighteenth or nineteenth century. He also is otherwise unknown.

20. *Muḥammad ibn al-Fattuḥ al-Khamā'irī* made No. 1 at Seville in 1236/7. This is the latest of his 13 known instruments, the earliest having been made in 1207/8. Eight instruments, including the Adler's, are listed in Mayer (1956) pp. 64-66; see also Turner (1985) 2 (Inv. 3407), and Gibbs and Saliba (1972) CCA No. 4001, pp. 184-187.

21. *Muḥammad ibn al-Mufaḍḍil ibn Kīrān;* made No. 9 in Morocco (he was muwaqqit at Wazzān) in 1847/8. Another astrolabe, Turner (1985) 11 (Inv. 430), was made by him in 1845.

22. *Muḥammad ibn Yūsuf ibn Ḥātim* made No. 2 in Spain in 1240/1. This seems to be his only known instrument; see Mayer (1956) p. 78.

23. *Muḥammad Amīn ibn Amīrzā Khān of Qumm* is known to have made an astrolabe in 1587/8 [Mayer (1956) p. 63]. Among the Adler's fakes is one, No. 42, that is attributed to a Muḥammad Amīn ibn Mīrzā, whose name seems to be a derivative of that of the genuine astrolabist. It is alleged that No. 42 was made in 1687 for a Shaykh 'Alīkhān Ḍifjāhī (?). This instrument seems to have been forged in India by the same person who made No. 43, which he claims to have been made by Muḥammad Amīn in 1723/4 for Ḥājjī Muḥammad Muḥsin.

1. Marjorie Webster gave the Webster astrolabe (acquisition number M316) to the Adler Planetarium. Unfortunately it was to late to include the instrument in the catalogue. For information about it contact the Adler Planetarium.

Again, there is a genuine Muḥammad Amīn ibn Muḥammad Ṭāhir (see Mayer (1956) pp. 63-64, Gibbs and Saliba CCA No. 31, pp. 68-69 and Gunther (1932) 31, pp. 145-146), who belonged to the workshop at Iṣfahān whose most prominent member was 'Abd al- A'immah. Either he or the other Muḥammad Amīn may have inspired the forger.

24. *Muḥammad Ibrāhīm al-Ḥasanī* made No. 25 in Iran in 1699/1700. He is otherwise unknown, but may be related to Ḥasan al-Ḥasanī al-Karānī al-Ḥusaynī, who made an astrolabe in 1677/8 (see Mayer (1956) p. 47). *Cf.* also the Mullā Ḥasanī whose name appears on the fake 'Abd al-'Alī astrolabe, No. 40.

25. *Muḥammad Mahdī al-khādim of Yazd,* a son of Muḥammad Amīn, made some 23 astrolabes and decorated others by 'Abd al- A'immah, Khalīl Muḥammad, and Muḥammad Muqīm al-Yazdī in the second half of the seventeenth century; see Mayer (1956) pp. 70-71 and 87, Turner (1985) 6 (Inv. 1849), Gibbs and Saliba (1972) CCA No. 25, pp. 65-68 and Gunther (1932) 25, pp. 139-140. Among these is one made in 1659/60 and described by Morley (1856) pp. 48-49. This inspired a forger to attribute No. 41 to Muḥammad Mahdī al-khādim and to claim that it also was made in 1659/60.

26. *Muḥammad Mu'min* made No. 31 in Iran in the eighteenth century. It appears that nothing further is known of him.

27. *Muḥammad Sharīf ibn Muḥammad* signed No. 34 as its maker. W. Hartner identified him with the Muḥammad Muḥsin ibn Muḥammad 'Alī al- Sharīf of Kirmān who engraved an astrolabe for Luṭf'alī ibn 'Abd al-Nabī in 1781 [Mayer (1956) p. 74]. Another astrolabe by the astrolabist is Turner (1985) 7 (Inv. 1436), made in 1780.

28. *Muḥammad Ṭabāṭabā'ī* made, but did not complete, No. 35 in Iran in 1792/3. Of this maker also nothing further is known.

Maghribi Astrolabes

1. Astrolabe

Muḥammad ibn al-Fattuḥ al-Khamā'irī
Seville, Spain
1236/7
"Muḥammad ibn al-Fattuḥ al-Khamā'irī made it in the city of Seville in the year A.H. 634."
24.8 x 19.3 x 4.5 cm.
Mater thickness 0.6 cm
Brass.
M-35

Face of No. 1

Astrolabe made by Muḥammad ibn al Fattuḥ al-Khamā'irī at Seville in A.H. 634 (A.D. 1236/7), according to the inscription on the back of the kursī: *ṣana'ahu muḥammad ibn al-fattuḥ al-khamā'irī bi-madīnat ishbīliyya fī sanah kh.l.d. al-hijrah.* Inscriptions in kūfic.

Consists of body, rete, uninscribed diopter with sights, pin, horse, handle, ring, and five tympans which have tabs at the top fitting into a corresponding slot in the body.

The rim of the front of the body is graduated for every degree; the numbers, running clockwise, are 5, 10, 15, ...360 (but after 100 the symbols for 100, 200, or 300 are omitted before numbers ending in 5). In the recess in the front is a projection for latitude 66°.

The upper half of the back of the body is graduated for every degree; the numbers, running clockwise from the left to the top and counter-clockwise from the right to the top, are 5, 10, 15, ...90. At the edge of the lower half are indicated the cotangents. There are four concentric rings indicating, from the outermost, the 30° in each zodiacal sign, the names of the zodiacal signs, the days of each Roman month, and the names of the Roman months. These names are: *yaniyyar, fabriyyar, mārs, abrīl,*

Reverse of No. 1

māyyah, yūniyyah, yūliyyah, aghusht, shatanbar, aktūbar, nawanber, and *dijanbar.* The vernal equinox is located at 15 March. In the center of these circles are, at the top, curves for the 24 hours of the day extending to 66° (see *Libros del saber,* vol. 3, p. 143) and, at the bottom, cosine lines for each part of a Radius of 60 parts.

On the rete the ecliptic is graduated for every 2°; the numbers for each zodiacal sign are 6, 12, 18, 24, 30. The names of the stars (some pointers are uninscribed) are listed. (★)

The five tympans have altitude circles for every 3°, azimuths for every 6°, unequal hour curves, and curves for prayer times. They indicate their latitudes. (■)

PROVENANCE — A.W.M. Mensing, Amsterdam, 1924; Max Adler, Chicago, 1930; AP gift, 1930.

ICA 153

REFERENCES — Combe, Sauvaget, and Wiet (1939), p. 138; Engelmann Catalogue (1924), p. 15; Gibbs and Saliba (1984), p. 187; Gunther (1932) 153, p. 300; Mayer (1956), p. 66; Renaud (1942b), p. 21; Stephenson, Bolt, and Friedman (2000), p. 65; Wiet (1935), p. 14.

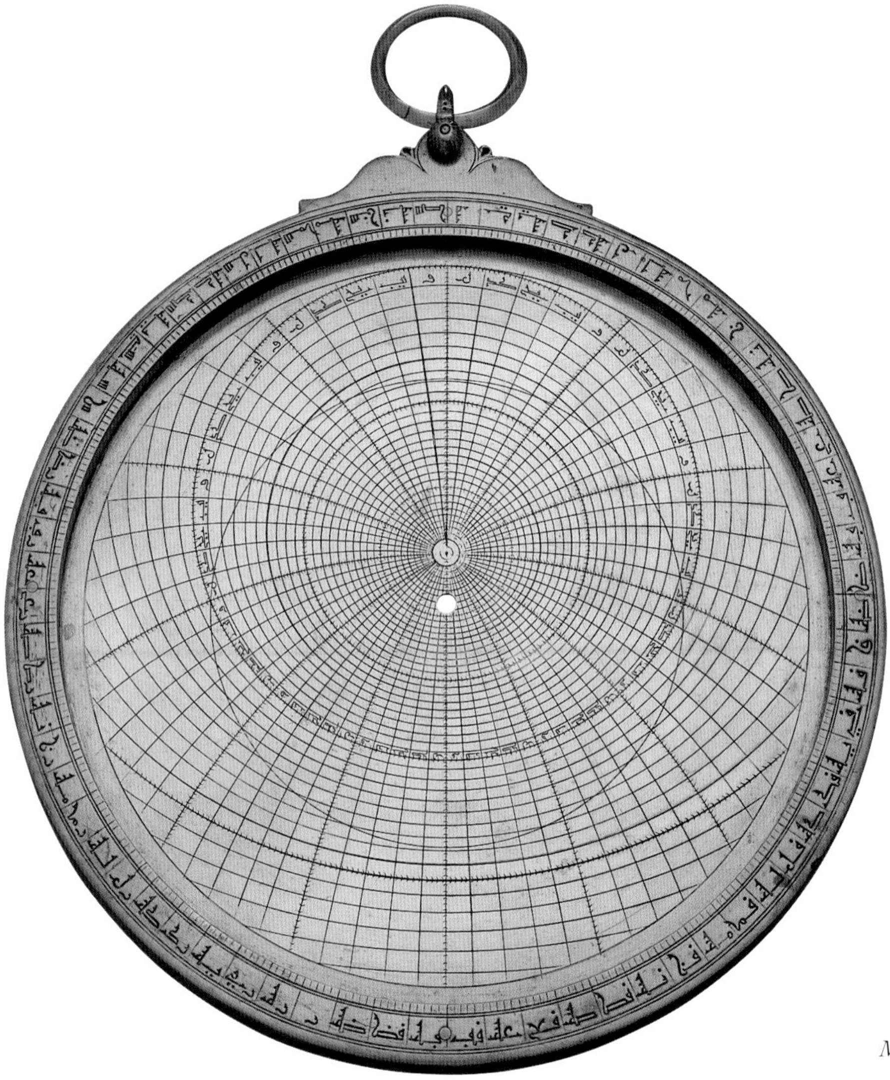

Mater of No. 1

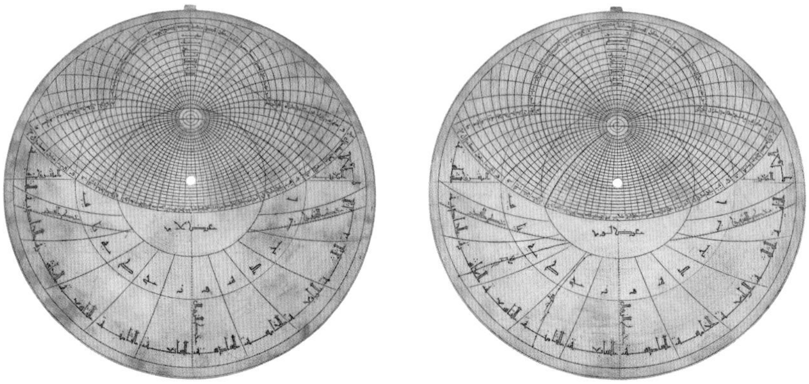

Tympan #1 of No. 1

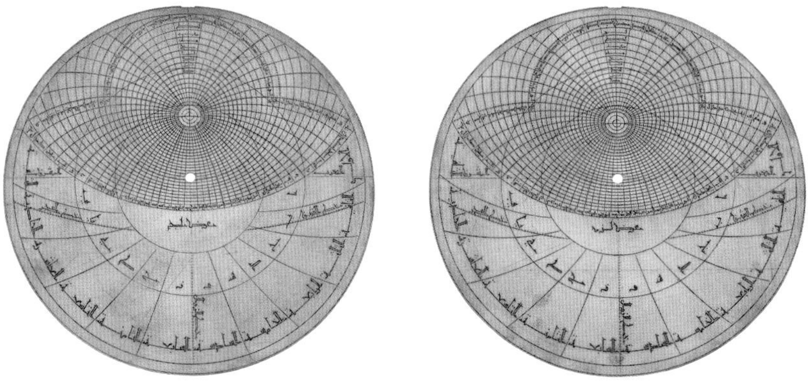

Tympan #2 of No. 1

Rete of No. 1

★ STAR LIST

Name	Identification
al-khaḍīb	β Cass.
dhanab qayṭūs	β Ceti.
ra's al-ḥawwā'	α Oph..
ra's al-ghūl	β Pers.
rijl al-dubb	μ Urs. Mai.
al-ridf	α Cygn.
sa'd nāshirah	γ Capr.
al-simāk al-a'zal	α Virg.
al-simāk al-rāmiḥ	α Boot.
al-shi'rā al-'abūr	α Can. Mai.
al-shi'rā al-ghumayṣā'	α Can. Min.
shafat al-saraṭān	α Canc.
'ayn al-thawr	α Tau.
al- 'ayyūq	α Aur.
al-fakkah	α Cor. .Bor.
al-qā'id	η Urs. Mai.
qadam al-jawzā'	β Ori.
qalb al-'aqrab	α Scorp.
mankib al-jawzā'	α Ori.
mankib al-faras	β Peg.
munīr al-shujā'	α Hydr.
al-nasr al-ṭā'ir	α Aquil.
al-wāqi'	α Lyr.
yad al-dubb	α Urs. Mai.

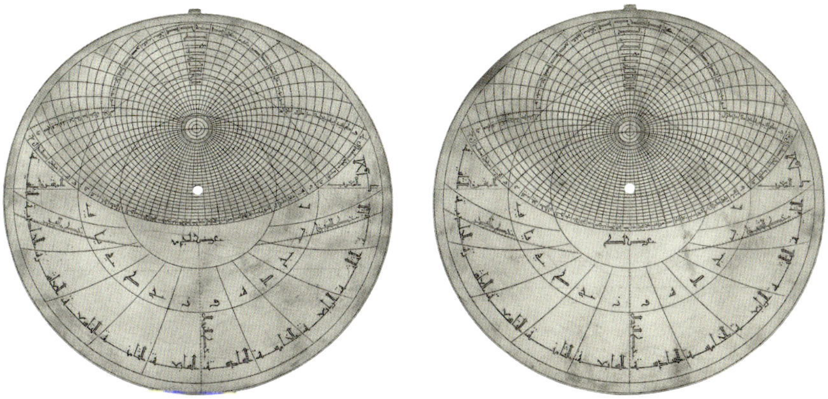

Tympan #3 of No. 1

Tympan #4 of No. 1

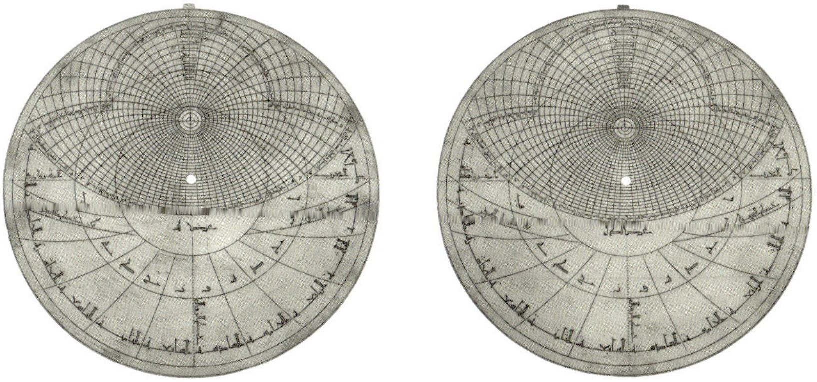

■ TYMPANS

Tympan	Latitude
Ia.	31;40°
Ib.	36;40°
IIa.	33°
IIb.	37;40°
IIIa.	33;40°
IIIb.	39°
IVa.	35°
IVb	39;30°
Va.	36°
Vb	40;30°

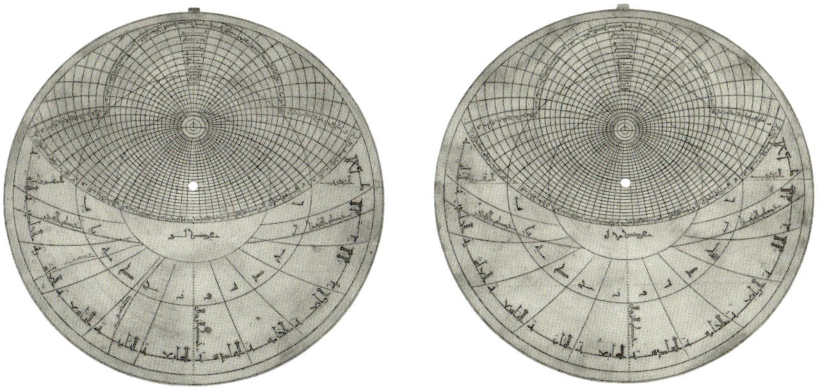

Tympan #5 of No. 1

2. Astrolabe

Muḥammad ibn Yūsuf ibn Ḥatim
Spain
1240/1
"Muḥammad ibn Yūsuf ibn Ḥatim. — may God forgive him — made this astrolabe. He finished it in the year 638."
22 x 15.3 x 2.8 cm
Mater thickness 0.6 cm
Brass
M-36

Face of No. 2

Astrolabe made by Muḥammad ibn Yūsuf ibn Ḥātim. in A.H. 638 (A.D. 1240/1) according to the inscription in the central square on the back: *ṣana'a hādhā al-asṭurlāb muḥammad ibn yūsuf ibn ḥātim 'afā allāh tammahu sanah kh.l.ḥ.* Inscriptions in kūfic.

Consists of body, rete, uninscribed diopter with sights, pin, washer, horse, handle, ring, and five tympans, of which the fifth is from a different instrument. The tympans have tabs at the top fitting into a corresponding slot in the body.

The rim of the front of the body is graduated for every degree; the numbers, running clockwise, are 5, 10, 5, 20, 5, ...90 in each quadrant. The recess in the front is uninscribed.

The upper half of the back is graduated for every degree; the numbers, running clockwise from the left to top and counter-clockwise from the right to the top, are 5, 10, 5, 20, 5, ...90. At the edge of the lower half are indicated the cotangents. There are five concentric circles indicating , from the outermost, the 30° in each zodiacal sign, the names of the zodiacal signs, the days of each Roman month, the names of the Roman months, and the names of the lunar mansions. The names of

Reverse of No. 2

the Roman months are: *yaniyyar, fabriyyar, mārs, abrīl, māyyah, yūniyyah, yūliyyah, aghusht, shitanbar, aktūbar, nawanbar*, and *dijanbar*. The vernal equinox is located at 12 March. Below the center is a shadow-square for 12 digits.

On the rete the ecliptic is graduated for every 6°; there are no numbers. Many pointers had silver knobs at their bases; some have fallen off. The names of the stars (some pointers are uninscribed) are listed. (★)

The five tympans have altitude circles for every 6°, azimuths for every 10°, unequal hour curves, and curves for prayer times. The four original tympans indicate their latitudes and longest daylights. (■)

The fifth, extraneous tympan also indicates its cities. It is said to be eleventh century in King (2004–05) II, p.885.(❑)

PROVENANCE — Raoul Heilbronner, Paris *ca.* 1900; French Government, Paris 1914; A.W.M. Mensing, Amsterdam, 1922; Max Adler, Chicago, 1930; AP gift, 1930.
ICA 154

EXHIBITIONS — *The Meeting of Two Worlds: The Crusaders and the Mediterranean Context,* University of Michigan Art Museum, Ann Arbor, MI, April – Oct., 1981

REFERENCES — Combe, Sauvaget, and Wiet (1939), p. 138; Engelmann Catalogue (1924), p. 15; Fox (1932), pp. 335-36; Fox (1933), pp. 35-36; Gunther (1932) 154, p. 300 (plate LXVII); Heilbronner Catalog (1922); King (1995), pp. 6, 10, 12; Mayer (1956), p. 78; Paris Universal Exhibition Catalogue (1900), p. 22 (lot. 28); Renaud (1942b), p. 21.

TYMPANS

Tympan	Latitude	Longest Daylight
Ia.	21;40°	13;27 hours
Ib.	25°	13;40 hours
IIa.	31°	14;4 hours
IIb.	32°	14;8 hours
IIIa.	33°	14;12 hours
IIIb.	34°	14;16 hours
IVa.	36°	14;24 hours
IVb.	37°	14;28 hours

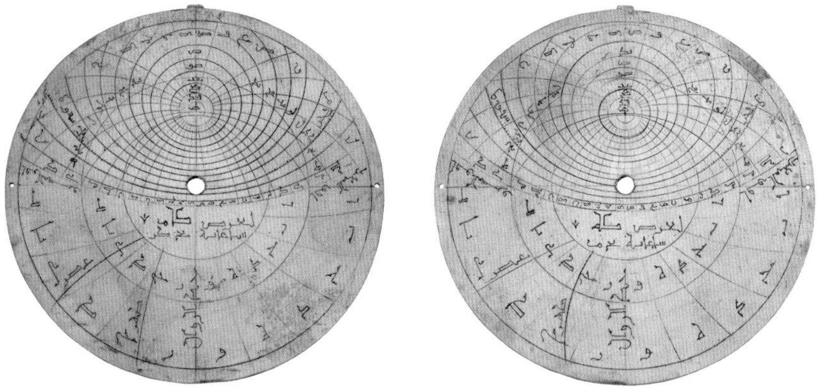

Tympan #1 of No. 2

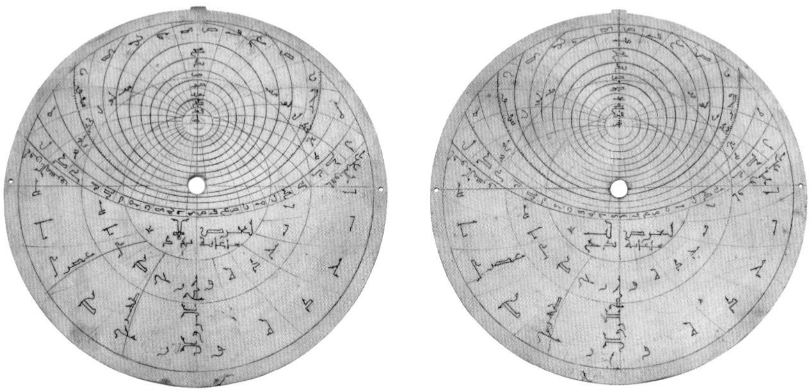

Tympan #2 of No. 2

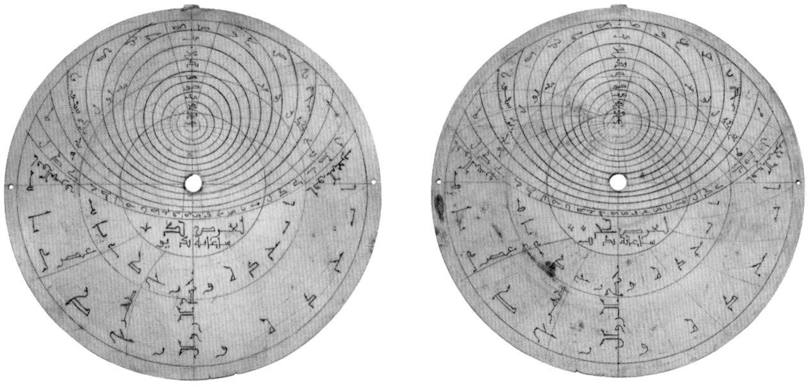

Tympan #3 of No. 2

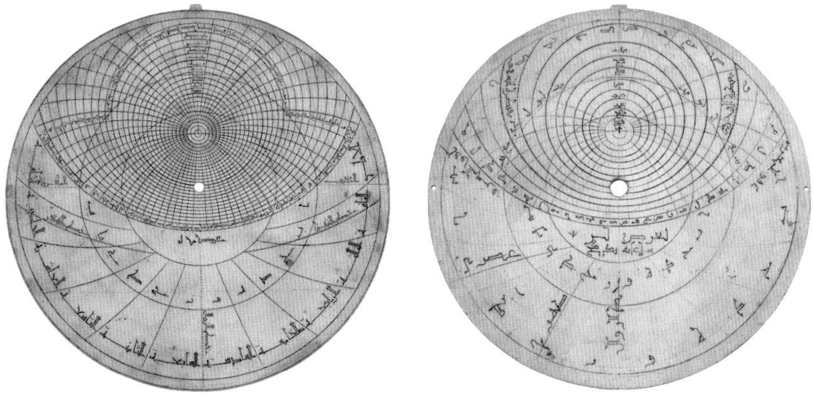

Tympan #4 of No. 2

★ STAR LIST

Name	*Identification*
a'zal	α Virg.
baṭn qayṭūs	ζ Ceti.
janāḥ al-ghurāb	γ Corv.
ḥawwā'	α Oph.
ḥayyah	α Serp.
khaḍīb	β Cass.
dubb	α Urs. Mai.
dabarān	α Tau.
dulfīn	ε Delph.
dhanab al-jady	δ Capr.
dhanab qayṭūs	ι Ceti.
rāmiḥ	α Boot.
rijl al-jawzā'	β Ori.
ridf	α Cygn.
ṭā'ir	α Aquil.
al-'abūr	α Can. Mai.
'ayyūq	α Aur.
ghumayṣā'	α Can. Min.
faras	β Peg.
fakkah	α Cor. Bor.
qā'id	η Urs. Mai.
qalb al-asad	α Leo.
qalb al-'aqrab	α Scorp.
mankib	α Ori.
munīr al-shujā'	α Hydr.
wāqi'	α Lyr.

Rete of No. 2

❑ TYMPANS

Tympan	*City*	*Latitude*	*Longest Daylight*
Va.	Ṭalīṭalah	40°	14;51 hours
Vb.	Burghush	42°	15;8 hours

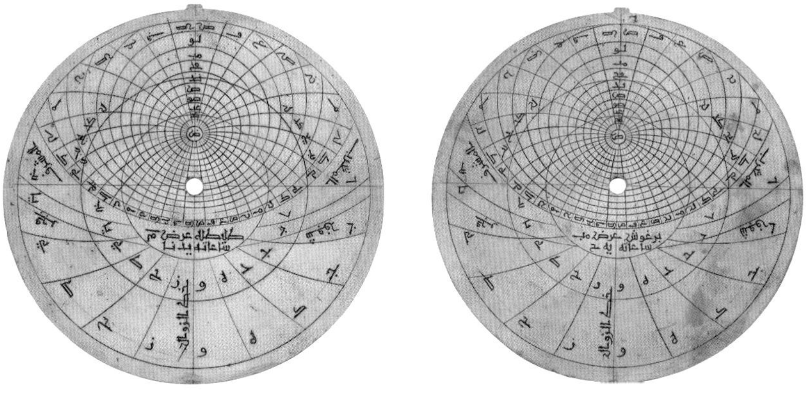

Tympan #5 of No. 2

3. Universal astrolabe

Maker unknown
Spain or North Africa
13th century
18.8 x 12 x 1.3 cm
Mater thickness 0.4 cm
Brass
A-71

Face of No. 3

A universal astrolabe, al-Shakkāziyya, of al-Zarqāllu. The location of the vernal equinox at 13 March indicates a date in the thirteenth century; the presence of the Roman months in the calendaric material shows that it was made in Spain or North Africa. Inscribed in kūfic.

Consists of a single plate with a graduated pointer with cursor on the front, throne to the east, handle, and ring.

The rim of the front is graduated for every degree; the numbers, running from the east and the west points clockwise and counter-clockwise to the north and south poles, are 6, 12, 18, ...90. The Zarqāllu projection has meridians and declination circles for every six degrees. The names of the stars (there are some star markers without names) are listed. (★)

The back is graduated for every degree; the numbers running clockwise from the left to the top and counter-clockwise from the right to the top, are 5, 10, 15, ...90. There are four concentric circles indicating the degrees of each zodiacal sign, the names of the zodiacal signs, the days of each Roman month, and the name of the Roman months. The last are: *janjar, fabriyyar, mārs, abrīl, māyyah, yūnī, yūlī, aghust, sabtabar, aktūbar, nawabar,* and *dijanbar*. The vernal equinox is located at 13 March. In the center is a shadow-square for 12 digits.

PROVENANCE — Jacob Zeitlin, Los Angeles, 1936; AP purchase 1936.
ICA 2544
EXHIBITIONS — Spertus Museum of Judaica, Chicago, IL, "In the Tradition of the Sephardim", Sept. 1992 – Jan. 1993

Reverse of No. 3

★ STAR LIST

Name	*Identification*
baṭn al-ḥūt	β Andr.
khaḍīb	β Cass.
dabarān	α Tau.
rāmiḥ	α Boot.
rijl	κ Ori.
ridf	α Cygn.
ṭā'ir	α Aquil.
'unq	α Hydr.
'ayyūq	α Aur.
ghumayṣā'	α Can. Min.
ghūl	β Pers.
fakkah	α Cor. Bor.
qā'id	η Urs. Mai.
mankib	β Peg.
nahr	θ Erid.
wāqi'	α Lyr.

4. Astrolabe

Maker unknown
North Africa
14th or 15th century
12.4 x 11.1 x 2.4 cm
Mater thickness 0.5 cm
Brass
A-396

Face of No. 4

Astrolabe probably made in the fourteenth or fifteenth century, as is indicated by the fact that the vernal equinox is located at March 13. Inscriptions in kūfic.

Consists of body, rete, uninscribed diopter with sights, pin, pointer, horse, handle, ring, cord, and two tympans. The typmans have tabs fitting into slots in the body.

The rim of the front of the body is graduated for every two degrees; the numbers, running clockwise from the top, are 10, 20, 30, …90 in each quadrant. In the recess in front are three concentric circles.

On the back are seven concentric rings. The outermost ring contains, in each quadrant, clockwise on the left and counter-clockwise on the right, the numbers 10, 20, 30, … 90. The second ring has lines dividing each 10° segment of the outermost ring into five equal parts. In the third ring are 36 segments labeled counter-clockwise 10, 20, 30| 10, 20, 30| … The fourth ring contains the names (Arabic) of the zodiacal signs associated with the twelve groupings of the third ring; *al-ḥamal* begins at the diameter bisecting the rings on the right. The fifth ring contains lines dividing each zodiacal sign into fifteen equal parts. The sixth ring indicates the number of days in each month according to the Julian calendar. Finally, the seventh ring gives the French names (in Arabic script) of the months.

The vernal equinox, as mentioned, is located at 13 March.

Below the center of the back is a shadow-square for 12 digits. The right and left sides are labeled *ma'kūs*, the two halves of the bottom together are labeled *mabsūṭ*.

On the rete the ecliptic is marked for every 6°; there are no numbers. The names of the zodiacal signs are too damaged to be read. There are twenty-six pointers, but the names of the stars are also illegible.

The two tympans have altitude circles for every 6°, and azimuths for

Reverse of No. 4

every 10°. They also indicate the seasonal hours (1–12) and the prayer times. Ia is for latitude 21;40°, Ib is latitude 37;30°. II is apparently by a different engraver. IIa is for the latitude of Mehnès, 31;30°, and IIb for latitude 37°. (■)

PROVENANCE — The Time Museum, Rockford, IL; AP purchase 1999.
EXHIBITIONS — Arab American National Museum, Dearborn, MI, Sept. 2006 – present
REFERENCES — Newbury (2004a); Turner, A. J. (1985), pp. 70-3, notes the similarity of the instrument to one acquired in Egypt by Jean-Joseph Marcel during the Napoleonic campaign, and described by Sedillot (1844).

Mater of No. 4

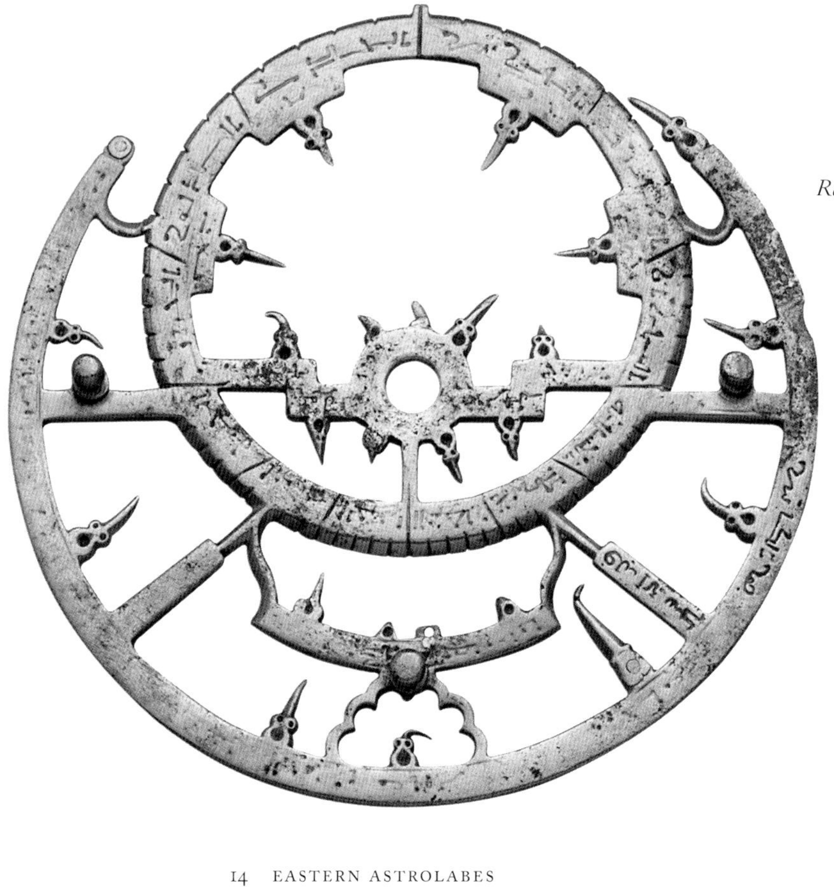

Rete of No. 4

■ TYMPANS

Tympan	*Latitude*
Ia.	21;40°
Ib.	37;30°
IIa.	31;30°
IIb.	37°

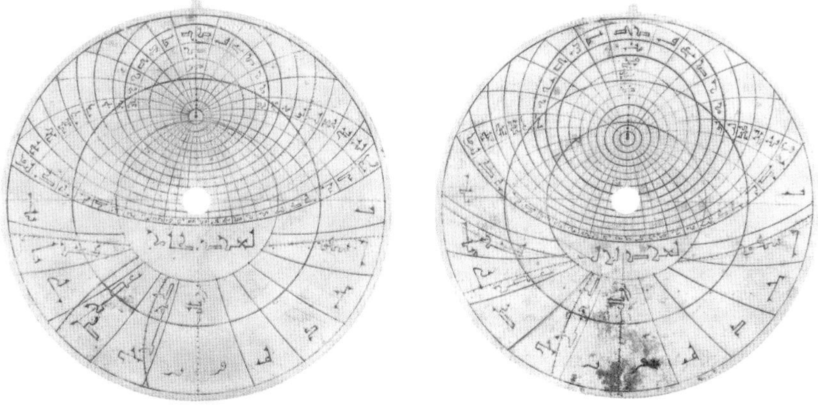

Tympan #1 of No. 4

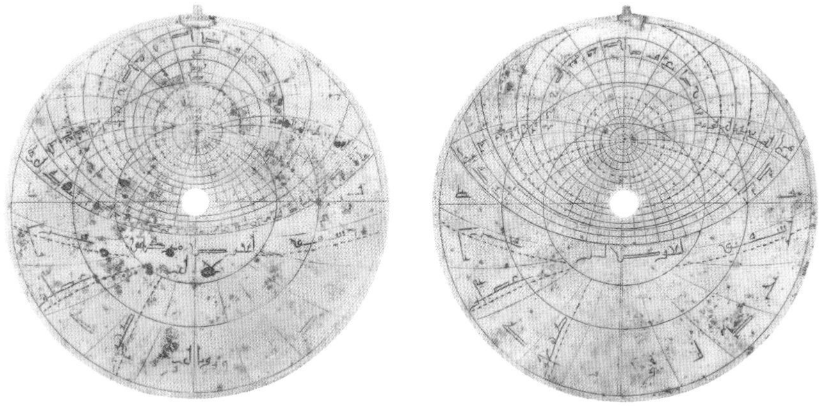

Tympan #2 of No. 4

Diopter, horse and pin of No. 4

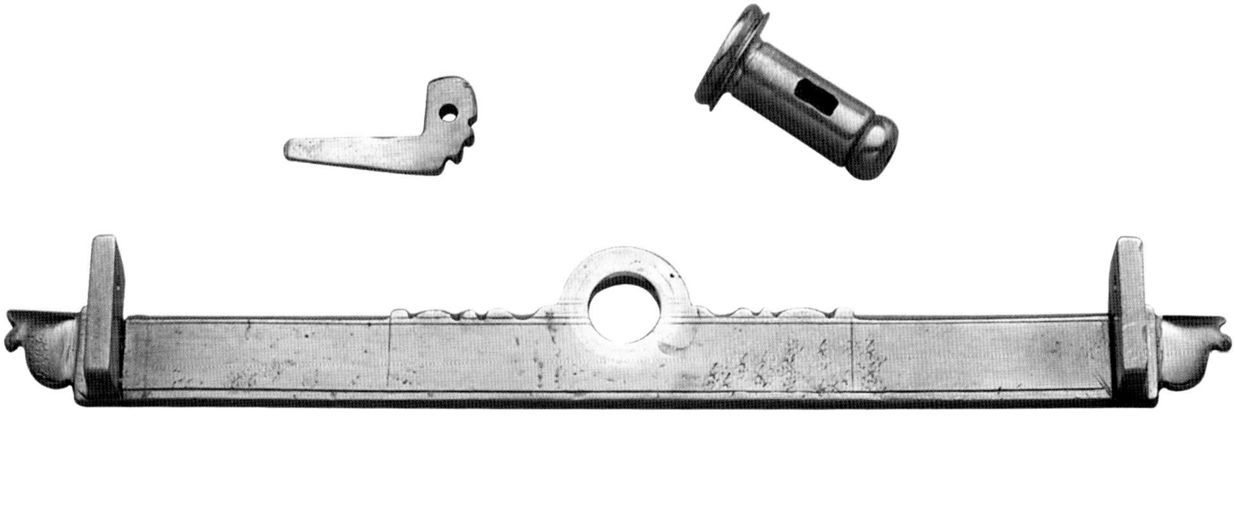

5. Astrolabe

Maker unknown
North Africa
1651 (?)
13.5 x 8.3 x 2.4 cm
Mater thickness 0.8 cm
Brass
M-37

Face of No. 5

Astrolabe made in A.D. 1651 if that number inscribed to the left on the rete is a date. Inscriptions in naskh and in Latin script.

Consists of body, rete, uninscribed diopter with sights, pin, pointer, horse, handle, ring, and three tympans (a piece of cardboard replaces a missing fourth tympan). The tympans have notches at the bottom fitting about a corresponding tab in the body. Into the back of the throne is set a compass.

The rim of the front of the body is graduated for every degree; the numbers, running clockwise, indicating every 15° or hour, are 1, 2, 3, ...12 from top to bottom and again from bottom to top. The recess in the front is uninscribed.

The back is graduated for every degree; the numbers (Arabic) for arcs of 6° are 1 to 15 clockwise from the left to the top, counter-clockwise from the right to the top. In the bottom half arcs of 10° within each zodiacal sign are marked 10, 20, 30. There are four concentric circles indicating: the Arabic names of the zodiacal signs, the days of the Roman months, the French names of the Roman months in Latin script, and the names of the Syriac months in Arabic script. The French and Syriac months are written as listed. (●)

The vernal equinox is located at 20 March (Gregorian). About the center of the back are inscribed the Arabic names of the months. Below the center is a shadow-square for 12 digits in Arabic.

On the rete the ecliptic is graduated for every 6°; there are no numbers. The names of the stars, written in Latin, in Latin script, are listed. (★)

The pointer on the front is marked in Latin script: *Lat. me. Lat. sept.* (*"southern latitude"* and *"northern latitude"*).

Reverse of No. 5

The three extant tympans have altitude circles for every 6°, azimuths for every 9°, and equal and unequal hour curves. Ia is for all horizons; there are no numbers. Ib is for a latitude of 18° (written in European); the other numbers are European as well. IIa is for a latitude of 30° (written in Arabic); there are no other numbers. IIb is for a latitude of 33° (written in Arabic); the other numbers are European. IIIa is for a latitude of 39° (written in Arabic); the other numbers are partly Arabic, partly European. IIIb is for a latitude of 42° (written in both Arabic and European); the other numbers are European. (■)

PROVENANCE A.W.M. Mensing, Amsterdam, 1924; Max Adler, Chicago, 1930; AP gift 1930.
ICA 115
REFERENCES — Engelmann (1924), p. 15; Gunther (1932), p. 247; Newbury (2004a).

MONTH NAMES

Syriac Months	*French Months*
kanūn 2	ianvier
shayyāt[1]	febvrier
adhār	mars
nīsān	apri
ayyādh[2]	mai
ḥazīrān[3]	ivin
tammūz	ivillet
ab	aovst
īlūl	september
tishrīn al-awwal	octobre
tishrīn al-akhīr	novembre
kanūn al-awwal	decembre

1. mistake for shubaṭ.
2. mistake for ayyar.
3. mistake for khazīran.

★ STAR LIST

Name	*Identification*
antares	α Scorp.
aqvila	α Aquil.
arctvrv	α Boot.
austr. cauda coeti	β Ceti
cignv	α Cygn.
cinrv[1]	α Canc.(?)
corvvs	γ Corv.
delphin	ε Delph.
hidra	α Hydr.
lira	α Lyr.
navis	ζ Puppis
oculus tavri	α Tau.
orion	α Ori.
proc.	α Can. Min.
serpent.	α Serp.
sirivs	α Can. Mai.
spica	α Virg.

1. perhaps mistake for cirrv = cirrus.

Rete of No. 5

Detail of pointer of No. 5

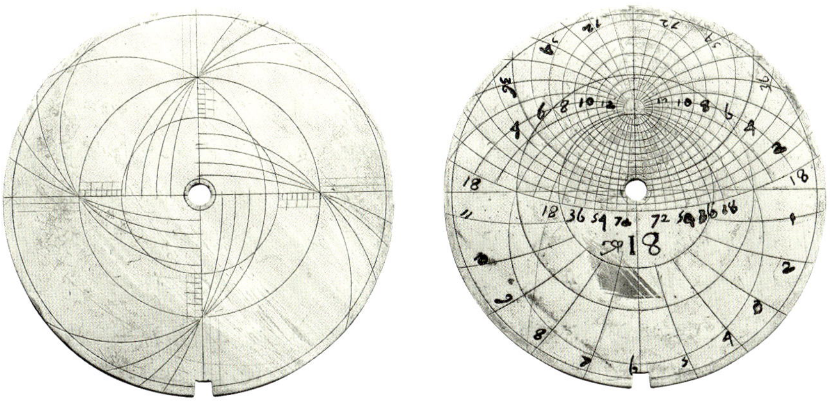

Tympan #1 of No. 5

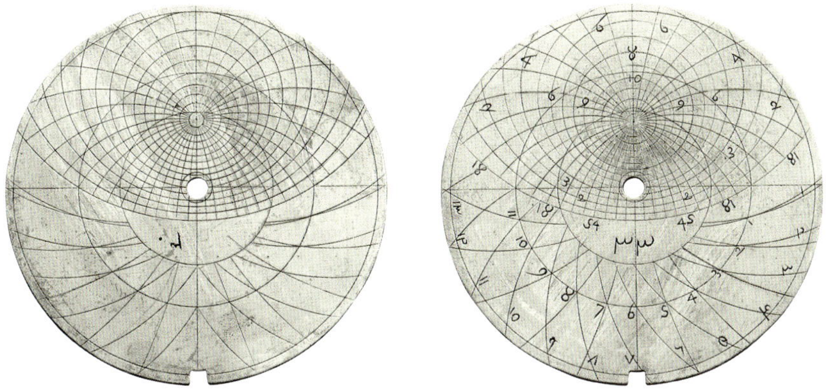

Tympan #2 of No. 5

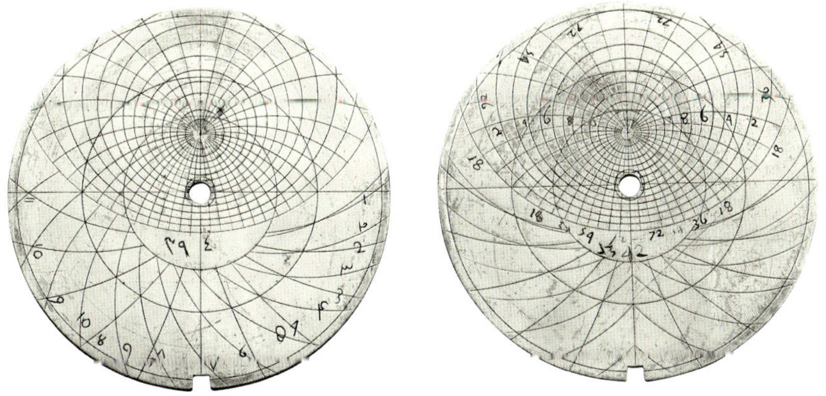

Tympan #3 of No. 5

TYMPANS

Tympan	*Latitude*
Ib.	18°
IIa.	30°
IIb.	33°
IIIa.	39°
IIIb.	42°

6. Astrolabe

Muḥammad ibn Aḥmad al-Baṭṭūṭī
Morocco
1720/1
"Praise to God. Muḥammad ibn Aḥmad al-Baṭṭūṭī made this in the year 1133".
28 x 19.9 x 4.3 cm
Mater thickness 0.9 cm
Brass
A-76

Face of No. 6

Astrolabe made by Muḥammad ibn Aḥmad al-Baṭṭūṭī in A.H. 1133 (A.D. 1720/1) according to the inscription in the center of the upper half of the back: *al-ḥamd allāh. wa ḥadā ṣana'ahu muḥammad ibn aḥmad al-baṭṭūṭī sanah* 1133 (1133 written in European numerals). Inscriptions in kūfic except for the date.

Consists of body, rete, uninscribed diopter with sights, pin, horse, handle, ring, and six tympans which have tabs at the top fitting into a slot in the body.

The rim of the front of the body is graduated for every degree; the numbers, running clockwise, are 5, 10, 15, ...360, except that , after 100, the symbols for 100, 200, or 300 are omitted in every number ending in 5.

In the recess of the front is a geographical table in two columns indicating the clime, the city, its longitude, and its latitude; the cities are arranged by increasing longitudes, with some confusion at the end. The list corresponds with that of Ibn al-Bannā' (see Kennedy and Kennedy (1987) p. 524) except for some mistakes in orthography and the substitution of ṭayyibah for yathrib. (■)(■■)

The back is graduated for every degree; the upper two quadrants are numbered 5, 10, 15, ...90 clockwise from the left to the top and counter-clockwise from the right to the top. There are six concentric circles indicating the degrees of each zodiacal sign, the names of the zodiacal signs, the days in each Roman month, the names of the Roman months, the names of the lunar mansions, and the names of the four seasons. The names of the Roman months are: *yaniyyar, fabrāyar, mārs, abrīl, māyyah, yūniyyah, yūliyyah, aghusht, shatanbar, aktūbar, nawanbar,* and *dijanbar*. The vernal equinox is located at 9 March. In the center of these circles are, at the top, curves for the 12 hours of the day and, at the bottom, a shadow-square for 12 digits.

Reverse of No. 6

On the rete the ecliptic is graduated for every 3°; there are no numbers. There are one or three silver knobs at the base of each pointer. The names of the stars are listed. (★)

The six tympans have altitude circles for every 3°. IIa, b, IVa, b, and VIa, b have azimuths for every 6° and Ia, b, IIIa, b, and Va, b for 5°. All tympans have unequal hour curves and curves for prayer-times. IVb was intended to be for a latitude of 30°, but the engraving was oriented 90° to the left and was not completed. The tympans (except for the last) indicate their cities and latitudes. (●)

PROVENANCE — Jacob Zeitlin, Los Angeles, 1936; AP purchase, 1936.
ICA 2549
EXHIBITIONS — Memphis Pink Palace Museum, "Egypt's Eternal Skies," March – June 1981; Newberry Library, 1974; Lakeview Museum, Peoria, IL, Aug. – Oct. 2000; Robert F. Longway Planetarium, Flint, MI, Sept. – Dec. 2002
REFERENCES — Gibbs and Saliba (1984), p. 225, n. 55; Newbury (2004a).

Mater of No. 6

COLUMN I

Clime	*City*	*Longitude*	*Latitude*	*Identification*
3	marrākush	21;0°	31;30°	Marrakech
5	shaqtrīn	23;40°	40;40°	Santarem
4	ṭanjah	24;0°	35;0°	Tangier
3	salā	24;0°	33;0°	Salé
3	muknāsat al-zaytūn	24;30°	34;0°	Meknès
3	fās	25;0°	33;40°	Fés
4	al-jazīrat al-khuḍr	25;30°	36;0°	Alger
4	sibtah	25;40°	35;20°	Ceuta
4	al-qaṣr	26;0°	35;0°	Alcazar
4	rubāṭ tārah	26;0°	34;30°	Rabat
3	sijilmāsah	27;0°	30;0°	Tafilalt
4	mālqah	26;30°	37;0°	Málaga
4	talqasān	27;0°	35;0°	Tlemcen
4	gharnāṭah	27;30°	37;30°	Granada
4	wahrān	31;0°	34;0°	Oran
4	yajāyah	33;0°	36;30°	Bejaia

★ STAR LIST

Name	*Identification*
al-aʿzal	α Virg.
baṭn qayṭūs	ζ Ceti.
janāḥ al-ghurāb	γ Corv.
al-ḥawwāʾ	α Oph.
al-ḥayyah	α Serp.
al-dabarān	α Tau.
al-dulfīn	ε Delph.
dhanab al-jady	δ Capr.
dhanab qayṭūs	ι Ceti
raʾs al-ghūl	β Pers.
al-rāmiḥ	α Boot.
rijl al-dubb	μ Urs. Mai.
al-ridf	α Cygn.
zubānā al-asad	ε (?) Leo.
al-shujaʿ	α Hydr.
al-ṭāʾir	α Aquil.
al-ʿabūr	α Can. Mai.
al-ʿayyūq	α Aur.
al-ghumayṣāʾ	α Can. Min.
al-fakkah	α Cor. Bor.
qadamā al-jawzāʾ	β Ori.
qalb al-asad	α Leo.
qalb al-ʿaqrab	α Scorp.
mankib al-jawzāʾ	α Ori.
mankib faras	β Peg.
al-wāqiʿ	α Lyr.

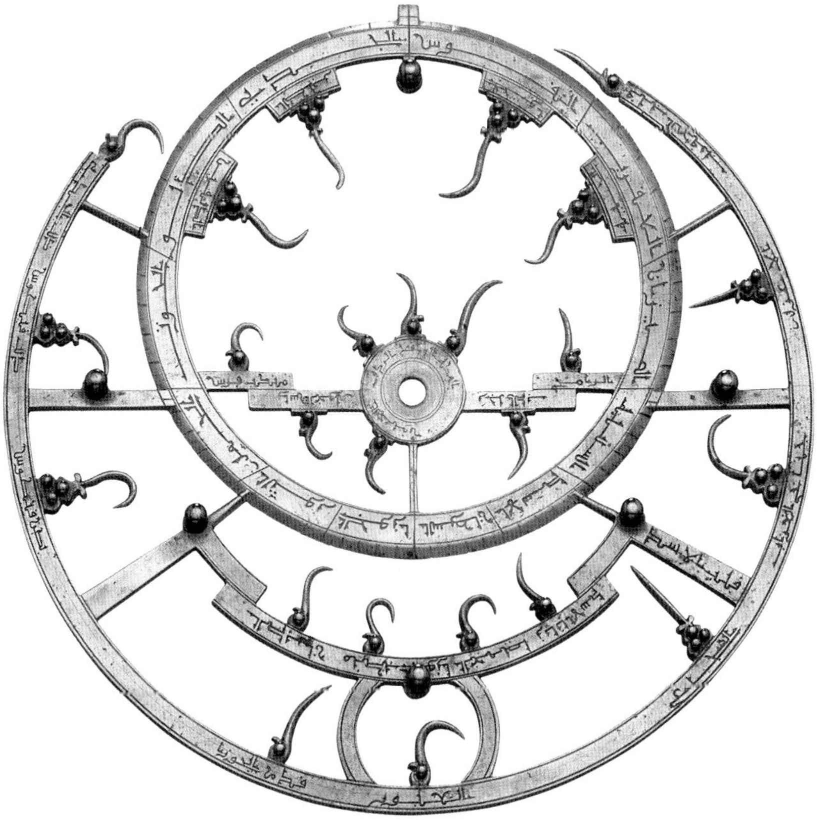

Rete of No. 6

■■ COLUMN II

Clime	*City*	*Longitude*	*Latitude*	*Identification*
3	al-qayrawān	33;45°	31;30°	Kairouan
3	tūnis	39;0°	33;0°	Tunis
4	al-mahdiyyah	40;0°	30;30°	Mahdia
3	tarābalas	41;40°	33;0°	Tripoli
3	barqah	43;0°	33;0°	Shahhat
3	al-iskandariyyah	61;0°	31;30°	Alexandria
3	miṣr	64;0°	30;0°	Cairo
3	bayt al-maqdis	67;0°	34;0°	Jerusalem
3	dimashq	71;0°	33;0°	Damascus
2	ṭayyibah	75;0°	25;30°	Medina
3	ʿasqalān	75;0°	33;0°	Ashqelon
2	makkah	77;0°	21;40°	Mecca
4	al-hind	90;0°	16;40°	(India)
2	al-ṣīn	84;0°	15;40°	(China)
1	samarqand	90;0°	0;40°	Samarkand
1	jurjān	84;0°	0;40°	Jurjān

TYMPANS

Tympan	*City*	*Latitude*
Ia.	makkah al-musharrafah	21;40°
Ib.	madīnat al-nabī	25;30°
IIa.	marrākush	31;30°
IIb.	bayt al-maqdis and wahrān	32°
IIIa.	dar'a	28;30°
IIIb.	madīnat Maknāsah	34°
IVa.	miṣr, sijilmāsah, and mahdiyyah	30°
Va.	madīnat fās	33;40°
Vb.	ṭanjah	35°
VIa.		36°
VIb.		37;30°

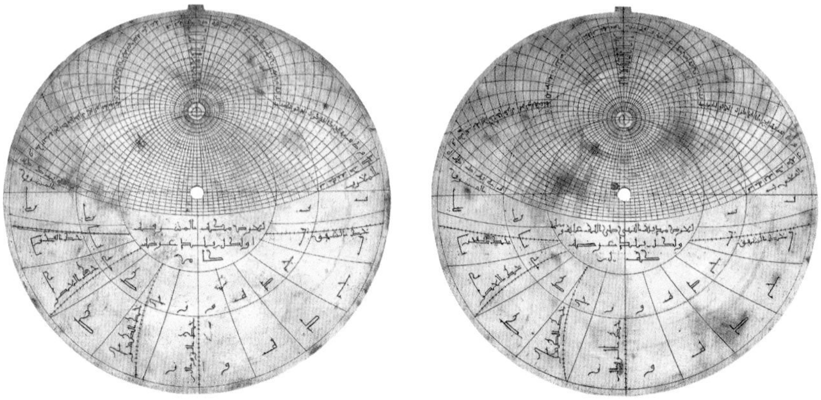

Tympan #1 of No. 6

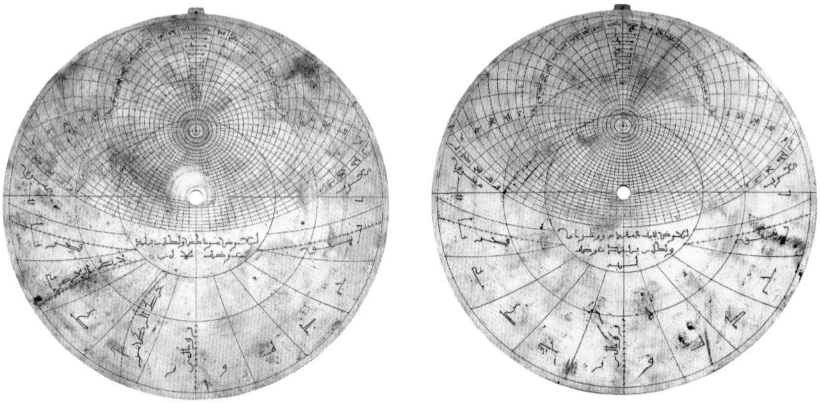

Tympan #2 of No. 6

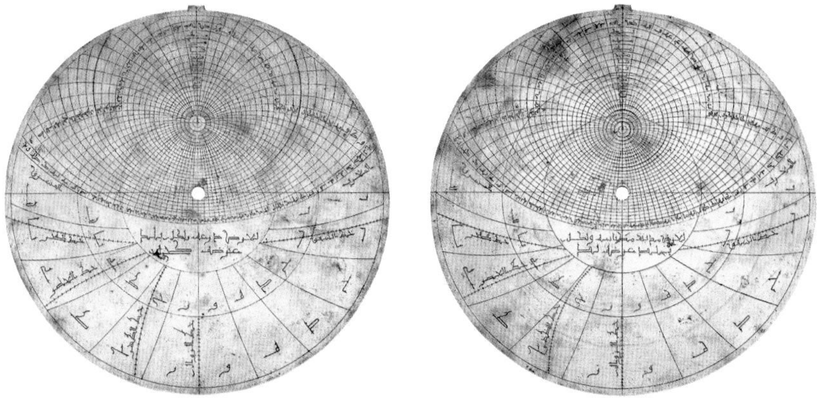

Tympan #3 of No. 6

Detail of reverse of No. 6

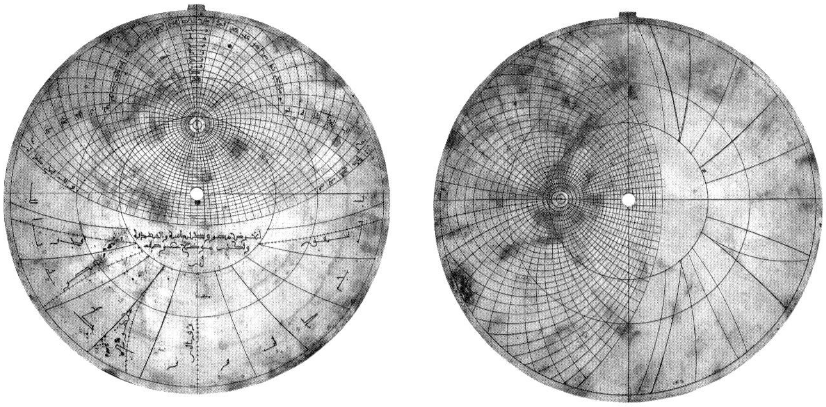

Tympan #4 of No. 6

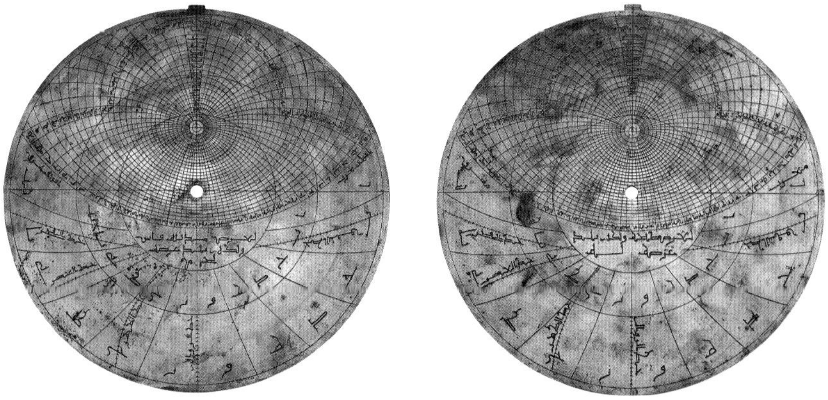

Tympan #5 of No. 6

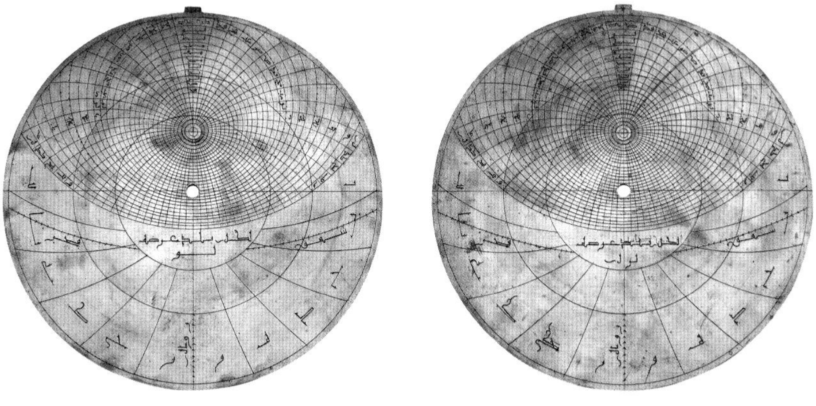

Tympan #6 of No. 6

7. Astrolabe

Muḥammad ibn Aḥmad al-Baṭṭūṭī
Morocco
1723/4
"Praise to God. Muḥammad ibn Aḥmad al-Baṭṭūṭī was its maker — may God forgive him and all Muslims. Amen. In the year 1136."
22.6 x 14.3 x 3.4 cm
Mater thickness 0.6 cm
Brass
A-87

Face of No. 7

Astrolabe made by Muḥammad ibn Aḥmad al-Baṭṭūṭī.in A.H. 1136 (A.D. 1723/4) according to the inscription within the shadow-square on the back: *al-ḥamd allāh. ṣāni'hi muḥammad ibn aḥmad al-baṭṭūṭī. ghafr allāh lihi wa li-jamī' al-muslimīn. amīn. sanah* 1136 (1136 written in European numerals). Inscribed in kūfic except for the date and a pious inscription in naskh on the side of the body *rāḥim (?) luṭf al-ghālib 'abdallāh abū ṭālib*.

Consists of body, rete, uninscribed diopter with sights, pin, horse, handle, ring, and six tympans which have tabs at the top which fit into a corresponding slot in the body.

The rim of the front of the body is graduated for every degree; the numbers, running clockwise, are, 5, 10, 15, ...360 (except that after 100 the symbols for 100, 200, or 300 are omitted from numbers ending in 5). The recess in the front is uninscribed except for traces of several concentric circles.

The back is graduated for every degree; the numbers are 5, 10, 15, ...90 in the upper quadrants, running clockwise from the left to the top and counter-clockwise from the right to the top. There are four concentric circles indicating the degrees of each zodiacal sign, the names of the zodiacal signs, the days of each Roman month, and the names of the Roman months. The last are: *yaniyyar, fabrāyar, mārs, abrīl, māyyah, yūniyyah, yūliyyah, aghusht, shatanbar, aktūbar, nawanbar,* and *dijanbar*. The vernal equinox is located at 8 March. In the lower part of the center is a shadow-square for 7 feet and 12 digits.

On the rete the ecliptic is graduated for every 6°; there are no numbers. The names of the stars are listed. (★)

The first five of the six tympans have altitude circles for every 6°, azimuths for every 10°, unequal hour curves, and curves for prayer times. The last tympan is on one side for

Reverse of No. 7

latitude 0° (*li-mawḍah lā ʿarḍ lihi*), on the other for all latitudes (*li-jamīʿ al-ʿurūḍ*); this second side has "altitude" circles whose centers are the North pole and the two equinoctial points. The other five tympans indicate their cities and latitudes. (■)

PROVENANCE — Acquired by A. Fletcher while in service in Russia, 1921; AP purchase, 1934.
ICA 1723
EXHIBITIONS — Purdue University, July 1961 – April 1962; Robert F. Longway Planetarium, Flint, MI, 1962; Lakeview Museum, Peoria, IL, Sept – Dec. 2002.
REFERENCES — Gibbs and Saliba (1984), p. 225, n. 55; Newbury (2004a).

Rete of No. 7

★ STAR LIST

Name	*Identification*
al-a'zal	α Virg.
baṭn qayṭūs	ζ Ceti
al-dabarān	α Tau.
al-dulfīn	ε Delph.
dhanab al-jady	δ Capr.
dhanab qayṭūs	ι Ceti
ra's al-ḥawwā'	α Oph.
ra's al-ghūl	β Pers.
al-rāmiḥ	α Boot.
rijl al-dubb	μ Urs. Mai.
ridf	α Cygn.
shujā'	α Hydr.
al-tā'ir	α Aquil.
al-'abūr	α Can. Mai.
'unq al-ḥayyah	α Serp.
'ayyūq	α Aur.
al-ghurāb	γ Corv.
al-ghumayṣā'	α Can. Min.
fakkah	α Cor. Bor.
qadamā al-jawzā'	β Ori.
qalb al-asad	α Leo.
qalb al-'aqrab	α Scorp.
mankib al-jawzā'	α Ori.
mankib faras	β Peg.
wāqi'	α Lyr.

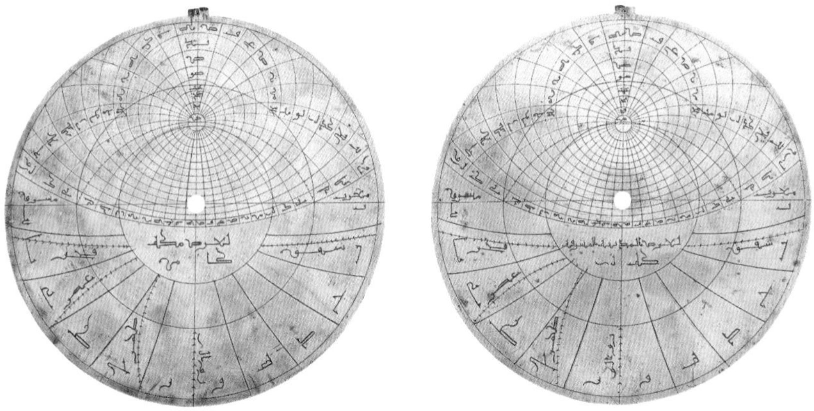

Tympan #1 of No. 7

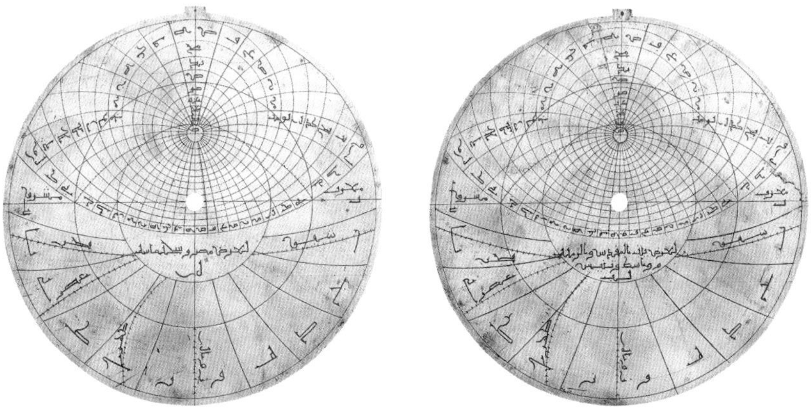

Tympan #2of No. 7

Tympan #3 of No. 7

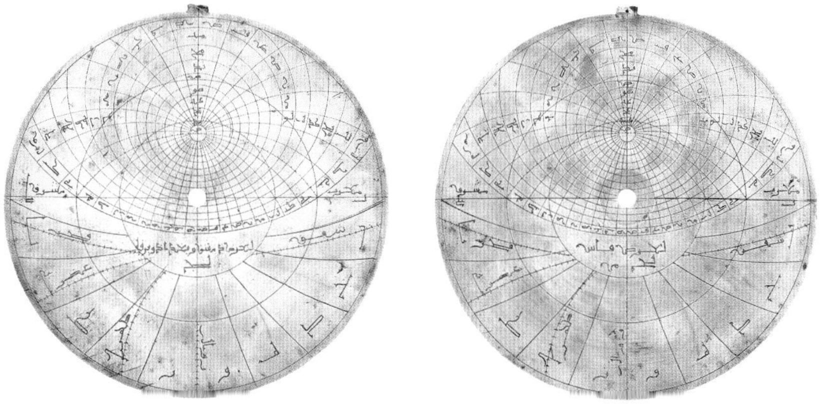

■ TYMPANS

Tympan	*City*	*Latitude*
Ia.	makkah	21;40°
Ib.	al-madīnat al-musharrafah	25;0°
IIa.	miṣr and sijilmāsah	30;0°
IIb.	bayt al-maqdis, al-ramlah, wāsiṭ, and tunīs	32°
IIIa.	dimashq, baghdād, and barqah	33°
IIIb.	fās	33;40°
IVa.	al-jazā'ir	35°
IVb.	ḥimṣah	38°
Va.	ḥalab, al-raqqah, and al-jazīrat al-khuḍrā	35;50°
Vb.	isṭanbūl	41°

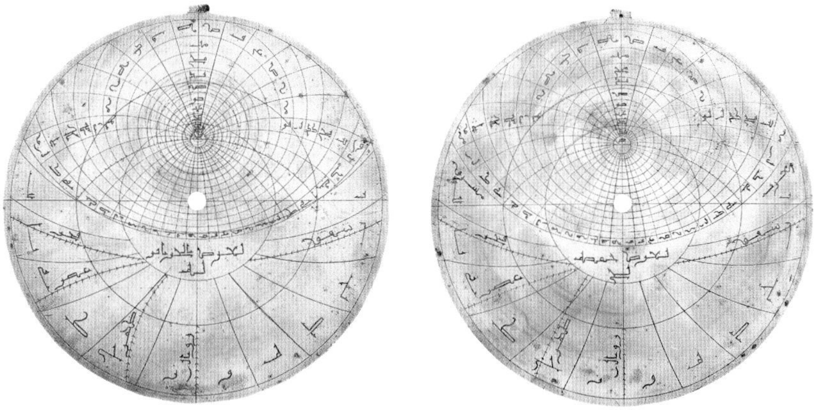

Tympan #4 of No. 7

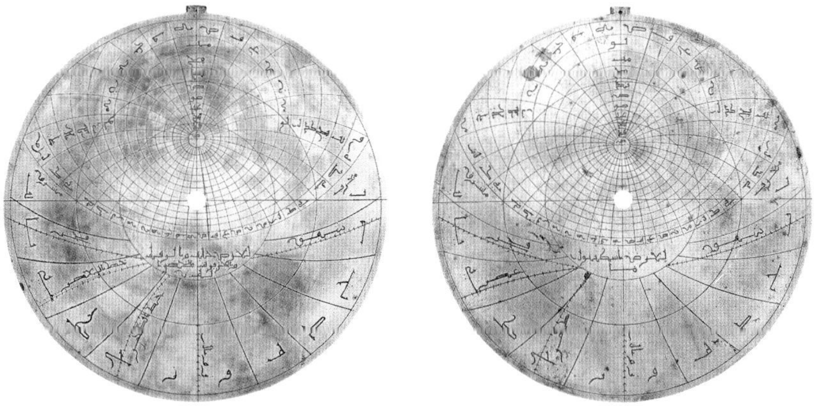

Tympan #5 of No. 7

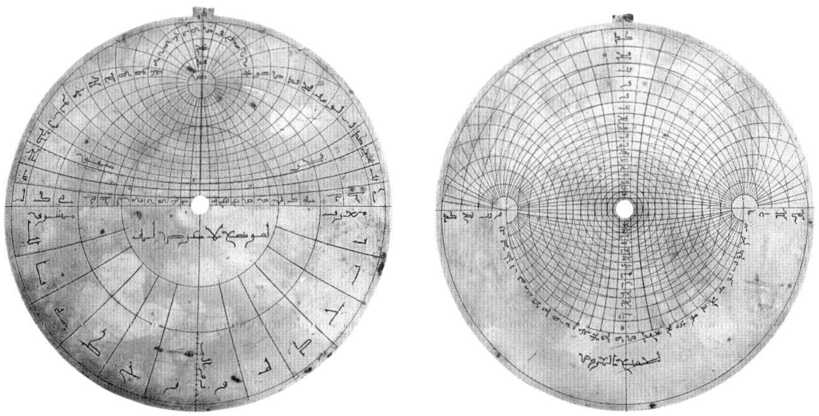

Tympan #6 of No. 7

8. Astrolabe

Maker unknown
North Africa
ca. 1750
25 x 15.2 x 2.6 cm
Mater thickness 0.7 cm
Brass
DPW-54

Mater of No. 8

Astrolabe made in North Africa in *ca.* 1750, as is indicated by the fact that the vernal equinox fell on 8 March. Inscriptions in kūfic.

Consists of body, rete, uninscribed diopter with sights, nut, bolt, handle, ring, and three tympans which have tabs at the top fitting into a corresponding slot in the body.

The rim of the front of the body is graduated for every degree; the numbers, running clockwise in each quadrant, are 5, 10, 15, ...90. The recess in the front is uninscribed.

The back is graduated for every degree; the numbers, running clockwise from the left to the top and from the right to the bottom and counter-clockwise from the left to the bottom and from the right to the top, are 5, 10, 15, ...90. There are four concentric circles indicating the degrees of each zodiacal sign, the names of the zodiacal signs, the days of the Roman months, and the names of the Roman months. The last are: *yaniyyar, fabriyyar, mārs, abrīl, māyyah, yūniyyah, yūliyyah, aghusht, shatanbar, aktūbar, nawanbar,* and *dijanbar*. The vernal equinox is located at 8 March. Above the center are curves for 1 to 12 hours. Below the center is a shadow-square for 12 digits.

On the rete the ecliptic is graduated for every 6°; there are no numbers. The star-pointers have silver knobs at their bases. The star names are listed. (★)

The three tympans have altitude-circles for every 5°, azimuths for every 6°, unequal hour curves, and curves for prayer times. They indicate their cities and latitudes. (■)

PROVENANCE — Chadenat, Paris, 1956; David P. Wheatland, Topsfield, MA, 1956; AP gift, 1985.

REFERENCES — Chadenat (1956), plate V, lot 6; Newbury (2004a) ; Stephenson, Bolt, and Friedman (2000), p. 108.

★ STAR LIST

Name	*Identification*
al-a'zal	α Virg.
baṭn qayṭūs	ζ Ceti
janāḥ al-ghurāb	γ Corv.
al-ḥawwā'	α Oph.
al-ḥayyah	α Serp.
al-dabarān	α Tau.
al-dulfin	ε Delph.
dhanab jady	δ Capr.
dhanab qayṭūs	ι Ceti
al-rāmiḥ	α Boot.
rijl al-jawzā'	β Ori.
rijlihi	μ Urs. Mai.
ridf	α Cygn.
al-shujā'	α Hydr.
al-ṭā'ir	α Aquil.
al-'abūr	α Can. Mai.
'ayyūq	α Aur.
al-ghumayṣā'	α Can. Min.
ghūl	β Pers.
fakkah	α Cor. Bor.
qalb 'aqrab	α Scorp.
mankib al-jawzā'	α Ori.
mankib faras	β Peg.
wāqi'	α Lyr.
yad al-dubb	α Urs. Mai.

Reverse of No. 8

Rete of No. 8

TYMPANS

Tympan	*City*	*Latitude*
Ia.	makkah	21°
Ib.	al-madīnah	25;30°
IIa.	marrākush	31;30°
IIb.	fās	38;40°
IIIa.	tilimsāh	35°
IIIb	al-jazā'ir	35°

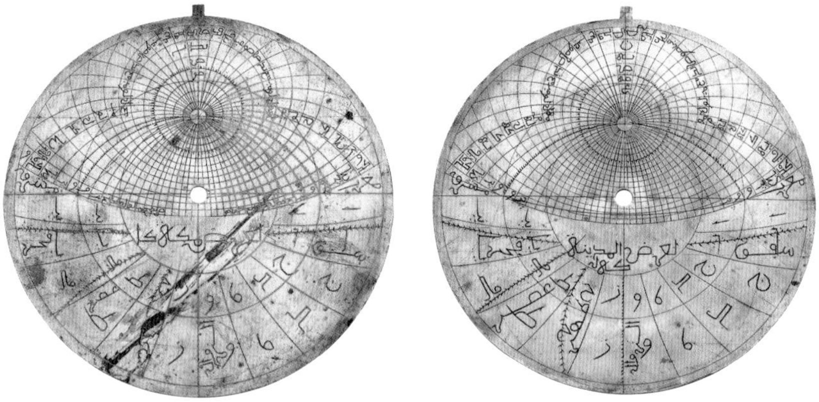

Tympan #1 of No. 8

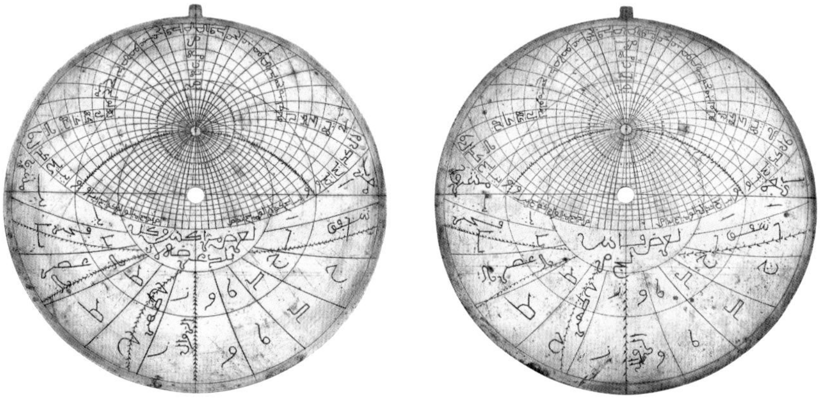

Tympan #2 of No. 8

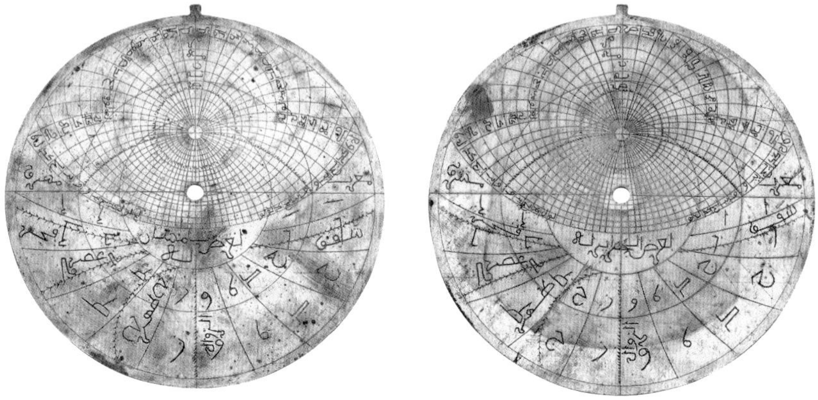

Tympan #3 of No. 8

9. Astrolabe

Muḥammad ibn al-Mufaḍḍil ibn Kīrān
Morocco
1847/8
"The work of Muḥammad ibn al-Mufaḍḍil ibn Kīrān. — may God befriend him."
"He made it in the year 1264."
12.2 x 7.8 x 1.7 cm
Mater thickness 0.6 cm
Brass (except for modern diopter, nut, and screw)
A-77

Face of No. 9

Astrolabe made by Muḥammad ibn al-Mufaḍḍil ibn Kīrān in A.H. 1264 (A.D. 1847/8) according to the inscriptions on the back of the throne: *'amal muḥammad ibn al-mufaḍḍil ibn kīrān. allāh walīhi;* and in the upper part of the center of the back: *ṣana'ahu sanah* 1264 (1264 written in European numerals). Inscribed in kūfic except for the date and for the numbers in the shadow-square. Some of the engraving is tinged red.

Consists of body, rete (without outer circle), handle, ring, and three tympans which have tabs at the top fitting into a corresponding slot in the body. The uninscribed diopter with sights, nut, and screw are later additions.

The rim of the front of the body is graduated for every 2°; the numbers, running clockwise, are 10, 20, 30,...100, 10, 20, ...200, 10, 20, ...300, 10, 20, ...60. The recess in the front is uninscribed.

The back is graduated for every 2°; the numbers in the upper half, running clockwise from the left to the top and counter-clockwise from the right to the top, are 10, 20, 30, ...90; those in the lower half are 10, 20, 30 for each zodiacal sign. There are four concentric circles indicating every 5° of each zodiacal sign, the names of the zodiacal signs, the days of each Roman month, and the names of the Roman months. The last are: *yanāyar, fabrāyar, mārs, abrīl, māyyah, yūniyyah, yūliyyah, ghusht, satanbar, aktabar, nawanbar*, and *dijanbar*. The vernal equinox is located at 8 March. In the lower part of the center is a shadow-square for 12 digits.

On the rete the ecliptic is graduated for every 6°; there are no numbers. The names of the stars are listed. (★)

The three rather crudely engraved tympans have altitude circles for every 6°, azimuths for every 10°, unequal hour curves, and curves for

★ STAR LIST

Name	*Identification*
ḥawwā'	α Oph.
ḥayyah	α Serp
dulfīn	ε Delph.
rāmiḥ	α Boot.
rijl	μ Urs. Mai.
ridf	α Cygn.
ṭā'ir	α Aquil.
'ayyūq	α Aur.
ghūl	β Pers.
faras	β Peg.
fakkah	α Cor. Bor.
wāqi'	α Lyr.
yad dubb (?)	α Urs. Mai.(?)

Reverse of No. 9

prayer-times. They indicate their cities and latitudes. (■)

PROVENANCE — Jacob Zeitlin, Los Angeles, 1940; AP purchase, 1940.
ICA 2550
REFERENCES — Turner, A. J. (1985), p. 108, note 259.

Rete of No. 9

TYMPANS

Tympan	*City*	*Latitude*
Ia.	Darʿah	24;30°
Ib.	Aslā	33;30°
IIa.	Fās	33;40°
IIb.	Rāwiyyah Wazzān	34;30°
IIIa.	Maknās	34°
IIIb.	Marrākush	31;30°

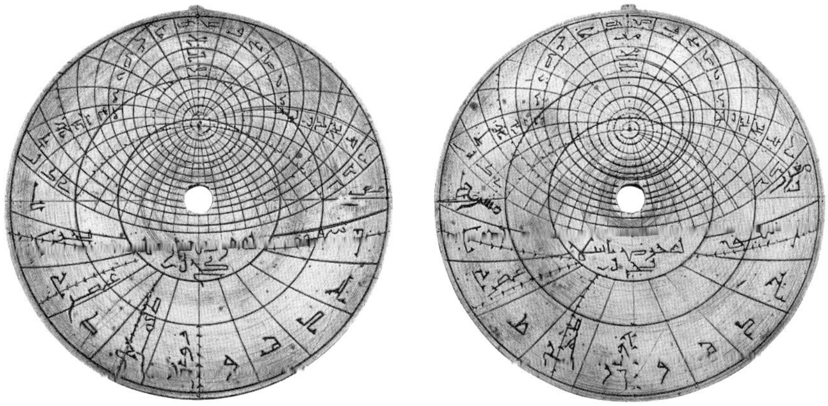

Tympan #1 of No. 9

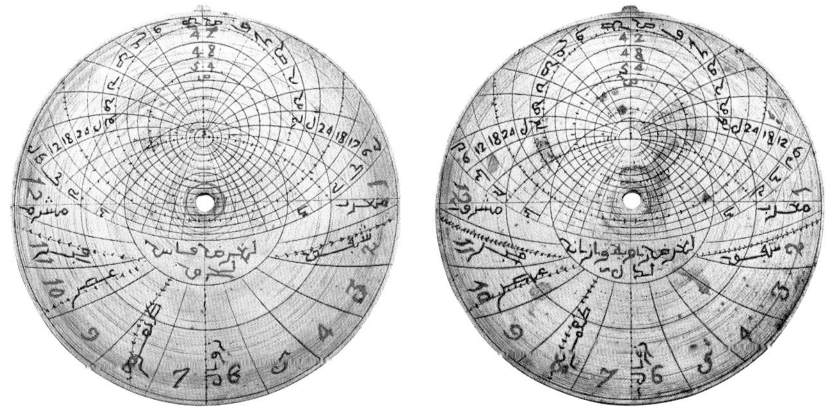

Tympan #2 of No. 9

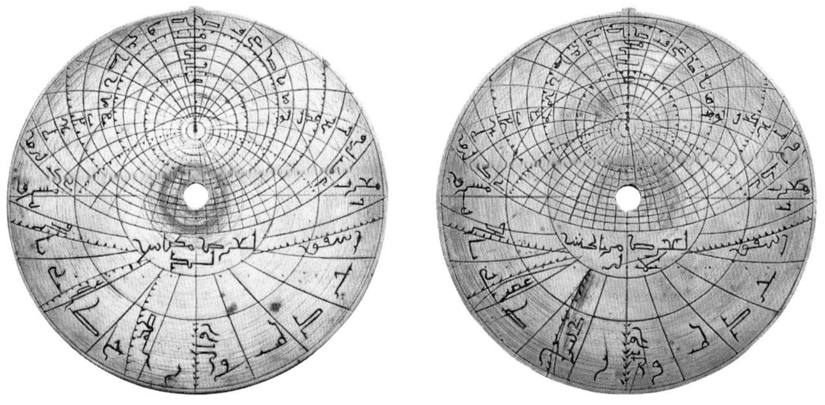

Tympan #3 of No. 9

Mashriqi
Astrolabes

10. Astrolabe

Badr <ibn 'Abdallāh>
Baghdād
1130/1.
"Badr, the *mawlā* (associate? freedman?) of the wonder of the age, Hibatallāh ibn al-Ḥusayn, the astrolabe-maker, in the beginning of the year 525." (A.H. 525 began on 4 December 1130).
"Belonging to the treasury of the exalted Sulṭān, the helper of the world and of religion, Abū al-Qasam Maḥmūd ibn Muḥammad ibn Malikshāh — may God make eternal his days."
"Document of the master, the mightiest king of kings and of sulṭāns. Order of Shaykh 'Alī ibn Malik. 7903 (793?). May God make eternal his reign."
17.6 x 13.3 x 1.2 cm
Mater thickness 0.5 cm
Brass
A-84

Face of No. 10

Astolabe made by Badr, the mawlā ("associate" or "freedman"?) of Hibatallāh ibn al-Ḥusayn, at the beginning of A.H. 525 (A.D. 1130/1) according to the inscription on the lower left edge of the back: *ṣana'ahu badr mawlā badī' al-zamān hibatallāh ibn al-ḥusayn al-asṭurlābī fī bidāyat sanah th.k.h.* The instrument belonged to the Saljuq Sulṭān Abū al-Qasam Maḥmūd ibn Muḥammad ibn Malikshāh, who ruled from 1118 till 1131, according to the inscription in the lower center of the back: *li-khizānat al-sulṭān al-mu'aẓẓam mughīth al-dunyā wa al-dīn abī al-qasam maḥmūd ibn muḥammad ibn malikshāh. khallad allāh ayyāmahu.* See No. 36 for an imitation of this inscription. Above Maḥmūd's inscription is another in naskh of doubtful import. It mentions the document of the mawlā (master), the king of kings and of sulṭāns, and the order of Shaykh 'Alī ibn Malik; if the final "7903" is read as "793", the note may be dated in A.H. 793 (1390/1), but this interpretation is uncertain. The inscription is: *rasm al-mawlā a'āẓim malik al-mulūk wa al-salāṭīn. amr shaykh 'alī ibn malik 7903. asad* (read *allāh*) *khallad mulkah.* Inscriptions in kūfic except for the preceding note in naskh.

Consists of a body, rete, handle, and ring. The three tympans, which have tabs at the bottom fitting into a corresponding slot in the body, and the pin and bent bar are later additions.

The rim of the front of the body is graduated for every degree; the numbers, running clockwise, are 5, 10, 15, ...360. The recess in the front is engraved as a tympan for the equator, with constant daylight of 12 hours. The front is a copy of an instrument by al-Khujandī (King (2004–05) II).

The upper half of the back is graduated for every degree; the numbers, running clockwise from

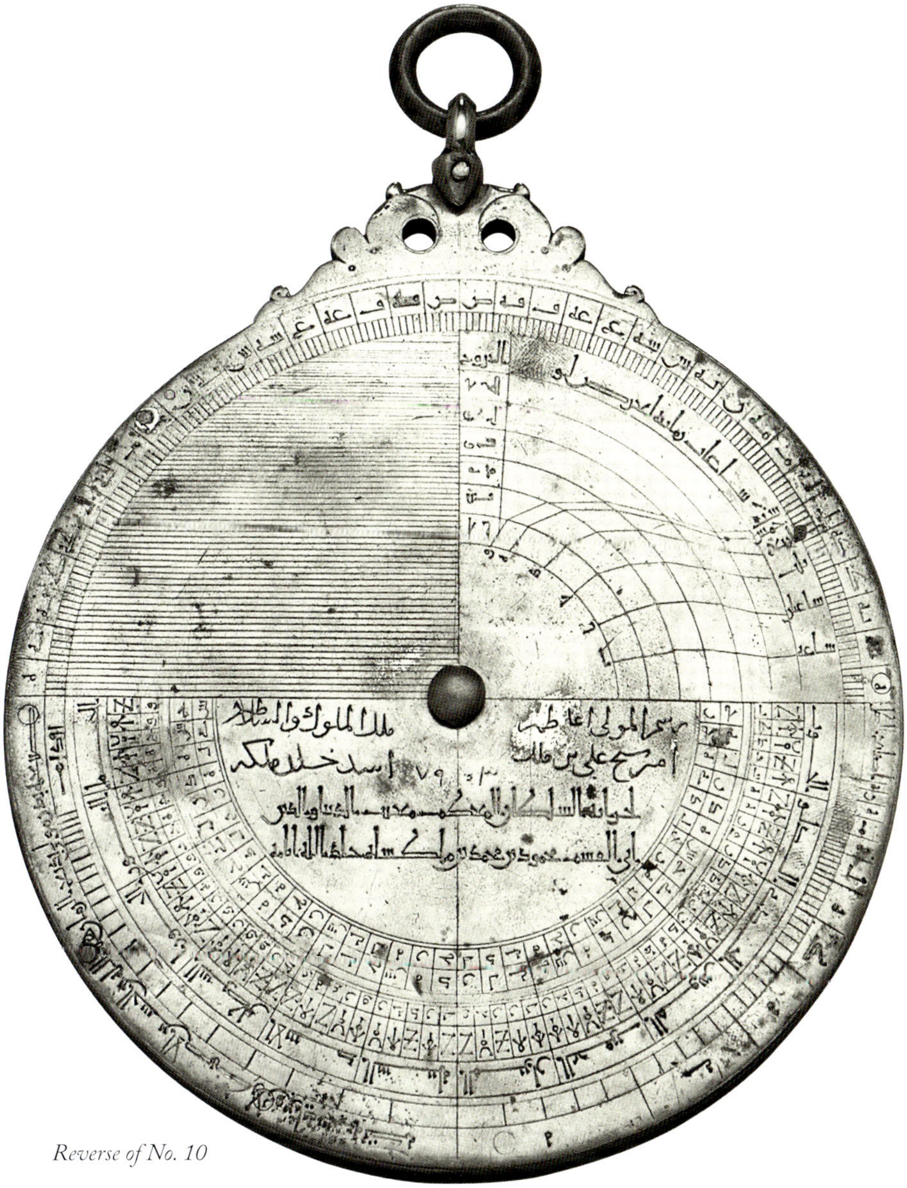

Reverse of No. 10

the left to the top and counter-clockwise from the right to the top, are 5, 10, 15, ...90. At the edge of the lower half are indicated the cotangents. In the upper left quadrant are the cosines of every degree from 1° to 90°. In the upper right quadrant are the altitude curves of the Sun for 1, 2, 3, 4, 5, and 6 hours at a latitude of 36°; at the edge is the inscription: *sā'āt zamāniyyah li-'arḍ l.w.* ("seasonal hours for latitude 36"). At the bottom are five semicircles indicating the zodiacal signs, the lords of the terms (represented by symbols found also in No. 36), the extents of the terms, and the lords of the decans according to the Greeks and to the Indians.

On the rete the ecliptic is graduated for every 3°; the numbers for each zodiacal sign are 6, 12, 18, 24, 30. The names of the stars are listed. (★)

Of the three tympans, the first is identical to the second. All three have altitude circles for every 3° and unequal hour curves (IIIb also has equal hour curves); none has azimuths. On each are indicated the latitude and longest daylight. (■)

PROVENANCE — Ali Kuli Khan of the House of Persian Art, New York, 1941; AP purchase, 1941.
ICA 2557

REFERENCES — Adler Planetarium (1970), pp. 17-18; Adler Planetarium (1973), pp. 21-23; Adler Planetarium (1977), cover; Adler Planetarium (1980), pp. 23-24; Architectural Digest (1973) ; Dor-Nev (1991), p. 20; Evans (1999), p. 255; King (1994), pp. 144, 182; King (1995), pp. 10, 12; King (2004-5), pp. 34, 54, 504–12; Newbury (2004a); Stephenson, Bolt, and Friedman (2000), p. 64.

★ STAR LIST

Name	*Identification*
janāḥ al-ghurāb	γ Corv.
al-dubb	α Urs. Mai.
dhanab al-jady	δ Capr.
dhanab al-dajājah	α Cygn.
dhanab al-dulfīn	ε Delph.
dhanab qayṭus	ι Ceti.
ra's al-ḥawwā'	α Oph.
ra's al-shirā'	ζ Puppis.
ra's al-ghūl	β Pers.
al-rāmiḥ	α Boot.
rijl al-jawzā'	β Ori.
al-simāk al-a'zal	α Virg.
al-sha'āmiyyah	α Can. Min.
al-ṭā'ir	α Aquil.
ṭaraf dhanab al-dubb	η Urs. Mai.
'unq al-ḥayyah	α Serp.
'unq al-shujā'	α Hydr.
'ayn al-thawr	α Tau.
al-'ayyūq	α Aur.
qalb al-asad	α Leo.
qalb al-'aqrab	α Scorp.
kaff al-khaḍīb	β Cass.
min jasad qayṭus	τ Ceti
mankib	β Peg.
al-munīr al-fakkah	α Cor. Bor.
al-wāqi'	α Lyr.
yad al-jawzā'	α Ori.
al-yamāniyyah	α Can. Mai.

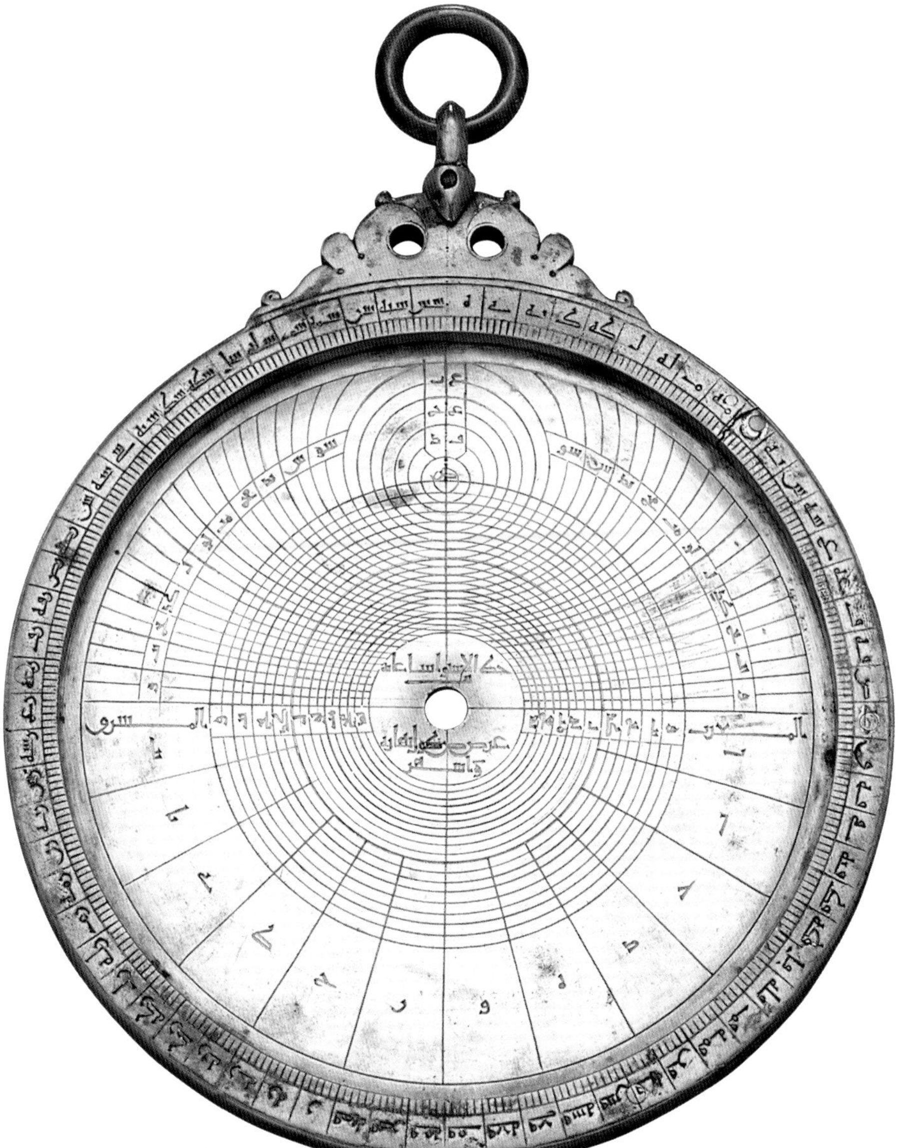

Mater of No. 10

Rete of No. 10

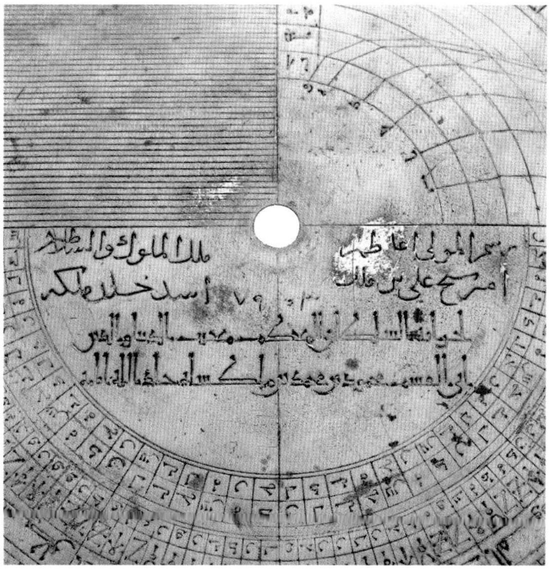

Detail of reverse of No. 10

■ TYMPANS

Tympan	*Latitude*	*Longest Daylight*
Ia = IIa.	22°	13;20 hours
Ib = IIb.	41;30°	15 hours
IIIa.	33°	14;13 hours
IIIb.	39°	14;53 hours

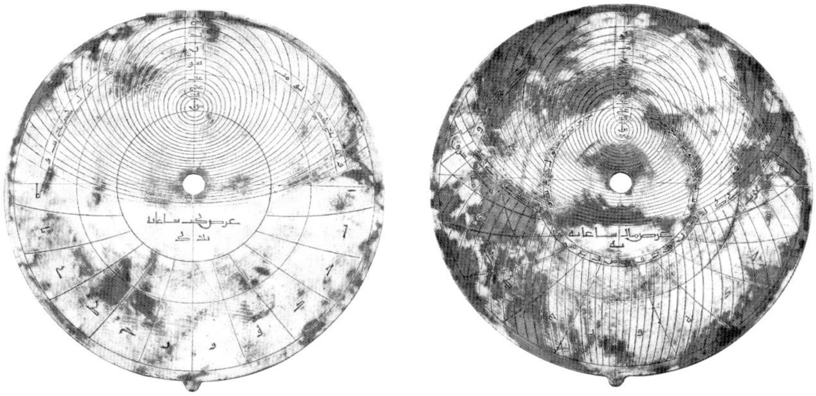

Tympan #1 of No. 10

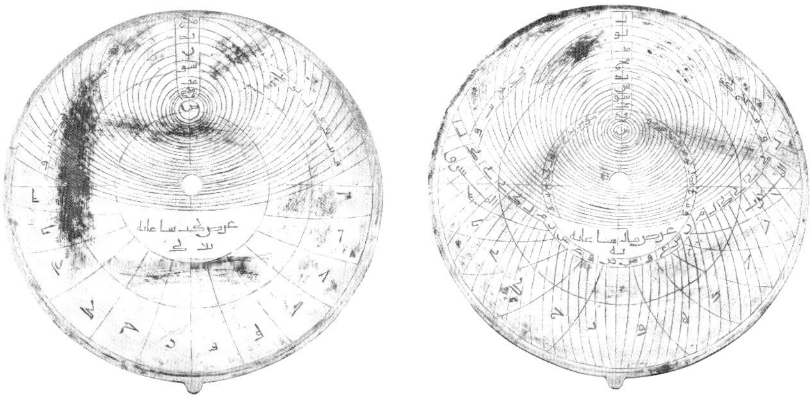

Tympan #2 of No. 10

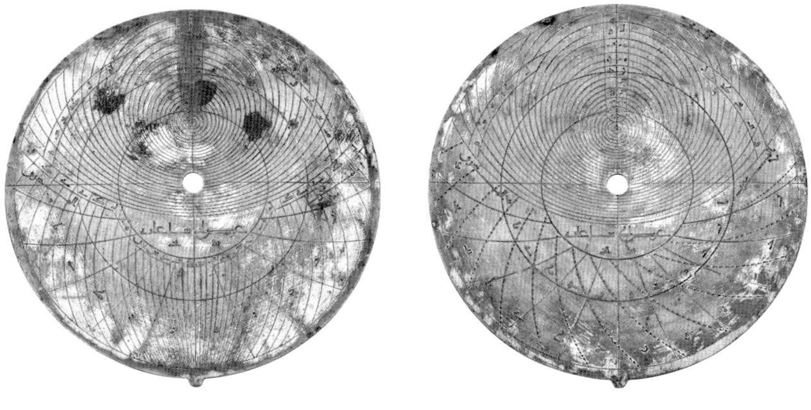

Tympan #3 of No. 10

11. Astrolabe

Ja'far ibn 'Umar ibn Dawlatshāh.
Iran
1388
"Ja'far ibn 'Umar ibn Dawlatshāh of Kirmān made it in the year 790 of the Hijra, 1699 of Alexander."
15.6 x 9.6 x 3.2 cm
Mater thickness 1 cm
Brass
N-70

Face of No. 11

Astrolabe made by Ja'far ibn 'Umar ibn Dawlatshāh of Kirmān in A.H. 790, Alexander 1699 (A.D. 1388), according to an inscription above and to the right of the table of the revolutions of the years on the back: *ṣana'ahu ja'far ibn 'umar ibn dawlatshāh al- kirmānī fī sanah dh.ṣ al-hijriyyah '. h. ṣ. ṭ. al-iskandariyyah.* Inscriptions in kūfic except that those on the rete and the tympans, which are later additions, are in naskh.

Consists of a body, graduated diopter, pin, washer, horse, handle, and ring; the rete and five tympans are later additions. The southern areas of the tympans have holes drilled through them to fit over a pillar erected in the body.

The rim of the face of the body is graduated for every degree; the numbers, running clockwise, are: 5, 10, 5, 20, 5, ...100, 5, 10, 5, ...200, 5, 10, 5, ...300, 5, 10, 5, ...60. The recess in the front has a geographical table in two circles, arranged in each separately according to increasing longitudes. (❍)(●)

The upper half of the back is graduated for every degree; the numbers, running clockwise from the left to the top and counter-clockwise from the right to the top, are 5, 10, 5, 20, 5, ...90. At the edge of the lower half are indicated the cotangents. In the upper left quadrant are the cosines for each of the sixty parts of the Radius. In the upper right quadrant is a table, inscribed upside-down, indicating the excess of revolution beyond 365 diurnal revolutions for 1 to 9 and 10 to 90 years; it is based on a tropical year containing 6, 5; 14, 32, 30 days. (❑)

In the lower part are eight semicircles indicating the names of the zodiacal signs; the lords of the terms and their extents according to Dorotheus; the Greek lords of the decans signified by the last letters of their names; the Indian lords of the decans signified by their symbols

Reverse of No. 11

(those for Mars and for Mercury are interchanged); the ends of the decans; and the diurnal and noctumal lords of the triplicities signified by the last letters of their names.

On the rete, which is a later addition, the ecliptic is not graduated. The names of the stars (not all pointers are inscribed) are written in nashk. (★)

None of the five tympans are original; the fourth was inscribed by a different hand and is made of a different metal. Those which are inscribed have altitude circles for every 6° and equal and unequal hour curves. The azimuth circles on tympan IV are for every 10°; those on the other tympans are too crudely executed to be determined. They indicate their latitudes and, sometimes, their longest daylights. The fifth tympan is not inscribed. (■)

PROVENANCE — Sotheby's, London, 1968; Kenneth Nebenzahl, Glencoe, IL, 1968; AP gift, 1986.

ICA 2605

EXHIBITIONS — University of British Columbia Museum of Anthropology, Vancouver, BC, *The Spirit of Islam*, Oct. 2001 – May 2002.

REFERENCES — Sotheby's (1968), p. 10, lot 23; Van Kleempoel (2005), p.215.

Mater of No. 11

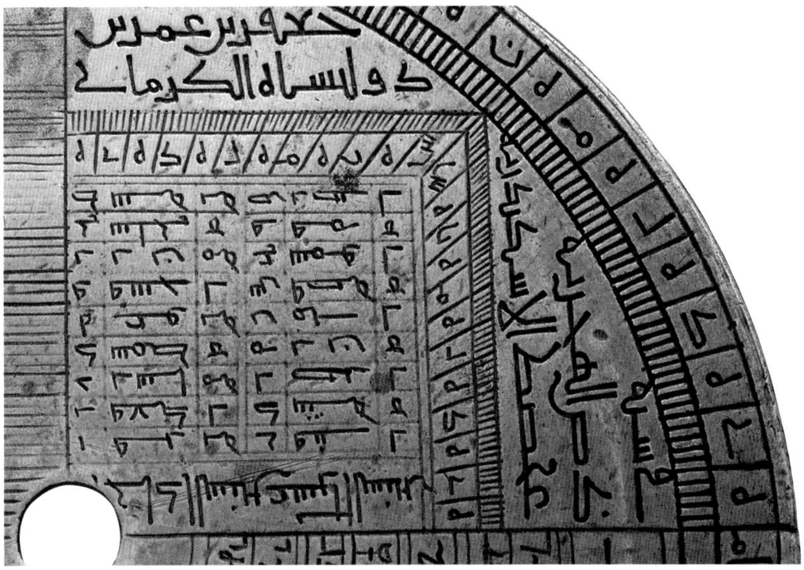

Detail of reverse of No. 11

EXCESS OF REVOLUTIONS

Year	Excess	Year	Excess
1	87;15°	10	152;30°
2	174;30	20	305;0°
3	261;45°	30	97;30
4	349;0°	40	250;0°
5	76;15°	50	42;30°
6	163;30°	60	195;0°
7	250;45°	70	347;30°
8	338;0°	80	140;0°
9	65;15	90	292;30°

OUTER CIRCLE

City	Longitude	Latitude	Inclination	Identification
baghdād	80;0°	33;25°	14;0°	Baghdād
marāghah	82;0°	37;20°	15;50°	Marāgheh
tabrīz	82;0°	38;0°	15;30°	Tabrīz
sulṭāniyyah	83;30°	36;36°	26;40°	Solṭānīyeh
hamadān	84;0°	35;10°	25;30°	Hamadan
qazwīn	85;0°	36;0°	24;30°	Qazvīn
sāwah	85;0°	35;40°	29;19°	Sāveh
qumm	85;40°	34;45°	32;13°	Qom
rayy	86;24°	35;35°	31;33°	Shahr Rey
kāshān	86;40°	34;0°	36;18° (?)	Kāshān
iṣfahān	86;40°	33;25°	40;23°	Eṣfahān
shīrāz	88;0°	29;30°	51;23°	Shīrāz
abarqūh	88;10°	31;30°	45;0°	Abarqū
yazd	89;0°	31;33°	47;57° (?)	Yazd
shabānkārah	89;0°	28;28°	33;28°	
kirmān	92;30°	30;5°	63;40°	Kerman
hurmūz	92;0°	25;0°	78;45°	Hormoz
ṭabas	92;0°	33;0°	52;0°	Ṭabas
ṭūs	92;30°	36;0°	45;45°	Ṭūs
nīshābūr	92;30°	36;25°	45;3°	Neyshābūr
harāh	94;4°	34;30°	33;50°	Herat
marw	97;0°	37;40°	50;45°	Mary
baṣrah	84;0°	30;0°	38;24° (?)	Basra
tustar	84;30°	31;30°	35;9°	Shustar

INNER CIRCLE

City	Longitude	Latitude	Inclination	Identification
miṣr	63;0°	30;20°	60;0°	Cairo
qūnya	65;0°	38;0°	43°	Konya
buyt al muqadda	66;30°	31;50°	29°	Jerusalem
dimashq	70;0°	33;0°	30;0°	Damascus
sūs	65;30°	22;0°	79°	Tarodannt
barbarah	65;0°	14;0°	39°	Berbara
ḥabashah	65;0°	9;30°	49°	(Ethiopia)
ʻadan	76;0°	11;0°	45°	Aden
sawmnāt	106;40°	10;31°	70°	Somnath
saradīb	130°	17;0°	66°	(Śrīlañkā)
jīn	130°	22°	75°	(China)
banjū	135;0°	21;0°	76°	(in China)
samarqand	98°	40;0°	49°	Samarkand
khujand	100;35°	41;16°	45°	Leninabad
balkh	101;0°	36;41°	54°	Wazirabad
kāshghar	106;30°	44;0°	54;30°	Kashgar

★ STAR LIST

Name	*Identification*
a'zal	α Virg.
janāḥ (?)	γ Corv.
dhanab qayṭus	ι Ceti
ra's al-ḥawwā'	α Oph.
rāmiḥ	α Boot.
rijl	β Ori.
sha'āmiyyah	α Can. Min.
ṭā'ir	α Aquil.
'ayn	α Tau.
'ayyūq	α Aur.
qā'id	α Crat.
fard	α Hydr.
kaff	β Cass.
nayyir	α Cor. Bor.
wāqi'	α Lyr.
yad	α Ori.

Rete of No. 11

Diopter, washer, pin, and horse of No. 11

TYMPANS

Tympan	*Latitude*	*Longest Daylight*
Ia.	22°	
Ib	33°	14;12 hours
IIa.	32°	
IIb.	34°	
IIIa.	66°	
IIIb.	<all horizons>	
IVa	30°	13;57 hours
IVb	33°	14;12 hours

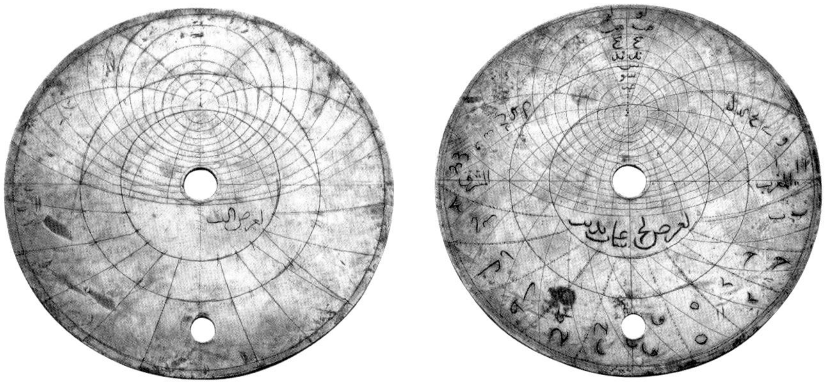

Tympan #1 of No. 11

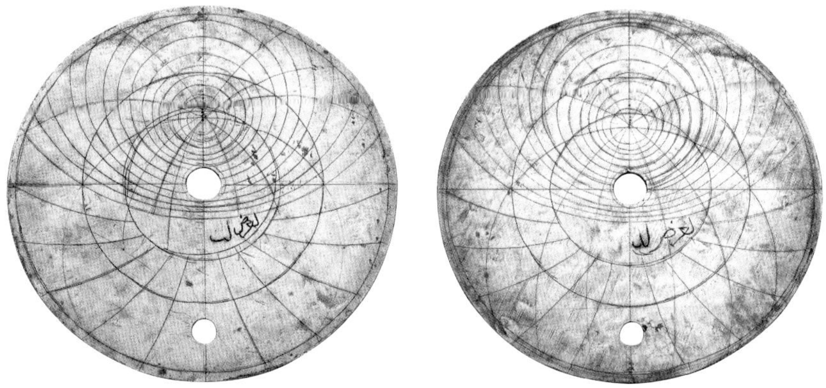

Tympan #2 of No. 11

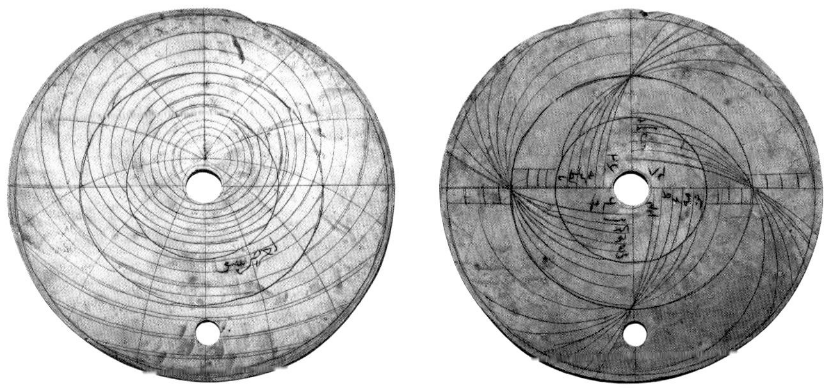

Tympan #3 of No. 11

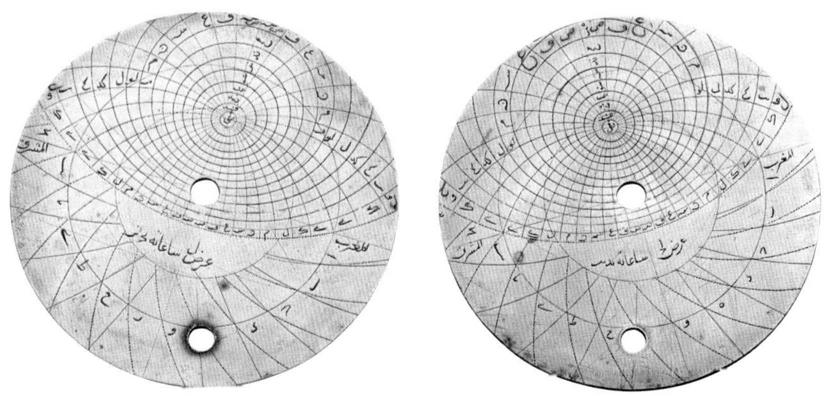

Tympan #4 of No. 11

12 Astrolabe

Maḥmūd ibn Jalāl ibn Jaʿfar.
Iran.
1433/4.
"Maḥmūd ibn Jalāl ibn Jaʿfar the astrolabe-maker made it"
"in the year 837 of the Hijra and 803 of Yazdijird."
13.5 x 8.9 x 1.1 cm
Mater thickness 0.5 cm
Brass
A-91

Face of No. 12

Astrolabe made by Maḥmūd ibn Jalāl ibn Jaʿfar al-asṭurlābī according to the inscription below the shadow-square on the back: *ṣanaʿahu maḥmū<d ib>n jalāl ibn jaʿfar al-asṭurlābī* (the letters in pointed brackets are covered by brass). To the left and right of the shadow table is written an inscription indicating that it was made in A.H. 837, A.Y. 803 (A.D. 1433/4): *fī sanah ḍ. l. z. al-hijriyyah wa ḍ. j. al-yazdijirdiyyah*. Inscription in kūfic.

Consists of a body, handle, and ring. The rete, inscribed and graduated diopter without sights, and three tympans are all fakes added later. The tympans have holes towards the south fitting over a brass pillar attached later as in No. 11. The horse is missing, being replaced by a bent bar.

The rim of the front of the body is graduated for every degree; the numbers, running clockwise, are 10, 20, 30, ...100, 10, 20, ...200, 10, 20, ...300, 10, 20, ...60. The recess in the front contains a geographical table in two circles. It is arranged by increasing longitudes except for the middle of the inner circle. (◯)(●)

The upper half of the back is graduated for every degree; the numbers, running clockwise from the left to the top and counter-clockwise from the right to the top, are 5, 10, 5, ...90. At the edge of the lower half are indicated the cotangents. In the upper left quadrant are the cosines for every 5° from 5° to 80°. In the upper right quadrant are altitude curves of the Sun for 1, 2, 3, 4, 5, and 6 hours for every latitude; at the edge is the inscription: *khuṭūṭ sāʿāt al-zamāniyyah li-kull ʿarḍ* ("lines of seasonal hours for every latitude"). In the lower part of the back is a shadow-square for 7 feet and 12 digits.

The rete is a fake; it is symmetrical, and has no star-names.

The three tympans are also fake. They have identical projections and meaningless inscriptions.

Reverse of No. 12

PROVENANCE — E. S. David, New York, 1943; AP purchase, 1943
ICA 2563
REFERENCES — Newbury (2004a).

Mater of No. 12

OUTER CIRCLE

City	*Longitude*	*Latitude*	*Inclination*	*Identification*
baghdād	80;0°	33;25°	13;0°	Baghdād
tabrīz	82;0°	38;0°	15;30°	Tabrīz
sulṭāniyyah	83;30°	36;36°	26;40°	Solṭānīyeh
qazwīn	85;0°	36;0°	27;30°	Qazvīn
qumm	85;40°	35;40°	32;43°	Qom
kāshān	86;0°	34;0°	36;8°	Kāshān
iṣfahān	86;40°	32;25°	40;23°	Eṣfahān
shīrāz	88;0°	29;30°	51;23°	Shīrāz
abarqūh	88;10°	30;0°	45;0°	Abarqū
yazd	89;0°	31;33°	47;17°	Yazd
hurmūz	92;0°	25;0°	78;45°	Hormoz
kirmān	92;30°	30;0°	63;2°	Kermān
ṭūs	92;30°	36;0°	45;45°	Ṭūs
nīshābūr	92;30°	36;21°	45;3°	Neyshābūr
harāh	94;4°	34;30°	53;50°	Herat
marw	97;0°	37;40°	50;45°	Mary

INNER CIRCLE

City	*Longitude*	*Latitude*	*Inclination*	*Identification*
bukhārā	97;30°	39;20°	47;32°	Bukhara
samarqand	98;0°	40;0°	52;0°	Samarkand
khujand	100;35°	41;56°	45;0°	Leninabad
miṣr	63;0°	30;20°	60;24°	Cairo
dimashq	70;0°	33;0°	30;17°	Damascus
madīnah	75;20°	25;0°	32;0°	Medina
ḥabashah	65;0°	9;0°	44;40°	(Ethiopia)
barbarah	65;0°	14;0°	39;10°	Berbara
ʿadan	76;0°	11;0°	5;50°	Aden
sawmnāt	106;0°	17;1°	66°	Somnath
sarandīb	130;0°	10;0°	70;0°	(Śrīlaṅkā)
jīn	135;0°	21;30°	76;15°	(China)

13. Astrolabe

‘Īsā ibn Allāhdād
Lahore
June 1601
“The weakest of the servants (of God), ‘Īsā the son of Allāhdād, the astrolabe-maker, of Lahore, the Humāyūnī, date: month Dhū al-ḥijjah
“year: 1009”
24.9 x 17 x 5.4 cm
Mater thickness 1.1 cm
Brass
N-69

Face of No. 13

Astrolabe made by ‘Īsā ibn Allāhdād in Dhū al-ḥijjah of A.H. 1009 (June 1601) according to the inscription written below the shadow-square on the back: *ṣana‘ahu aḍa‘af al-‘ibād ‘īsā wald allāhdād asṭurlābī lāhūrī hamāyūnī tā’rīkh shahr dhī al-ḥijjah.* Above this is written : sanah 1009. Inscriptions in naskh.

Consists of body, rete, uninscribed diopter missing one sight, handle, ring, modern screw, washer, and nut, and three genuine with two extraneous tympans which have notches at the bottom to fit about a corresponding tab in the body.

The rim of the face of the body is graduated for every degree; the numbers, running clockwise, are 5, 10, 15, ...360. The recess in the front has a geographical table in two circles, arranged more or less by region. (❍)(●)

The upper half of the back is graduated for every 0;30°; the numbers, running clockwise from the left to the top and counter-clockwise from the right to the top, are 5, 10, 15, ...90. At the edge of the lower half are indicated the cotangents. In the upper left quadrant are the cosines for each of the 60 parts of the Radius. In the upper right quadrant are the altitude curves of the Sun for 1, 2, 3, 4, 5, and 6 hours, and the curve of the noon altitude of the Sun for a latitude of 31;50° (i.e, for Lahore) inscribed: *khaṭṭ zawāl ‘arḍ l.a.n.* (“line of noon; latitude 31;50”). In the lower part is the shadow-square for 7 feet and 12 digits; below this are two semicircles indicating the names of the zodiacal signs and of the lunar mansions.

On the rete the ecliptic is graduated for every 2°; the numbers in each zodiacal sign are 6, 12, ...30. The names of the stars are very badly preserved; some I have been unable to read or to identify. (★)

The three genuine tympans have altitude circles for every 3°, equal and

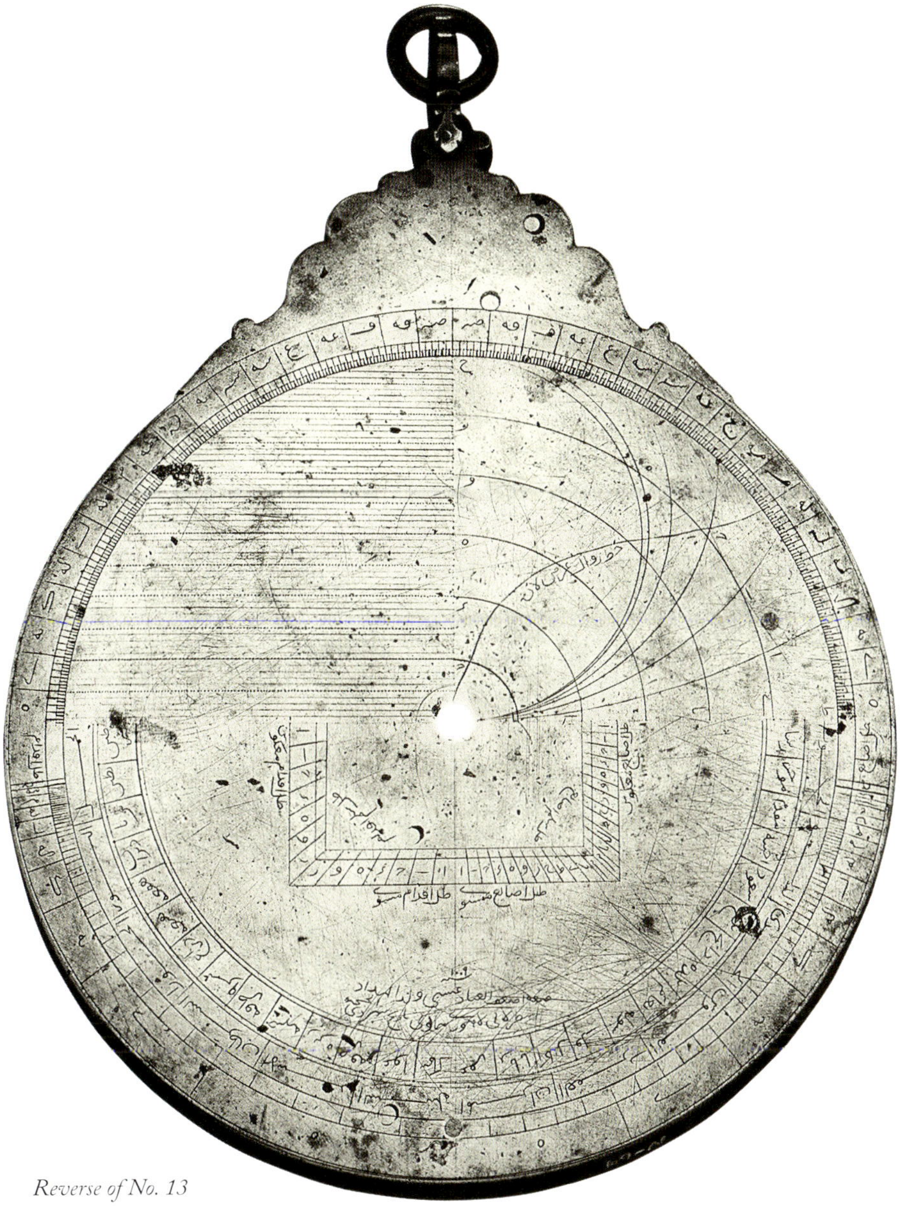

Reverse of No. 13

unequal hour curves, and no azimuths. The first two indicate their latitudes and longest daylights. (■)

One side of the third tympan has horizons for latitudes 9°,12°, 15°, 18°, 21°, 24°, 27°, 30°, 33°, 36°, 39°, 42°, 45°, 48°, 51°, 54°, 57°, 60°, 63°, 66°, 69°, 72°, 75°, 78°, and 81°. The other side has a projection for latitude 66°; it is labeled: *ṣafīḥat mīzān al-ʿankabūt* ("tympan for balancing the rete"). The two extraneous tympans are uninscribed.

PROVENANCE — Sotheby's, London, 1968, Kenneth Nebenzahl, Glencoe, IL, 1968; AP gift, 1986.

REFERENCES — Gunther (1932), no. 68, p. 187; Newbury (2006); Sarma (1994); Sotheby's (1968), p. 10, lot 22.

Mater of No. 13

OUTER CIRCLE

City	*Longitude*	*Latitude*	*Identification*
makkah mubārak	77;10°	21;40°	Mecca
madīnat rusūl	75;20°	25;0°	Medina
iskandariyyah	61;54°	30;58°	Alexandria
miṣr	63;20°	30;20°	Cairo
yaman	77;0°	14;30°	Ṣan'ā'
ṭā'if	77;30°	21;20°	Tā'if
bayt al-muqaddas	66;30°	31;50°	Jerusalem
dimashq	70;0°	33;15°	Damascus
anṭākiyyah	71;26°	35;30°	Antakya
ḥalab	72;10°	35;50°	Aleppo
qusṭanṭiyyah	59;50°	45;0°	Istanbul
anīth madīnat al-ḥukamā'	65;40°	37;20°	Athinai
mawṣil	77;0°	34;30°	Mosul
ardabīl	82;30°	38;0°	Ardabīl
marāghah	82;0°	37;33°	Marāgheh
tabrīz	82;0°	38;0°	Tabrīz
bulghār	90;0°	49;30°	(Bulgars)
kūfah	79;30°	31;30°	Kūfah
baṣrah	84;0°	30;0°	Basra
baghdād	80;30°	33;25°	Baghdād
'askar mukram	54;30°	31;15°	Band Qir
kāzirūn	87;0°	29;55°	Kāzerūn
sābūr	86;55°	30;0°	Shāpūr
shīrāz	88;0°	29;36°	Shīrāz
yazd	89;0°	32;30°	Yazd
hamadān	83;0°	35;10°	Hamadan
qazwīn	85;0°	36;55°	Qazvīn
iṣfahān	86;40°	32;25°	Eṣfahān
kāshān	86;0°	34;0°	Kāshān
rayy	86;20°	35;0°	Shahr Rey
ṭāliqān	85;45°	36;10°	Taliq-an
astarābād	89;30°	36;50°	Gorgān
jurjān	92;0°	36;50°	Jurjān
basṭām	89;30°	36;10°	Bastam
sabzawar	91;30°	36;5°	Sabzevār
nīshābūr	92;30°	36;21°	Neyshābūr
ṭūs	94;30°	37;0°	Ṭūs
harāt	94;33°	34;30°	Herat
sarakhs	94;30°	37;0°	Sarakhs
fāryāb	99;0°	36;45°	Doulatabad
balkh	101;0°	36;41°	Wazirabad
badakhshān	104;24°	37;10°	Faizabad
khuwārizm	94;30°	42;17°	Kunyaurgench
bukhārā	97;30°	39;50°	Bukhara
samarqand	99;16°	37;37°	Samarkand
khujand	105;30°	41;55°	Leninabad
kāshghar	106;30°	44;0°	Kashgar
hurmuz	92;0°	25;0°	Hormoz
kābul	104;40°	34;30°	Kabul
qaṣr sind mansūrah	105;0°	27;40°	Mansūra
sawmnāt	106;0°	17;0°	Somnath
mūltān	107;35°	29;40°	Multan
qandahār	106;40°	33;0°	Kandahar
lahāwar	109;20°	31;50°	Lahore
qanawj	115;50°	26;35°	Kannauj

INNER CIRCLE

City	*Longitude*	*Latitude*	*Identification*
bānārasī	117;20°	26;50°	Varanasi
sarāndīb	130;0°	10;0°	(Śrīlaṅkā)
tibat	110;0°	40;0°	(Tibet)
kashmīr	103;0°	35;0°	Srinagar
dihlī	114;48°	28;17°	Delhi
akrah	114°	27;13°	Agra
kwāliyār	115°	26;29°	Gwalior
ajmīr	111;5°	24;0°	Ajmer
sulṭān kūt	109;4°	33;0°	
kawil	114;19°	28;4°	Aligarh
sunnām	110;25°	30;30°	Sunam
pānī pat	113;20°	28;52°	Panipat
tānīsar	112;35°	30;10°	Thanesar
budāūn	114;39°	27;32°	Budaun
sanbhal	115;0°	28;1°	Sambhal
kurah	117;6°	26;36°	Manikpur
aḥmadābād	108;40°	23;15°	Ahmedabad
chānpānīr	108;45°	22;30°	Chāmpāner
kanbāyat	109;4°	22;20°	Cambay
bījābūr	105;30°	17;20°	Bijapur
ujayn	102;0°	22;30°	Ujjain
amarkūt	105°	25°	Umarkot
jawnpūr	119;6°	26;36°	Jawnpur
mānikpūr	118;6°	26;49°	Manikpur
hānsī	102;25°	29;15°	Hansi

★ STAR LIST

Name	*Identification*
baṭn al-ḥūt	β Andr.
tālī masāf al-nahr	θ Erid.
janāḥ al-ghurāb	γ Corvi
janāḥ al-faras	γ Peg.
.... al-janūbī	
dhanab al-jady	δ Cap.
dhanab al-dulfīn	ε Delph.
dhanab qayṭus	β Ceti
ra's al-ḥawwā'	α Oph.
ra's al-ghūl	β Pers.
rijl al-jawzā' yumnā	κ Ori.
rijl al-yusrā	β Ori.
rukbat al-ḥawwā'	ζ Oph.
zubānā (?) al-asad	ε Leo.
sāq sākib al-mā'	δ Aquar.
surrat al-faras	δ Peg.
simāk rāmiḥ	α Boot.
simāk a'zal	α Virg.
shi'rā yamāniyyah	α Can. Mai.
shi'rā sha'āmiyyah	α Can. Min.
ṭaraf al-sufīnah	ρ Puppis
ẓahr al-asad	δ Leo.
ẓahr al-dubb	α Urs. Mai.
'iṣābat al-rāmī	ρ (?) Sag.
'ayn al-tinnīn	α Drac.
'ayn al-thawr	α Tau.
'ayyūq	α Aur.
fard al-shujā'	α Hydr.
.....al-furūdayn	ζ (?) Can. Mai.
fam al-faras	ε Peg.
fam al-qayṭus	γ Ceti
qā'idat al-bāṭiyyah	α Crat.
qalb al-asad	α Leo.
qalb al-'aqrab	α Scorp.
kaff al-khaḍīb	β Cass.
kiffah janūbī	α Lib.
kiffah shamālī	β Lib.
mu'akhkhar (?) shujā'	σ (?)Hydr.
al-mutaqaddim	ε Virg.
minqār al-dajājah	β Cyg.
mankib al-faras	β Peg.
muqaddam	
nasr ṭā'ir	α Aquil.
al-nasr al-wāqi'	α Lyr.
nayyir al-fakkah	α Cor. Bor.
<yad> al-jawzā' yusrā	γ Ori.
yad al-jawzā' yumnā	α Ori.
yad al-ḥawwā'	δ Oph.

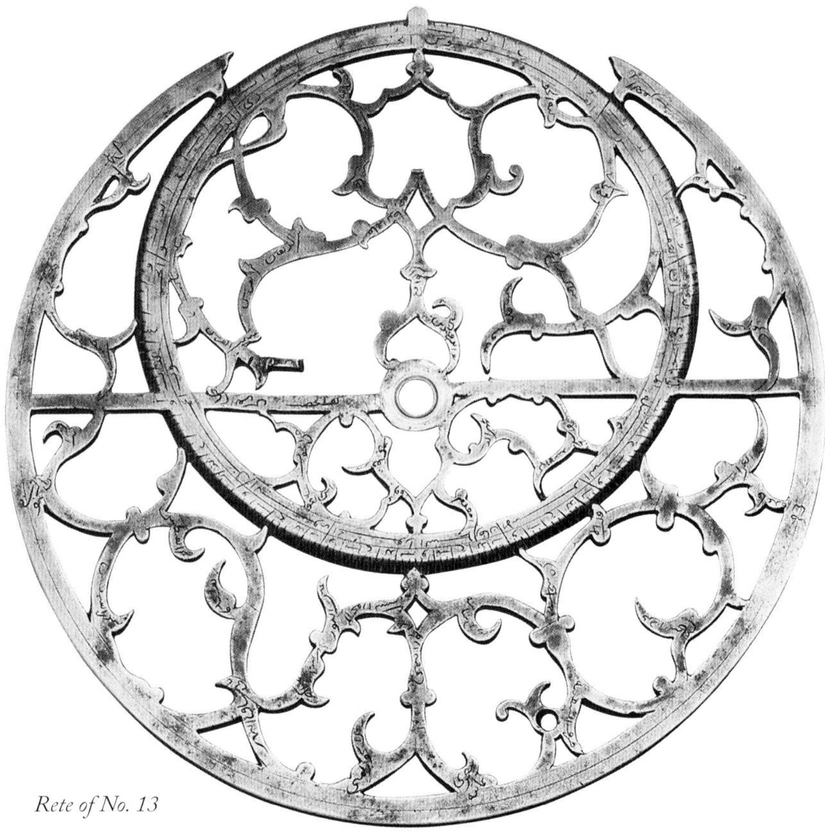

Rete of No. 13

TYMPANS

Tympan	*Latitude*	*Longest Daylight*
Ia.	22°	13;22 hours
Ib.	25°	13;25 hours
IIa.	27°	13;46 hours
IIb.	32°	14;8 hours

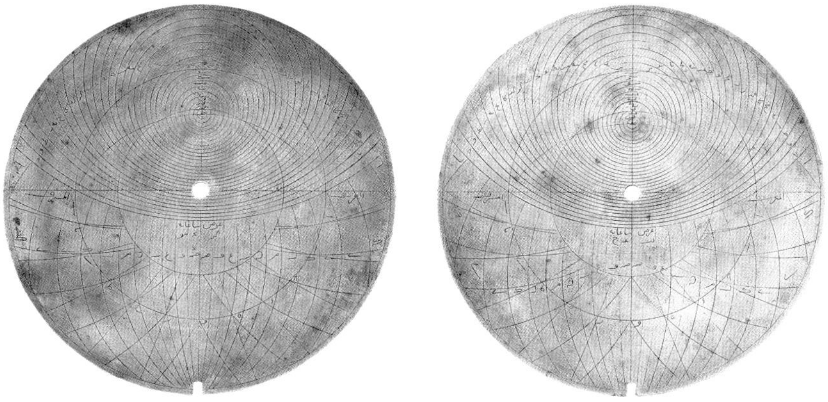

Tympan #1 of No. 13

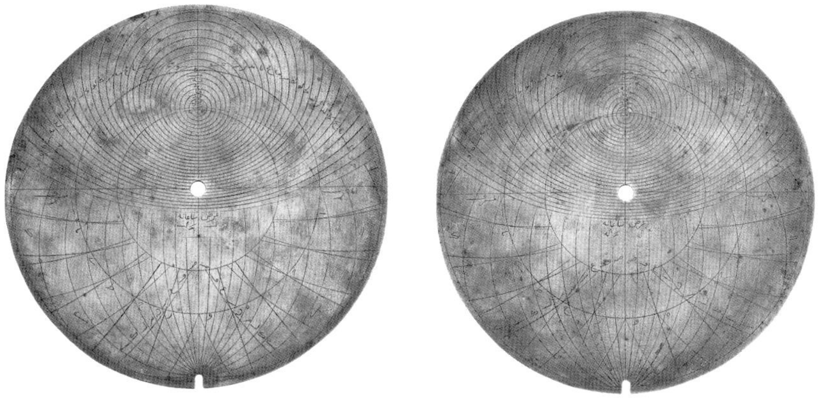

Tympan #2 of No. 13

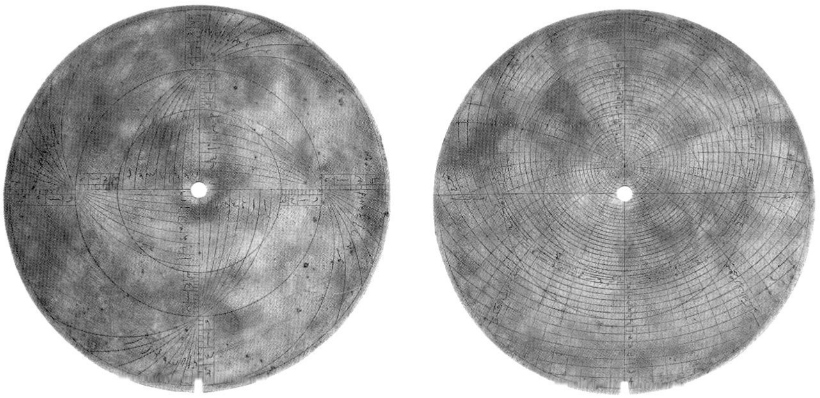

Tympan #3 of No. 13

14. Astrolabe

ʿĪsā ibn Allāhdād
Lahore
1604/5
"The weakest of the servants (of God), ʿĪsā the son of Allāhdād, made it in the year 1013."
18.5 x 12 x 4.2 cm
Mater thickness 1 cm
Brass
A-79

Face of No. 14

Astrolabe made by ʿĪsā ibn Allāhdād in A.H. 1013 (A.D. 1604/5) according to the inscription written upside down in the shadow-square on the back: *ṣanaʿahu aḍaʿaf al-ʿibād ʿīsā wald allāhdād sanah 1013*. Inscriptions in naskh.

Consists of a body, rete (later addition), uninscribed diopter with sights, pin, horse, handle, ring, and three genuine tympans which have notches at the bottom fitting about a corresponding tab in the body. There are also five extraneous tympans which also have notches at the bottom.

The rim of the face of the body is graduated for every degree; the numbers, running clockwise, are 5, 10, 15, ...360. The recess in the front has a geographical table in two circles, arranged more or less by region. (❍)(●)

The upper half of the back is graduated for every degree; the numbers, running clockwise from the left to the top and counter-clockwise from the right to the top, are 5, 10, 15, ...90. At the edge of the lower half are indicated the cotangents. In the upper left quadrant are the cosines for each of the 60 parts of the Radius. In the upper right quadrant are the altitude curves of the Sun for 1, 2, 3, 4, 5, and 6 hours, and the curve of the noon altitude of the Sun for a latitude of 31;50° (i.e., for Lahore) inscribed: *khaṭṭ niṣf al-nahār ʿarḍ l.a.n.* ("line of noon; latitude 31;50"). In the lower part is the shadow-square for 7 feet and 12 digits; below this are two semicircles indicating the names of the zodiacal signs and of the lunar mansions.

On the rete, which is a later addition, the ecliptic is graduated for every 6°; there are no numbers. The names of the stars (one pointer — probably for β Pers. — has no inscription) are listed (★).

The three genuine tympans have altitude circles for every 3°, equal and

Reverse of No. 14

unequal hour curves, and no azimuths. They indicate their latitudes and longest daylights. (■)

The other five tympans include three which have altitude circles for every 6°, azimuths for every 10° (except that there are no azimuths on VIa or VIb), and equal and unequal hour curves (except for Vb). They also indicate their altitudes and longest daylights. (■)

VIIa has horizons for latitudes 4°, 12°, 20°, 28°, 36°, 44°, 52°, and 60°; for 10°, 18°, 26°, 34°, 42°, 50°, 58°, and 66°; for 8°, 16°, 24°, 32°, 40°, 48°, 56°, and 64°; and for 6°, 14°, 22°, 30°, 38°, 46°, 54°, and 62°. VIIb has horizons for latitudes 3°, 11°, 19°, 27°, 35°, 43°, 51°, and 59°; for 9°, 17°, 25°, 33°, 41°, 49°, 57°, and 65°; for 7°, 15°, 23°, 31°, 39°, 47°, 55°, and 63°; and for 5°, 13°, 21°, 29°, 37°, 45°, 53°, and 61°.

The last tympan bears on one side a projection for a latitude of 90°; it has altitude circles for every 6° and azimuths for every 10°. On the other side is a combined projection, resulting in S-shaped curves, for the latitudes 72° and 18°; this is similar to a projection on a tympan belonging to an astrolabe in the Benaki Museum in Athens (Inv. No. 13178).

PROVENANCE — Purchased by E.S. David in Paris, 1935, AP purchase, 1935.

ICA 1096

REFERENCES — Abbot (1937); Gibbs and Saliba (1984), p. 225, n. 52; Nadvi (1937) ; Newbury (2004a); Newbury (2006); Sarma (1994).

Mater of No. 14

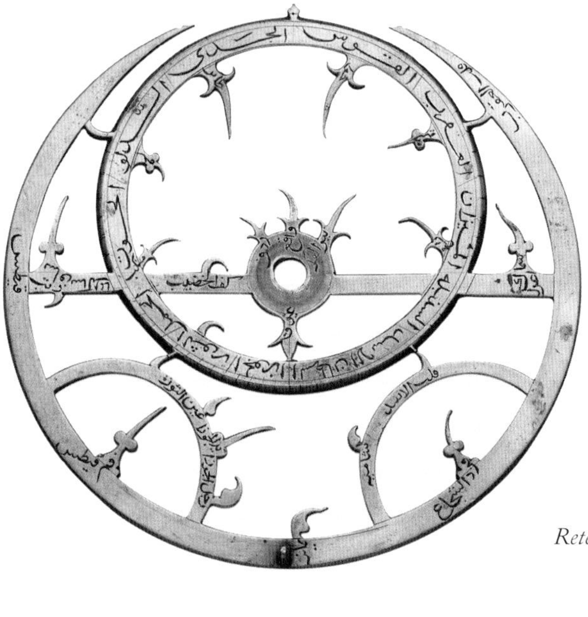

Rete of No. 14

★ STAR LIST

Name	*Identification*
ḥawwā'	α Oph.
ḥayyah	α Serp.
dhanab qayṭus	ι Ceti
rāmiḥ	α Boot.
rijl al-jawzā'	β Ori.
ridf	α Cygn.
sāq sākib al-mā'	δ Aquar.
simāk a'zal	α Virg.
sha'āmiyyah	α Can. Min.
ṭā'ir	α Aquil.
'ayn al-thawr	α Tau.
'ayyūq	α Aur.
fard al-shujā'	α Hydr.
fam	ε Peg.
fam qayṭus	γ Ceti
qalb al-asad	α Leo.
qalb al-'aqrab	α Scorp.
kaff al-khaḍīb	β Cass.
mankib	β Peg.
nayyir al-fakkah	α Cor. Bor.
wāqi'	α Lyr.
yad al-hawwā'	α Ori.
yamāniyyah	α Can. Mai.

OUTER CIRCLE

City	*Longitude*	*Latitude*	*Identification*
makkah mubārak	77;10°	21;40°	Mecca
madīnat rusūl	75;20°	25;0°	Medina
ṭā'if	77;30°	21;20°	Tā'if
tihāmah	82;30°	23;0°	Yamāma
hajar baḥrayn	83;0°	25;15°	Manāmah
dimashq	70;0°	33;15°	Damascus
ba'albak	70;15°	33;15°	Ba'albek
ḥimṣ	70;15°	34;0°	Homs
ḥalab	72;10°	35;50°	Aleppo
ṭarsūs	68;40°	36;42°	Tarsus
rūmiyya kubrā	45;27°	41;50°	Roma
madīnat al-ḥukamā'	65;40°	37;20°	Athinai
qusṭanṭiyyah	59;50°	45;0°	Istanbul
mawṣil	77;0°	34;30°	Mosul
nakhjawān	81;15°	38;0°	Nakhichevan
marāghah	82;0°	37;20°	Marāgheh
tabrīz	82;0°	38;0°	Tabrīz
ardabīl	82;30°	38;0°	Ardabīl
mūqān	83;0°	38;0°	
bāb al-abwāb	85;0°	43;0°	Darbent
bulghār	90;0°	59;30°	(Bulgars)
kūfah	79;30°	31;30°	Kūfah
madā'in	72;0°	33;10°	Ctesiphon
baghdād	80;30°	33;25°	Baghdād
wāsiṭ	81;30°	30;20°	Wasit
baṣrah	84;0°	30;0°	Basra
'abbādān	84;30°	30;0°	Abādān
kāzirūn	86;0°	39;15°	Kāzerūn
shīrāz	88;0°	29;36°	Shīrāz
iṣṭakhr	88;30°	30;0°	Takhṭ-e Jamshīd
yazd	89;0°	32;30°	Yazd
shahrūr	82;20	32;30	Shahrazūr
nihāwand māh al-baṣrah	83;15°	34;30°	Nahāvand
zankān	83;40°	36;30°	Zanjan
hamadān	83;0°	35;10°	Hamadan
sāwah	85;0°	35;0°	Sāveh
karaj	84;45°	30;0°	Karaj
qazwīn	85;0°	36;55°	Qazvīn
iṣfahān	86;40°	32;25°	Eṣfahan
kāshān	86;0°	34;0°	Kāshān
qumm	86;40°	34;45°	Qom
rayy	86;20°	35;0°	Shahr Rey
ṭāliqān	85;41°	36;10°	Taliq-an
hawsam az jīlān	85;10°	37;0°	
astarābād	89;30°	36;50°	Gorgān
jurjān	90;0°	37;50°	Jurjān
simnān	88;0°	36;0°	Semnān

INNER CIRCLE

City	*Longitude*	*Latitude*	*Identification*
basṭām	81;30°	36;10°	Bastam
sabzawār	91;30°	36;5°	Sabzevār
nīshābūr	92;30°	36;21°	Neyshābūr
ṭūs	94;30°	37;0°	Ṭūs
qāyin	93;13°	33;40°	Qāyen
zūzan	93;30°	35;20°	Zūzan
harān	94;13°	34;30°	Herat
sarakhs	94;30°	36;0°	Sarakhs
marw	97;0°	37°	Mary
balkh	101;0°	36;41°	Wazirabad
badakhshān	104;24°	37;10°	Faizabad
bukhārā	97;30°	39;50°	Bukhara
samarqand	99;16°	39;37°	Samarkand
kābul	104;40°	34;30°	Kabul
kāshghar	106;30°	44;0°	Kashgar
khutan	107;0°	42;0°	Khotan
hurmūz	92;0°	25;0°	Hormoz
khabīṣ	93;0°	31;0°	Shadab
daybul hiyya tāttah	102;7°	25;10°	Tatta
multān	107;35°	29;40°	Multan
qandahār	106;40°	33;0°	Kandahar
lahāwar	109;20°	31;50°	Lahore
kashmīr	108;0°	35;0°	Srinagar

TYMPANS

Tympan	*Latitude*	*Longest Daylight*
Ia.	28°	13 hours
Ib.	29°	13 hours
IIa.	31°	14 hours
IIb.	33°	14 hours
IIIa.	34°	14 hours
IIIb.	35°	14 hours
IVa.	27°	13;27 hours
IVb.	32°	14;18 hours

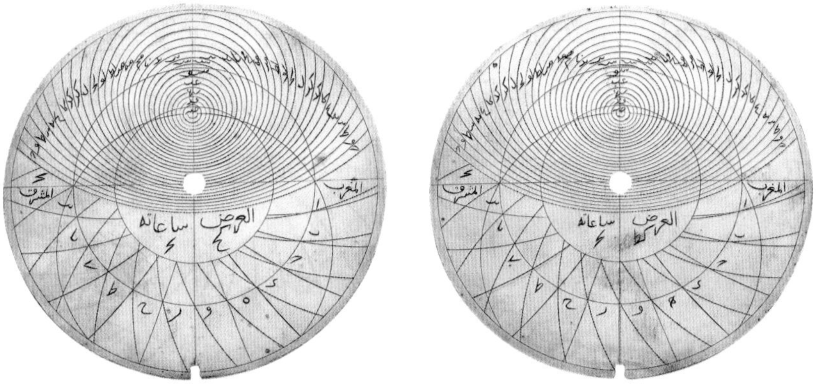

Tympan #1 of No. 14

Tympan #2 of No. 14

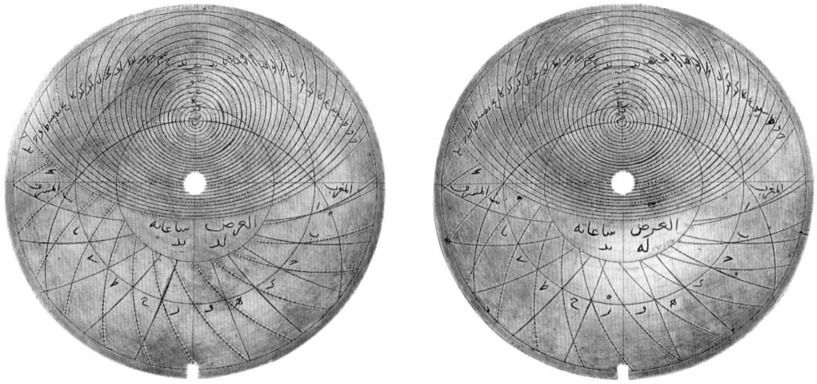

Tympan #3 of No. 14

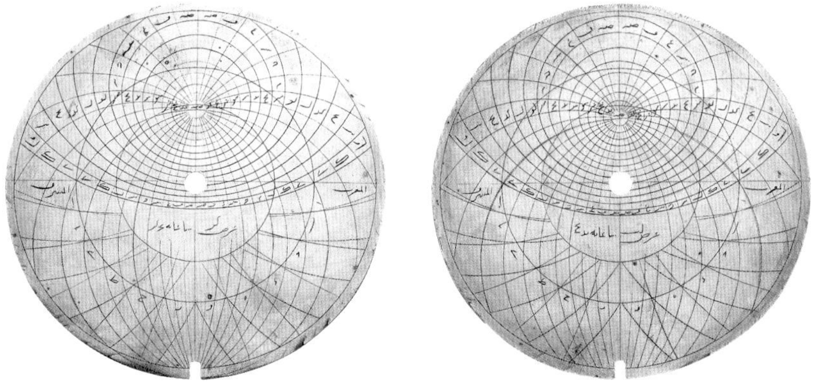

Tympan #4 of No. 14

TYMPANS

Tympan	Latitude	Longest Daylight
Va.	36°	14;30 hours
Vb.	66°	not recorded
VIa.	40°	14;54 hours
VIb.	45°	15;30 hours.

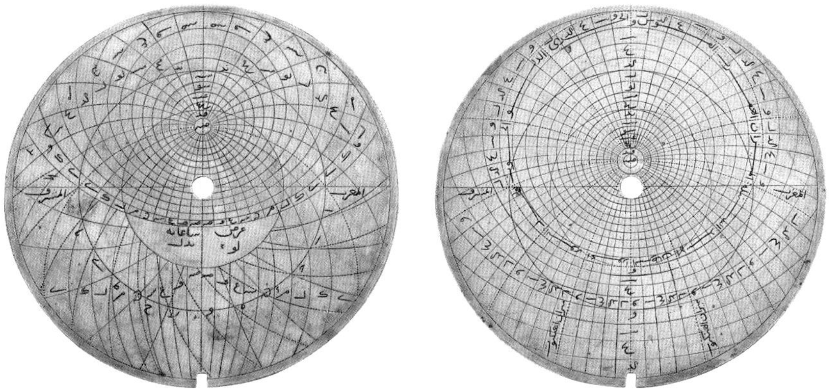

Tympan #5 of No. 14

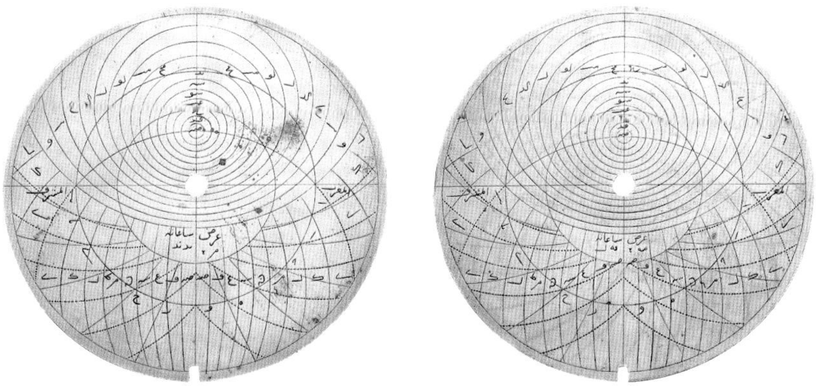

Tympan #6 of No. 14

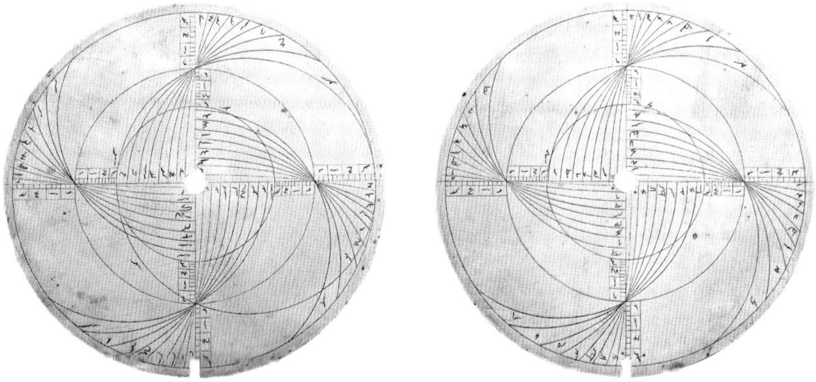

Tympan #7 of No. 14

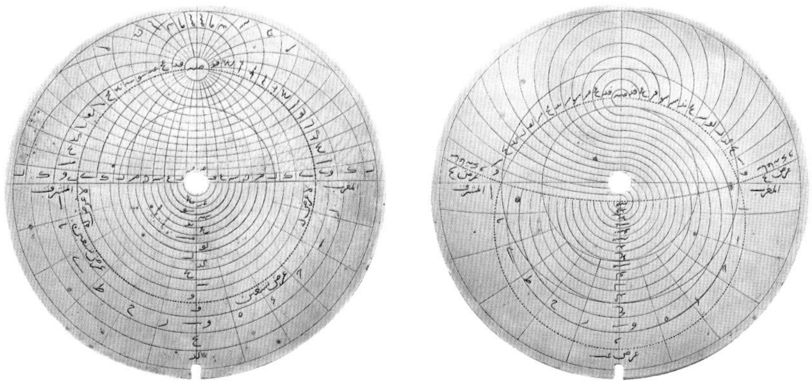

Tympan #8 of No. 14

15. Astrolabe

Ḍiyā' al-Dīn Muḥammad
Lahore
1637/8
"The work of Ḍiyā' al-Dīn Muḥammad ibn Qā'im Muḥammad ibn Mullā 'Īsā ibn Mullā Allāhdād, the astrolabe-maker, the Humāyūnī, the resident of Lahore, in the year 1047 of the Hijra."
17.1 x 11.7 x 4.2 cm
Mater thickness 0.8 cm
Body and rete of brass; diopter, pin and horse of a darker metal; tympans of copper (engraver apparently the same)
A-86

Face of No. 15

An astrolabe made by Ḍiyā' al-Dīn Muḥammad, the son of Qā'im Muḥammad, the son of 'Īsā, (the maker of Nos. 13 and 14), the son of Allāhdād, of Lahore in A.H. 1047 (A.D. 1637/8) according to the inscription written upside down below the center on the back: *'amal ḍiyā' al-dīn muḥammad ibn qā'im muḥammad ibn mullā 'īsā ibn mullā allāhdādī asṭurlābī humāyūnī lāhawarī fī sanah 1047 hijrī*. Inscribed in naskh.

Consists of body, rete, graduated diopter with scales and sights, pin, horse, handle, ring, and four tympans with notches at the bottom fitting about a corresponding tab on the body.

The rim of the front of the body is graduated for every degree; the numbers, running clockwise, are 6, 12, 18,...360. In the recess in the front is a geographical table in two circles. (❍)(●)

The upper half of the back is graduated for every degree; the numbers, running clockwise from the left to the top and counter-clockwise from the right to the top, are 6, 12, 18, ...90. In the lower half near the edge are indicated the cotangents. In the upper left quadrant are drawn the cosines for every 2 of the 60 parts of the Radius. In the upper right quadrant are drawn the curves for the noon altitude of

Reverse of No. 15

the Sun (labeled *khuṭṭ niṣf al-nahār* [line of noon]) for latitudes 27° and 32°. In the lower half is a shadow-square for 7 feet and 12 digits, inside of which are two semicircles giving the names of the zodiacal signs and of the lunar mansions.

On the rete the ecliptic is graduated for every 6°; the numbers are, for each zodiacal sign, 6, 12, 18, ...30. The names of the stars are listed. (★)

Three of the four tympans have altitude circles for every 6° and equal and unequal hour curves; only Ib and IIIb have azimuths, for every 10°. They indicate their latitudes and longest daylights. (■)

One side of the fourth tympan has horizons for latitudes 4°, 14°, 24°,34°, 44°, 54°, and 64°; for 6°, 16°, 26°, 36°, 46°, 56°, and 66°; for 8°, 18°, 28°, 38°, 48°, and 58°; and for 10°, 20°, 30°, 40°, 50°, and 60°. The other side has a projection for latitude 66°; it is labeled: *ṣafiḥat mīzān al-ʿankabūt* ("tympan for balancing the rete").

PROVENANCE — Ali Kuli Khan; AP purchase, before 1945.

ICA 2558

REFERENCES — Chughtai (1994); Gibbs and Saliba (1984), p. 134; Newbury (2004a); Newbury (2006); Sarma (1994).

Mater of No. 15

Rete of No. 15

★ STAR LIST

Name	*Identification*
janāḥ al-ghurāb	γ Corv.
dhanab dajājah	α Cygn.
dhanab qayṭus	ι Ceti
ra's al-ḥawwā'	α Oph.
rijl al-yusrā	β Ori.
sāq ayman	δ Aquar.
simāk a'zal	α Virg.
simāk rāmiḥ	α Boot.
sha'āmī	α Can. Min.
shi'rā yamāniyyah	α Can. Mai.
ṣadr	π Ceti
ẓahr al-dubb	α Urs. Mai.
'unq al-ḥayyah	α Serp.
'ayn al-thawr	α Tau.
'ayyūq	α Aur.
ghūl	β Pers.
fard al-shujā'	α Hydr.
fam al-faras	ε Peg.
qā'idah	α Crat.
qalb al-'aqrab	α Scorp.
kaff al-khaḍīb	β Cass.
masāfat al-nahr	θ Erid.
mankib al-faras	β Peg.
nasr ṭā'ir	α Aquil.
nasr wāqi'	α Lyr.
nayyir fakkah	α Cor. Bor.
yad yumnā	α Ori.

OUTER CIRCLE

City	Longitude	Latitude	Identification
madīnat rusūl	75;20°	25;0°	Medina
makkah mubārak	77;10°	21;40°	Mecca
ḥalab	72;10°	35;50°	Aleppo
sumyāṭ	72;35°	40;0°	Samsat
arbīl	77;20°	35;0°	Arbīl
kūfah	79;30°	31;30°	Kūfah
baṣrah	84;0°	30;0°	Basra
shīrāz	88;0°	29;36°	Shīrāz
yazd	89;0°	32;30°	Yazd
iṣfahān	86;40°	32;25°	Eṣfahān
qumm	85;40°	35;45°	Qom
rayy	86;20°	35;0°	Shahr Rey
ṭūs	92;30°	36;0°	Ṭūs
samarqand	99;56°	39;37°	Samarkand
hurmūz	92;0°	25;0°	Hormoz
kābulistān	104;20°	33;35°	Kabul
tattah	102;7°	25;10°	Tatta
mūltān	107;35°	29;40°	Multan
qandahār	107;40°	33;0°	Kandahar
kashmīr	108;0°	35;0°	Srinagar
tibat	110;0°	40;0°	(Tibet)
lāhawar	109;20°	31;50°	Lahore
siyālkūt	109;10°	33;0°	Sialkot
dimashq	70;0°	33;15°	Damascus
miṣr	63;20°	30;20°	Cairo
tabrīz	82;0°	38;0°	Tabrīz
madā'in	72;0°	33;10°	Ctesiphon
baghdād	80;30°	33;25°	Baghdād
balkh	101;0°	36;41°	Wazirabad
badakhshān	104;24°	37;30°	Faizabad
bukhārā	97;30°	39;50°	Bukhara

INNER CIRCLE

City	Longitude	Latitude	Identification
shunām	114;0°	30;30°	Sunam
bānī batah	113;20°	28;52°	Panipat
tānīsar	112;26°	30;10°	Thanesar
dihlī	113;35°	28;20°	Dehli
akrah	114;0°	27;13°	Agra
qanawj	116;50°	26;36°	Kannauj
jawnbūr	119;6°	26;36°	Jaunpur
bānārasī	117;20°	26;55°	Varanasi
ajmīr	111;5°	25;0°	Ajmer
kwāliyār	115;0°	26;29°	Gwalior
aḥmadābād	108;40°	23;15°	Ahmedabad
kanbāyat	109;10°	22;20°	Cambay
ujayn	102;0°	22;30°	Ujjain
mandū	107;10°	22;13°	Mandu
burhānbūr	109;0°	20;31°	Burhanpur
jiddah	76;0°	21;0°	Jiddah
bījābūr	105;30°	17;20°	Bijapur
dawlatābād	111;0°	20;30°	Daulatabad
aḥmudānakar	107;0°	19;0°	Ahmednagar

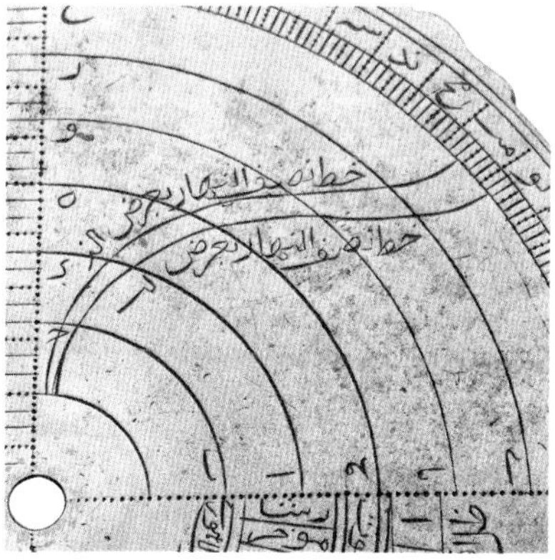

Detail of reverse of No. 15

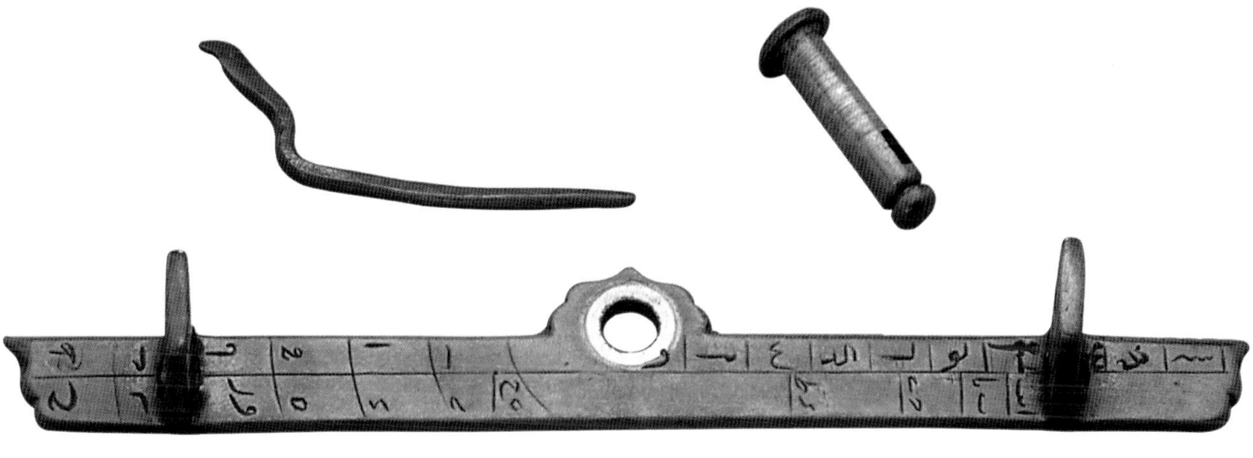

TYMPANS

Tympan	*Latitude*	*Longest Daylight*
Ia.	27°	13;43 hours
Ib.	32°	14;8 hours
IIa.	30°	13;58 hours
IIb.	34°	14;12 hours
IIIa.	36°	14;32 hours
IIIb.	40°	14;54 hours

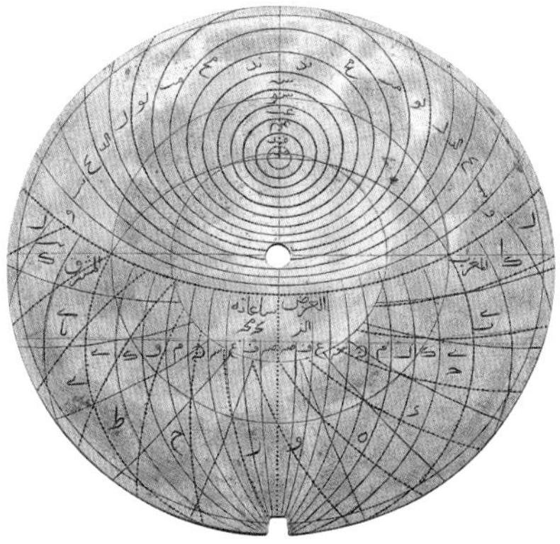
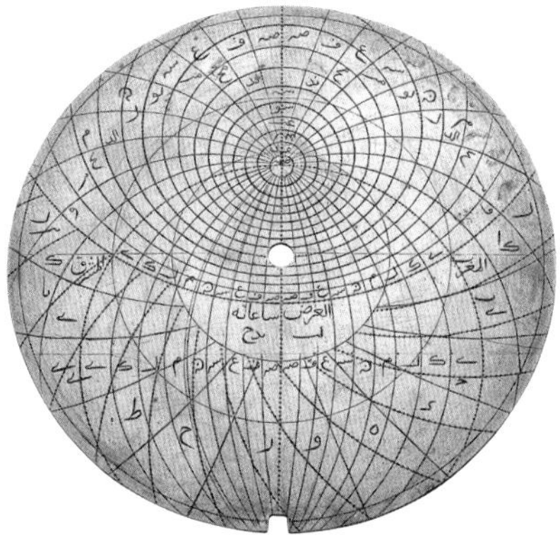

Tympan #1 of No. 15

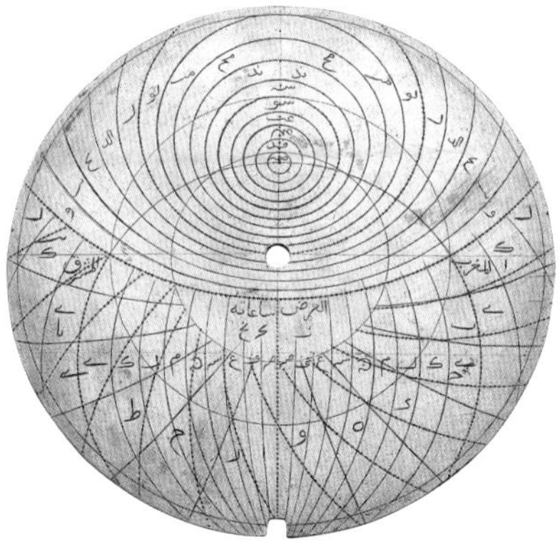
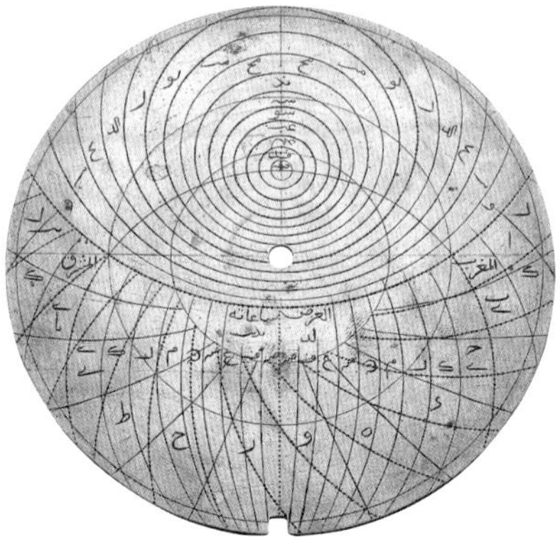

Tympan #2 of No. 15

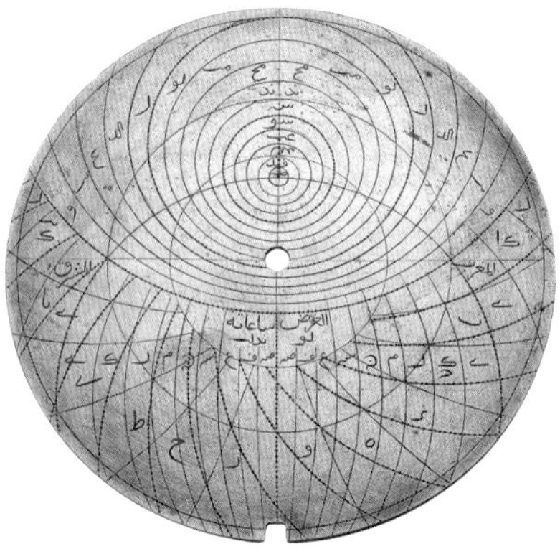
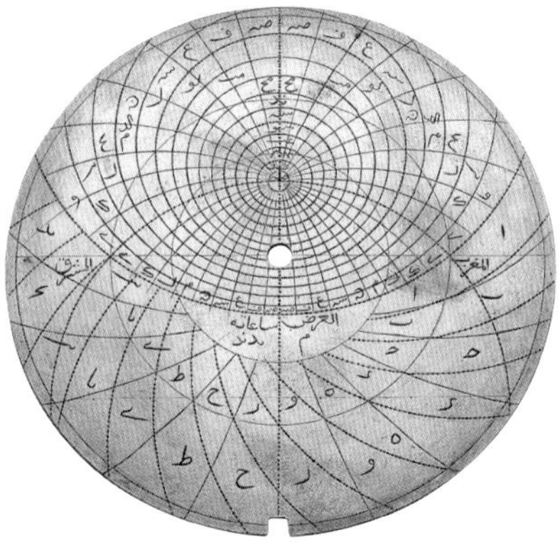

Tympan #3 of No. 15

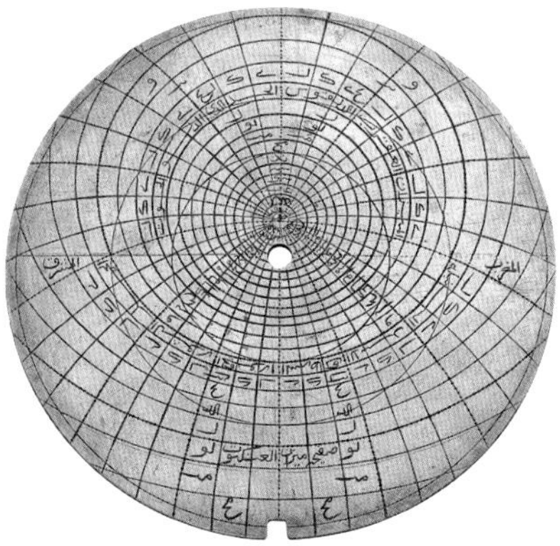

Tympan #4 of No. 15

16. Astrolabe

Ḍiyā' al-Dīn Muḥammad
Lahore
1647/8
"The work of Ḍiyā' al-Dīn Muḥammad ibn Mullā Qā'im Muḥammad ibn Mullā 'Īsā ibn Mullā Allāhdād, the astrolabe-maker, the Humāyūnī, the resident of Lahore, in the year 1057 of the Hijra."
20.3 x 12.1 x 3.9 cm
Mater thickness 0.9 cm
Brass, but tympans copper (?)
A-70

Face of No. 16

Astrolabe made by Ḍiyā' al-Dīn Muḥammad, the son of Qā'im Muḥammad, the son of 'Īsā, the son of Allāhdād, of Lahore in A.H. 1057 (A.D. 1647/8) according to the inscription below the center of the back: *'amal ḍiyā' al-dīn muḥammad ibn mullā qā'im muḥammad ibn mullā 'īsā ibn mullā allāhdād asṭurlābī humāyūnī lāhawarī sanah 1057 hijrī*. Inscriptions in naskh.

Consists of body, rete, graduated diopter with scales and sights, pin, handle, ring, and four tympans with notches at the bottom fitting about a corresponding tab in the body; the horse is missing.

The rim of the front of the body is graduated for every degree, the numbers, running clockwise, are 5, 10, 15, ... 360. In the recess in the front is a geographical table in two circles. (❍)(●)

The upper half of the back is graduated for every degree; the numbers, running clockwise from the left to the top and counter-clockwise from the right to the top, are 5, 10, 15, ...90. At the edge of the lower half are indicated the cotangents. In the upper left quadrant are the cosines for every 2 of the 60 parts of the Radius. In the upper right quadrant are the curves for the noon altitude of the Sun (labeled *khaṭṭ nisf al-nahār*) for latitudes 27° and 32°. In the lower part is the shadow-square for 7 feet and 12 digits, within which are two semicircles indicating the names of the zodiacal signs and of the lunar mansions.

On the rete the ecliptic is graduated for every 6°; the numbers are, for each zodiacal sign, 6, 12, 18, ...30. The names of the stars are listed. (★)

Three of the four tympans have altitude circles for every 3°, azimuths for every 30° (except on III), and equal and unequal hour curves; the first two tympans are identical. They

★ STAR LIST

Name	Identification
al-a‘zal	α Virg.
baṭn al-ḥūt	β Andr.
tālī masāf al-nahr	θ Erid.
janāḥ al-ghurāb	γ Corv.
dhanab al-jady	δ Capr.
dhanab al-dhālfīn	ε Delph.
dhanab al-qayṭus al-shamālī	ι Ceti
dhanab al-qayṭīs al-janūbī	β Ceti
ra’s al-jāthī	α Herc.
rāmiḥ	α Boot.
rijl al-jawzā’ al-yusrā	β Ori.
rijl al-jawzā’ al-yumnā	κ Ori
rijl al-musalsalah	γ Andr.
sāq sākib al-mā’i janūbī	δ Aquar.
shi‘rā sha’āmiyyah	α Can. Min.
ṣadr al-qayṭus	π Ceti
ẓahr al-dubb al-a‘ẓam	α Urs. Mai.
al-‘anāq	ζ Urs. Mai.
‘unq al-ḥayyah	α Serp.
‘ayn al-tinnīn	α Drac.
‘ayn al-thawr	α Tau.
‘ayyūq	α Aur.
kaff al-khaḍīb	β Cass.
kiffah janūbī	α Libr.
<fa>rd al-shujā‘	α Hydr.
al-faras	β Peg.
fam al-faras (?)	ε Peg. (?)
fam al-qayṭus	γ Ceti
qā‘idat al-bāṭiyyah	α Crat.
nasr ṭā’ir	α Aquil.
wāqi‘	α Lyr.
yad al-jawzā’ al-yusrā	γ Ori.
yad al-ḥawwā’	δ Oph.

Reverse of No. 16

indicate their latitudes and longest daylights. (■)

On one side of the fourth tympan are horizons for latitudes 4°, 12°, 20°, 28°, 36°, 44°, 52°, and 60°; for 10°, 18°, 26°, 34°, 42°, 50°, 58°, and 66°; for 8°, 16°, 24°, 32°, 40°, 48°, 56°, and 64°; and for 6°, 14°, 22°, 30°, 38°, 46°, 54°, and 62°. On the other side is a projection for latitude 66° labeled: *ṣafīḥat mīzān al-‘ankabūt* (“tympan for balancing the rete”).

PROVENANCE — E.S. David, 1940; AP purchase, 1940.
ICA 1095
EXHIBITIONS — Michigan State University, April – Nov. 1977.
REFERENCES — Abbot (1937); Gibbs and Saliba (1984), p. 134; Newbury (2004); Newbury (2006); Sarma (1994); Stephenson, Bolt, and Friedman (2000), p. 64.

OUTER CIRCLE

City	*Longitude*	*Latitude*	*Identification*
makkah muʿaẓẓamah	77;10°	21;40°	Mecca
madīnah musharrafah	75;20°	25;0°	Medina
mahdiyyah maghrab	42;0°	32;30°	Mahdia
iskandariyya	61;57°	30;58°	Alexandria
miṣr	63;20°	30;20°	Cairo
āmid	73;40°	33;0°	Diarbekir
marāghah	82;0°	36;20°	Marāgheh
tabrīz	82;0°	38;0°	Tabrīz
ardabīl	82;30°	38;0°	Ardabīl
bulghār	90;0°	49;30°	(Bulgars)
kūfah	79;30°	31;30°	Kūfah
madāʾin	72;0°	33;10°	Ctesiphon
baghdād	80;30°	33;25°	Baghdād
baṣrah	84;0°	30;0°	Basra
iṣfahān	85;0°	34;45°	Eṣfahān
sabzawār	91;30°	36;5°	Sabzevār
nīshābūr	92;30°	36;21°	Neyshābūr
ṭūs	94;30°	37;0°	Ṭūs
harāh	94;33°	34;30°	Herat
balkh	101;0°	36;41°	Wazirabad
badakhshān	104;24°	37;10°	Faizabad
bukhārā	97;30°	39;50°	Bukhara
samarqand	99;56°	39;37	Samarkand
kāshghar	106;30°	44;0°	Kashgar
hurmūz	92;0°	25;0°	Hormoz
aḥmadābād	108;40°	23;15°	Ahmedabad
ujayn	112;0°	22;30°	Ujjain
amarkūt	105;0°	25;0°	Umarkot
kābul	104;40°	34;30°	Kabul
mīwāt	117;10°	34;50°	
barāwan	117;59°	27;32°	Baraum
hānsī	102;25°	29;15°	Hansi
tānīsar	112;33°	30;10°	Thanesar
pānīpatah	113;20°	28;52°	Panipat
dihlī	113;35°	28;20°	Dehli

INNER CIRCLE

City	*Longitude*	*Latitude*	*Identification*
ajmīr	111;5°	24;0°	Ajmer
kwāliyār	114;0°	26;29°	Gwalior
kashmīr	108;0°	35;0°	Srinagar
tibat	110;0°	40;0°	(Tibet)
sarandīp	130;0°	10;0°	(Śrīlaṇkā)
bānārasī	117;20°	26;50°	Varanasi
qanawj	115;50°	26;50°	Kannauj
lahāwar	109;20°	31;50°	Lahore
qandahār	107;40°	33;0°	Kandahar
mūltān	107;35°	29;40°	Multan
sanbhal	115;0°	28;10°	Sambhal
kurah	117;6°	26;16°	
chānpānīr	108;45°	22;30°	Chāmpāner
kambāyat	109;10°	22;20°	Cambay
jiddah	76;0°	21;0°	Jiddah
bījābūr	105;30°	17;20°	Bijapur
dawlatābād	111;0°	20;30°	Daulatabad
burhānpūr	109;0°	20;31°	Burhanpur
jawnpūr	119;6°	26;36°	Jaunpur
mānikpūr	118;10°	26;49°	Manikpur
awadah	118;6°	27;22°	Ayodhya
darband jīn	178;40°	23;0°	Canton
pūrusawar	115;15°	31;10°	Peshawar

Rete of No. 16

Mater of No. 16

TYMPANS

Tympan	*Latitude*	*Longest Daylight*
Ia.= IIa	27°	13;43 hours
Ib.= IIb	32°	14;8 hours
IIIa.	36°	14;32 hours
IIIb.	40°	14;58 hours

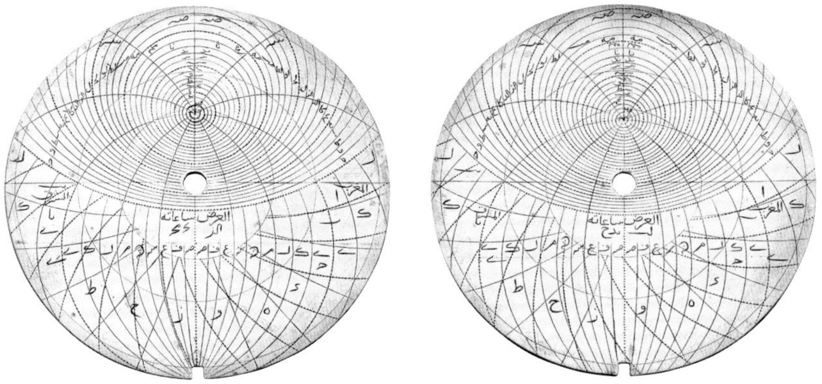

Tympan #1 of No. 16

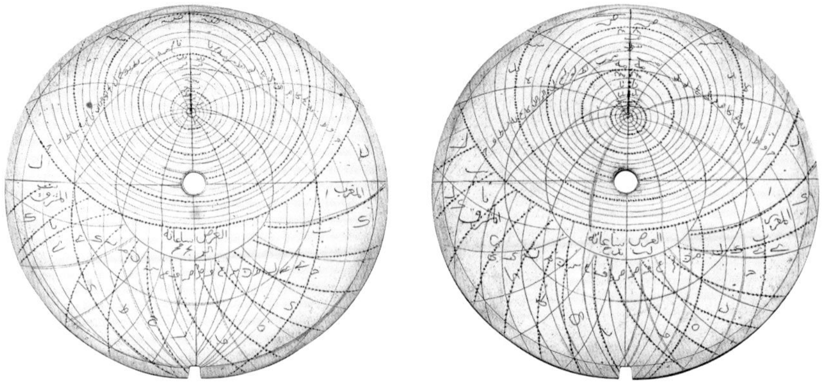

Tympan #2 of No. 16

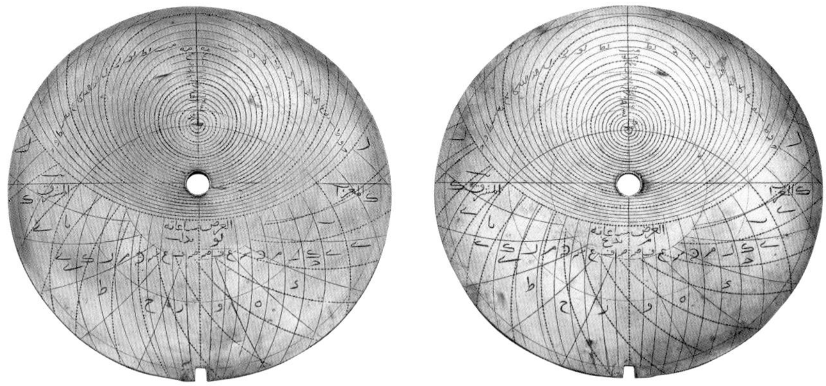

Tympan #3 of No. 16

Tympan #4 of No. 16

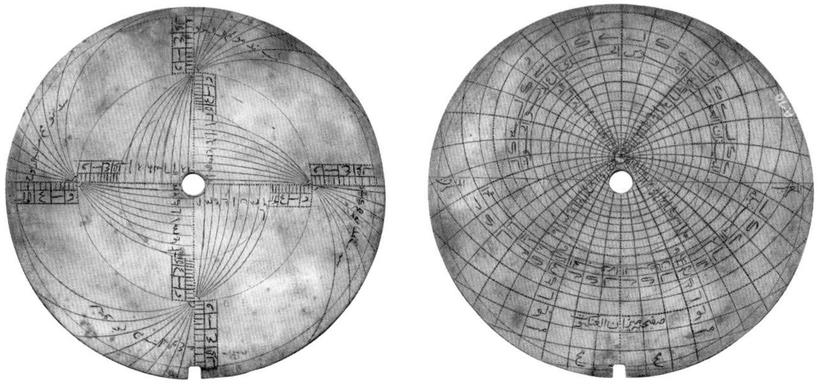

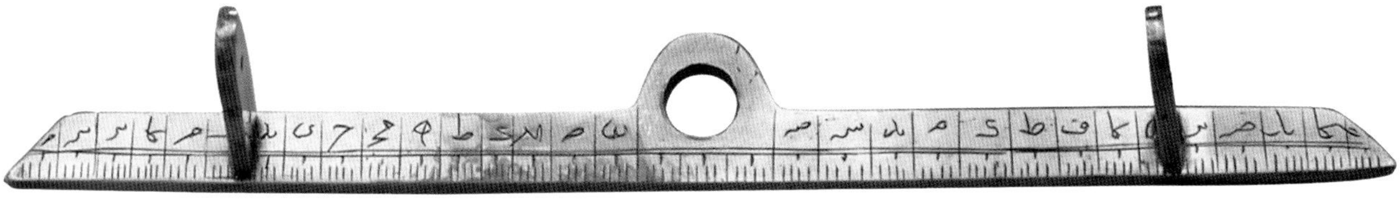

17. Astrolabe

Ḍiyā' al-Dīn Muḥammad
Lahore
1660/1
"The work of the least of the servants (of God), Ḍiyā' al-Dīn Muḥammad ibn Qā'im Muḥammad ibn Mullā 'Īsā in the year 1071 of the Hijra — the Humāyūnī, the resident of Lahore."
12.7 x 8.3 x 2.1 cm
Mater thickness 0.6 cm
Brass for body, diopter, pin, washer, and horse, copper for rete and tympans
A-81

Face of No. 17

Astrolabe made by Ḍiyā' al-Dīn Muḥammad, the son of Qā'im Muḥammad, the son of 'Īsā, of Lahore in A.H. 1071 (A.D. 1660/1) according to the inscription in the shadow-square on the back: *humāyūnī lāhawarī 'amal aqall al-i'bād ḍiyā' al-dīn muḥammad ibn qā'im muḥammad ibn mullā 'īsā fī sanah 1071 hijrī.* Inscriptions in naskh.

Consists of body, rete, uninscribed diopter with sights, pin, washer, horse, handle, ring, and five tympans which have notches at the bottom fitting about a corresponding tab in the body.

The rim of the front of the body is graduated for every 2°; the numbers, running clockwise in each quadrant, are 6, 12, 18, ...90. In the recess in the front is a geographical list in two circles. (❍)(●)

The upper half of the back is graduated for every 2°; the numbers, running clockwise from the left to the top and counter-clockwise from the right to the top, are 6, 12, 18, ...90. At the edge in the lower half are indicated the cotangents. In the upper left quadrant are the cosines for every 2 of the 60 parts of the Radius. In the upper right quadrant are the curves for the noon altitude of the Sun (labeled *khaṭṭ niṣf al-nahār*) for latitudes 24° and 32°. Below the center is the shadow-square for 7 feet and 12 digits. Below this are two semicircles indicating the names of the zodiacal signs and of the lunar mansions.

On the rete the ecliptic is graduated for every 6°; there are no numbers. The names of the stars (some pointers have no inscriptions) are listed. (★)

The first four tympans have altitude circles for every 6°, azimuths for every 30° (except for every 10° on IVb), and equal and unequal hour curves. They indicate their latitudes and longest daylights. (■)

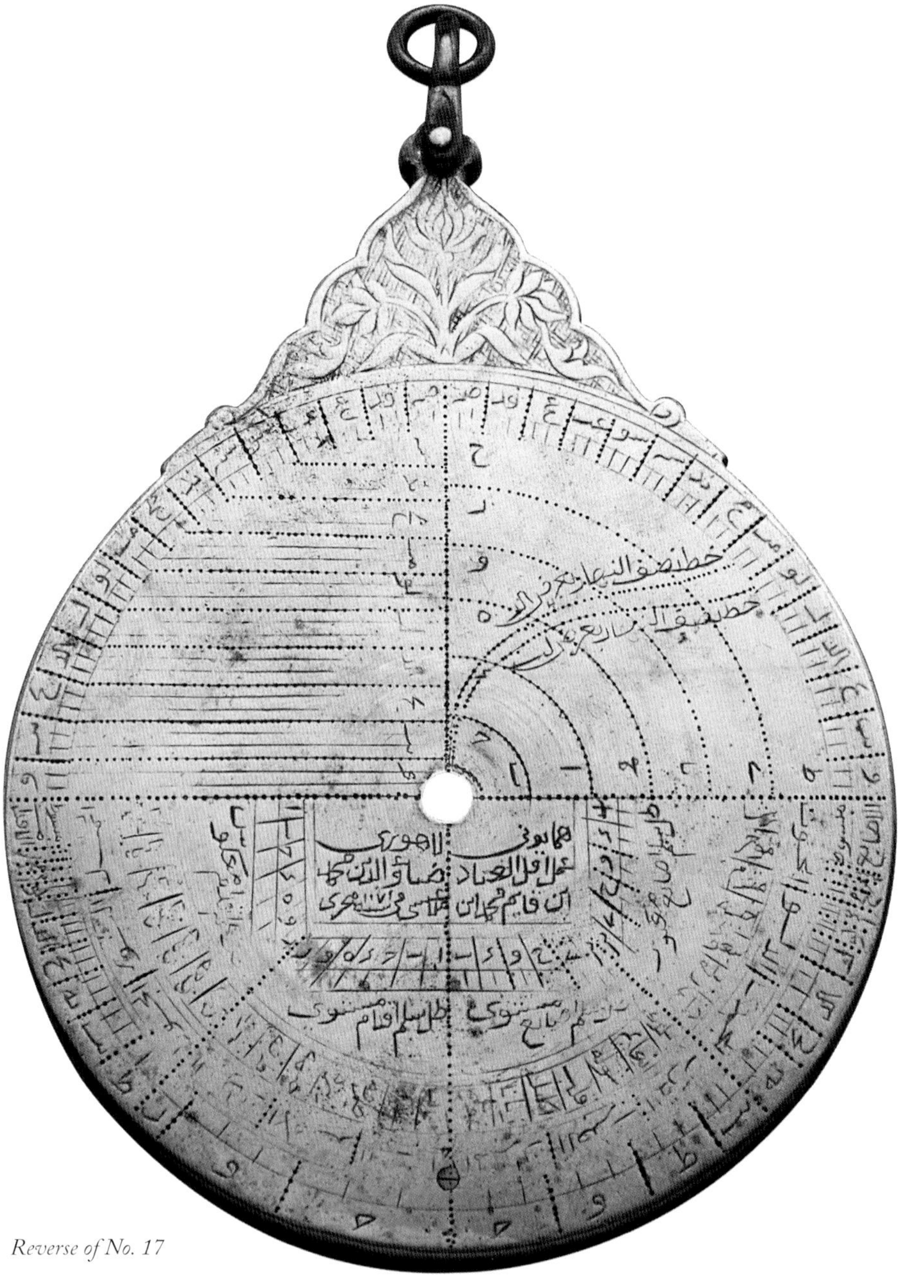

Reverse of No. 17

On one side of the fifth tympan are horizons for latitudes 6°, 16°, 26°, 36°, 46°, 56°, and 66°; for 8°, 18°, 28°, 38°, 48°, and 58°; for 10°, 20°, 30°, 40°, 50°, and 60°; and for 12°, 22°, 32°, 42°, 52°, and 62°. On the other side is a projecton for latitude 66° labeled: *ṣafīḥat mīzān al-ʿankabūt* ("tympan for balancing the rete").

PROVENANCE — Ali Kuli Khan of the House of Persian Art, New York, 1941; AP purchase, 1941.
ICA 2554
REFERENCES — Gibbs and Saliba (1984), p. 134; Newbury (2004a); Newbury (2006); Sarma (1994).

OUTER CIRCLE

City	Longitude	Latitude	Identification
makkah mu'aẓẓamah	77;10°	21;40°	Mecca
madīnah musharrafah	75;20°	25;0°	Medina
miṣr	63;20°	30;20°	Cairo
ḥalab	72;10°	35;50°	Aleppo
tabrīz	82;0°	38;0°	Tabrīz
ardabīl	82;30°	38;0°	Ardabīl
bulghār	90;0°	49;30°	(Bulgars)
kūfah	79;30°	31;30°	Kūfah
madā'in	82;0°	33;10°	Ctesiphon
baghdād	80;30°	33;25°	Baghdād
baṣrah	84;0°	30;0°	Basra
shīrāz	88;0°	29;36°	Shīrāz
yazd	88;0°	32;30°	Yazd
hamadān	83;0°	35;10°	Hamadan
iṣfahān	86;40°	32;25°	Eṣfahān
sabzawār	91;30°	36;5°	Sabzevār
nīshāpūr	92;30°	36;21°	Neyshābūr
balkh	101;0°	36;41°	Wazirabad
badakhshān	104;24°	37;10	Faizabad
bukhārā	97;30°	39;50°	Bukhara
samarqand	99;56°	39;37°	Samarkand
kābul	104;40°	34;30°	Kabul
qandahār	107;40°	33;0°	Kandahar

INNER CIRCLE

City	Longitude	Latitude	Identification
multān	107;35°	29;40°	Multan
lāhawar	109;20°	31;50°	Lahore
shāhjahānābād	113;35°	28;39°	Delhi
akbarābād	114;0°	26;43°	Agra
ajmīr	111;5°	24;0°	Ajmer
kwāliyār	115;0°	26;29°	Gwalior
bījāpūr	105;30°	17;20°	Bijapur
dawlatābād	111;0°	21;40°	Daulatabad
aḥmadābād	108;40°	23;15°	Ahmedabad
kulkundah	115;39°	18;10°	Golconda
dahākah bankālah	130;0°	27;0°	Dacca
burhānpūr	109;0°	20;31°	Burhanpur
patnah	120;45°	27;40°	Patna
kashmīr	103;0°	35;0°	Srinagar
tibat	110;0°	40;0°	(Tibet)

Mater of No. 17

★ STAR LIST

Name	*Identification*
janāḥ al-ghurāb	γ Corv.
jawzā' yumnā	κ Ori.
al-ḥayyah	α Serp.
dulfīn	ε Delph.
dhanab al-jady	δ Capr.
dhanab al-dajājah	α Cygn.
dhanab al-qayṭus	ι Ceti
ra's al-jāthī	α Herc.
ra's al-ḥawwā'	α Oph.
rāmiḥ	α Boot.
sāq sākib al-mā'ī	δ Aquar.
simāk a'zal	α Virg.
shi'rā sha'āmiyyah	α Can. Min.
shi'rā yamāniyyah	α Can. Mai.
ṣadr al-qayṭus	π Ceti
ẓahr dubb	α Urs. Mai.
'ayn al-thawr	α Tau.
fard al-shujā'	α Hydr.
fam al-faras	ε Peg.
fam al-qayṭus	γ Ceti
qā'idat al-bāṭiyyah	α Crat.
qalb al-asad	α Leo
qalb al-'aqrab	α Scorp.
kaff al-khaḍīb	β Cass.
kiffah janūbī	α Libr.
mankib faras	β Peg.
nasr ṭā'ir	α Aquil.
nasr wāqi'	α Lyr.
nayyir fakkah	α Cor. Bor.
yad jawz al-yusrā	γ Ori.

Rete of No. 17

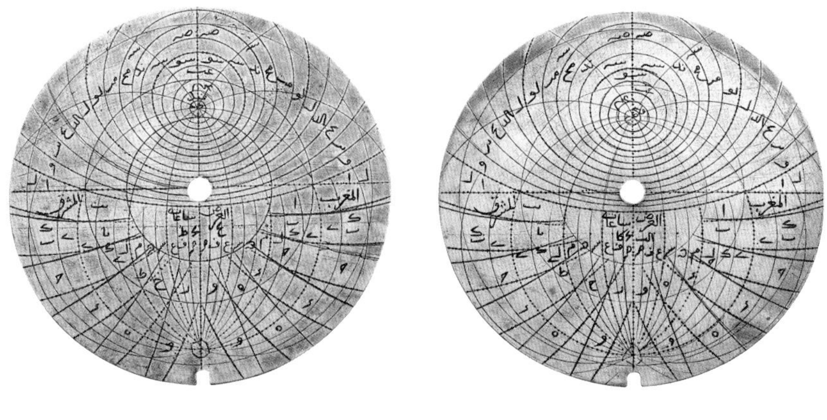

Tympan #1 of No. 17

TYMPANS

Tympan	*Latitude*	*Longest Daylight*
Ia.	18°	13;9 hours
Ib.	22°	13;21hours
IIa.	24°	13;32 hours
IIb.	27°	13;43 hours
IIIa.	29°	13;52 hours
IIIb.	32°	14;8 hours
IVa.	34°	14;20 hours
IVb.	36°	14;32 hours

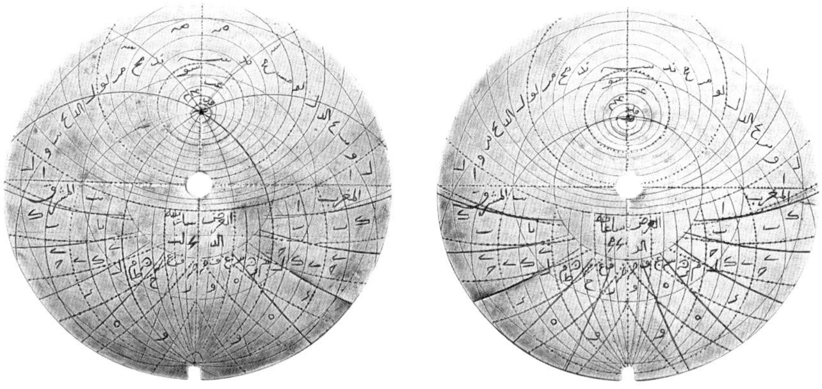

Tympan #2 of No. 17

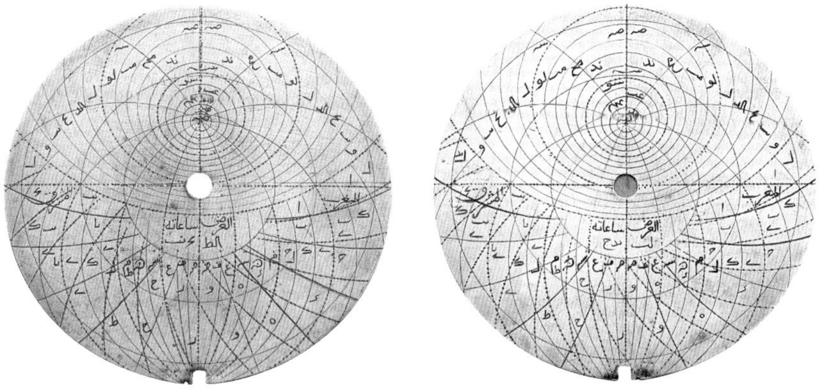

Tympan #3 of No. 17

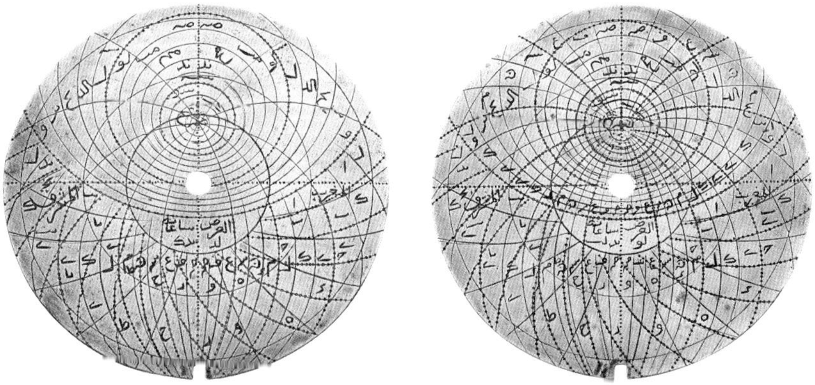

Tympan #4 of No. 17

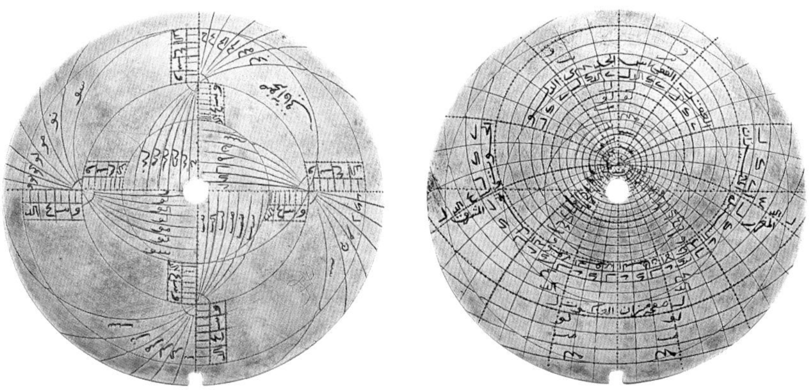

Tympan #5 of No. 17

18. Astrolabe

Ḍiyā' al-Dīn Muḥammad
Lahore
1662/3
"The work of Ḍiyā' al-Dīn Muḥammad in the year 1073 of the Hijra."
15.2 x 9.5 x 3.6 cm
Mater thickness 0.7 cm
Brass
A-78

Face of No. 18

Astrolabe made by Ḍiyā' al-Dīn Muḥammad in A.H.1073 (A.D. 1662/3) according to the inscription within the shadow-square on the back: *amal ḍiyā' al-dīn muḥammad sanah 1073 hijrī.* Inscriptions in naskh.

Consists of body, rete, uninscribed diopter with sights, pin, horse, handle, ring, and five tympans with notches fitting about a tab in the body.

The rim of the front of the body is graduated for every 2° (with a dot after each graduation line); the numbers, running clockwise in each quadrant, are 6, 12, 18, ...90. In the recess in the front is a geographical table in two circles. (❍)(●)

The upper half of the back is graduated for every 2°; the numbers, running clockwise from the left to the top and counter-clockwise from the right to the top, are 6, 12, 18, ...90. At the edge of the lower half are indicated the cotangents. In the upper left quadrant are the cosines for every 2 of the 60 parts in the Radius. In the upper right quadrant are two curves for the noon altitude of the Sun; toward the top of the quadrant and upside down is the inscription: *khaṭṭ nisf al-nahār bānah (?) 'arḍ l.w. y.ḥ* ("line of noon ...(?) latitude 36;18"). Below the center is the shadow-square for 7 feet and 12 digits. Below this are five concentric semicircles indicating the bounds of the terms, the planetary lords of the decans (designated by their final letters) according to the Greek system, the names of the zodiacal signs, the planetery lords of the terms (designated by their final letters), and the names of the lunar mansions.

On the rete the ecliptic is graduated for every 6°; the numbers for each zodiacal sign are 6, 12, 18, ...30. The names of the stars are listed. (★)

The first four tympans have altitude circles for every 6°; equal and unequal hour curves; azimuths on Ia (every 10°) and on IIa (every 45°). They indicate their latitudes and longest daylights. (■)

★ STAR LIST

Name	*Identification*
tālī masāfat al-nahr	θ Erid.
janāḥ al-ghurāb	γ Corv.
dhanab al-jady	δ Capr.
dhanab al-dajājah	α Cygn.
dhanab dulfīn	ε Delph.
dhanab al-qayṭus	β Ceti
dhanab al-qayṭus	ι Ceti
ra's ḥawwā'	α Oph.
ra's ghūl	β Pers.
rijl jawzā' yusrā	β Ori.
rijl jawzā' yumnā	κ Ori.
sāq sākib al-mā'	δ Aquar.
simāk a'zal	α Virg.
simāk rāmiḥ	α Boot.
shi'rā sha'āmī	α Can. Min.
ṣadr <a>l-qayṭus	π Ceti
ẓahr	δ Leo.
ẓahr dubb	α Urs. Mai.
'unq	α Serp.
'ayn al-thawr	α Tau.
fard al-shujā' (on 2 pointers)	α Hydr.
fam al-faras	ε Peg.
fam al-qayṭus	γ Ceti
qā'idat al-bāṭiyyah	α Crat.
qalb al-'aqrab	α Scorp.
kaff al-khaḍīb	β Cass.
mu'akhkhar al-shujā'	σ Hydr.
mankib al-faras	β Peg.
nasr ṭā'ir	α Aquil.
nasr wāqi'	α Lyr.
nayyir fakkah	α Cor. Bor.
yad jawzā' yusrā	γ Ori.
yad al-jawzā' yumnā	α Ori.

Reverse of No. 18

On one side of the fifth tympan are the horizons for latitudes 4°, 12°, 20°, 28°, 36°, 44°, 52°, 60°, and 68°; for 6°, 14°, 22°, 30°, 38°, 46°, 54°, 62° and 70°; for 8°, 16°, 24°, 32°, 40°, 48°, 56°, 64°, and 72°; for 10°, 18°, 26°, 34°, 42°, 50°, 58°, 66°, and 74;30°. On the other side is the projection for latitude 66° labeled: *ṣafīḥat mīzān al-'ankabūt* ("tympan for balancing the rete").

PROVENANCE — Sotheby's, London, 1938; Percy Webster, London, 1938; AP purchase, 1939.
ICA 2551
REFERENCES — Newbury (2004a); Newbury (2004b); Newbury (2006); Sarma (1994).

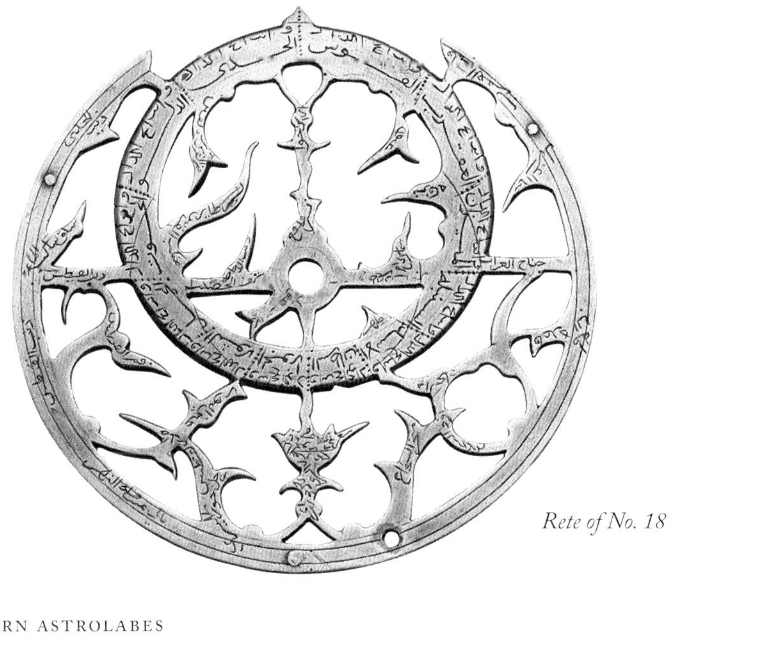

Rete of No. 18

OUTER CIRCLE

City	Longitude	Latitude	Identification
makkah mu'aẓẓamah	77;10°	21;40°	Mecca
madīnah musharrafah	75;20°	25;0°	Medina
miṣr	63;20°	30;20°	Cairo
dimashq	70;0°	33;15°	Damascus
ḥalab	72;10°	35;50°	Aleppo
al-rūm	77;0°	39;40°	Erzurum
mawṣil	77;0°	34;30°	Mosul
tabrīz	82;0°	38;0°	Tabrīz
ardabīl	82;30°	38;0°	Ardabīl
bulghār	90;0°	49;30°	(Bulgars)
kūfah	79;30°	31;30°	Kūfah
madā'in	72;0°	33;10°	Ctesiphon
baghdād	80;30°	33;25°	Baghdād
baṣrah	84;0°	30;0°	Basra
shīrāz	88;0°	29;36°	Shīrāz
yazd	89;0°	32;30°	Yazd
iṣfahān	86;40°	32;25°	Eṣfahān
sabzawār	91;30°	36;5°	Sabzevār
nīshāpūr	92;30°	36;21°	Neyshābūr
balkh	101;0°	36;41°	Wazirabad
badakhshān	104;24°	37;10°	Faizabad
bukhārā	97;30°	39;50°	Bukhara
samarqand	99;56°	39;37°	Samarkand
ṭūs	94;30°	37;0°	Ṭūs
tūn	92;30°	32;30°	Ferdows
harāh	94;33°	34;30°	Herat
sarakhs	94;30°	32;0°	Sarakhs
khujand	105;30°	41;55°	Leninabad
kāshghar	106;30°	44;0°	Kashgar
hurmūz	92;0°	25;0°	Hormoz
kashmīr	103;0°	35;0°	Srinagar

INNER CIRCLE

City	Longitude	Latitude	Identification
kābul	104;40°	34;30°	Kabul
qandahār	107;40°	33;0°	Kandahar
multān	107;35°	29;40°	Multan
lāhawar	109;20°	31;50°	Lahore
shāhjahānābād	113;35°	28;39°	Delhi
akbarābād	114;0°	26;43°	Agra
ajmīr	111;5°	24;0°	Ajmer
kwāliyār	115;0°	26;29°	Gwalior
bījāpūr	105;30°	17;20°	Bijapur
dawlatābād	111;0°	21;40°	Daulatabad
aḥmadābād	108;40°	23;15°	Ahmedabad
kulkundah	114;39°	18;10°	Golconda
dahākah bankālah	130;0°	27;0°	Dacca
burhānpūr	109;0°	20;31°	Burhanpur
patnah	120;45°	26;40°	Patna

Mater of No. 18

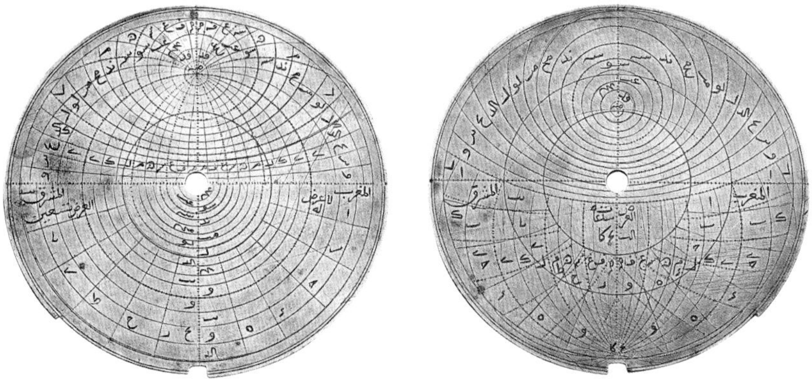

Tympan #1 of No. 18

■ TYMPANS

Tympan	*Latitude*	*Longest Daylight*
Ia.	90°	
Ib.	22°	13;21hours
IIa.	24°	13;32 hours
IIb.	27°	13;43 hours
IIIa.	29°	13;52 hours
IIIb.	32°	14;8 hours
IVa.	36°	14;32hours
IVb.	40°	14;54 hours

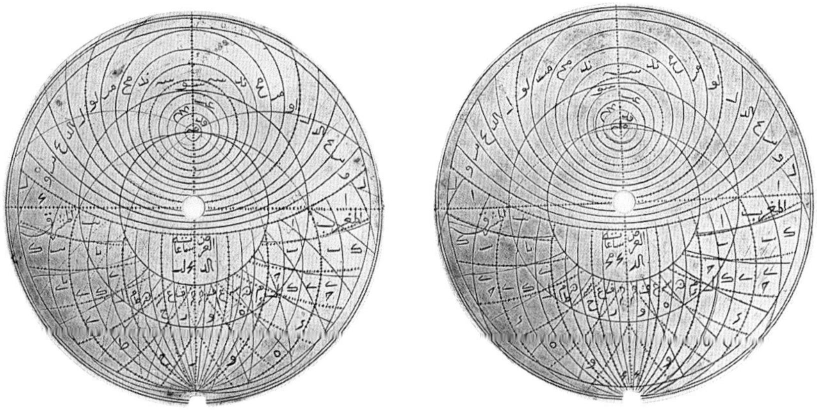

Tympan #2 of No. 18

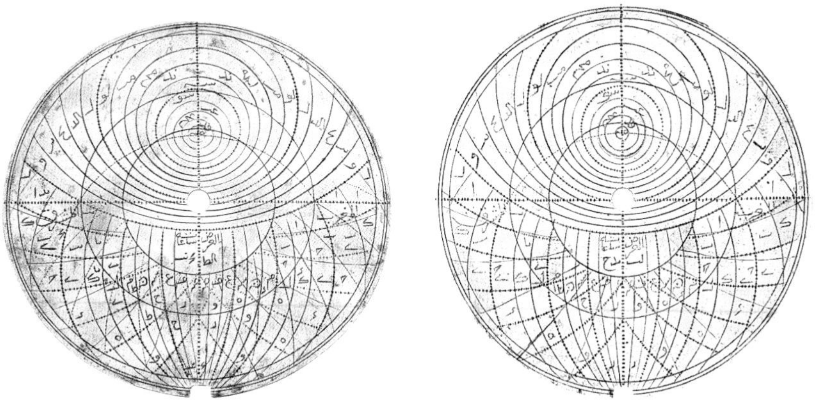

Tympan #3 of No. 18

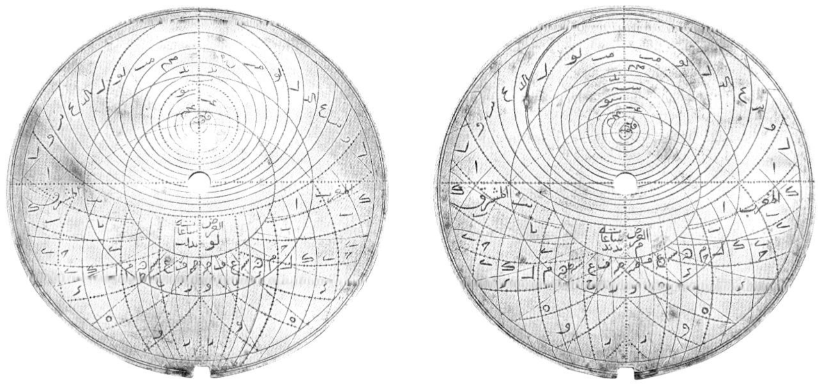

Tympan #4of No. 18

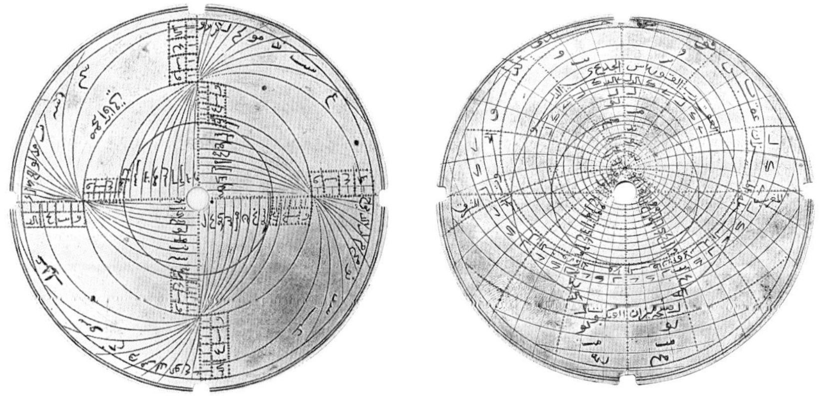

Tympan #5 of No. 18

19. Astrolabe with Projections from South and North Poles

Maker unknown
India (Lahore?)
17th century
17.3 x 12.8 x 2.8 cm
Mater thickness 0.8 cm
Brass
L-100

Northern face of No. 19

Astrolabe with a rete on each side of the body, one representing a projection from the South Pole, providing a map of the northern hemisphere; the other one from the North Pole, providing a map of the sourthern hemisphere. Made in India (perhaps Lahore), probably in the late seventeenth century. Inscriptions in naskh.

Consists of body (a ring), two retes, ruler, pin, horse (in form of bird), ring, chain, and seven tympans which have tabs fitting into slots at the top and bottom of the body. On each side are four rotating lugs that hold the retes and tympans in place. There are no diopters, but their purpose is served by sighting holes on either side of the East-West bar on each rete.

Each side of the body-ring is graduated for 1°; the numbers, running clockwise, in each quadrant are 6, 12, 18, …90.

The ecliptic in each rete has its rim graduated for every 1°. In the projection from the South Pole the zodiacal signs are listed counter-clockwise beginning at the North with Capricorn, in the projection from the North Pole beginning with Cancer. The graduations corresponding to each sign are divided into 5 equal parts labeled 6, 12, 18, 24, 30. The names of the stars on each rete are listed. (★)

There are seven tympans. The first five are for given latitudes, their altitude-circles are for every 2° (Ia; Ib; IIa; IIb; IIIa; IIIb; IVa; IVb) or every 3° (Va top and bottom). They indicate their latitudes and longest daylights. (■) There are seasonal and equal hour curves.

Tympan VIa has horizons for latitudes 5, 9, 13, 17, 21, 25, 29, 33, 37, 41, 45, 49, 53, 57, 61, and 65; for 8, 12, 16, 20, 24, 28, 32, 36, 40, 44, 48, 52, 56, 60, and 64; for 7, 11, 15, 19, 23, 27, 31, 35, 39, 43, 47, 51, 55, 59, 63, and 67; and for 6, 10, 14, 18, 22, 26, 30, 34, 38, 42,

Southern face of No. 19

46, 50, 54, 58, 62, and 66. Tympan VIb has a projection for 66° labeled: *ṣafīḥat mīzān al-ʿankabūt* ("tympan for balancing the rete").

Tympan VII serves as the back of the body. On one side is a geographical table in three circles abbreviated from the list found on No. 33. (☽)(●)(●)

The bottom of the other side has several semicircles in which are written the names of the lunar mansions, the names of the zodiacal signs, the planetary lords of the terms (indicated by their final letters), and the degrees in each term; within these semicircles are a sine table to the left, a shadow square for 12 digits and 7 feet to the right. On the upper half are many more semicircles and to the right, the noon-altitudes of the Sun for latitudes 19°, 27°, and 32°.

PROVENANCE — The Time Museum, Rockford, IL; City of Chicago Purchase, 1999; AP long-term loan from the City of Chicago, Public Art Program, 1999.
ICA 3917

EXHIBITIONS — Time Museum, Rockford, IL ; Adler Planetarium, "Episodes in the Story of Time", April – Sept. 2001.
REFERENCES — Newbury (2004a); Newbury (2006); Turner, A. J. (1985), pp. 74-83.

★ STAR LIST PROJECTION FROM SOUTH POLE

Name	*Identification*
baṭn	β Andr.
tālī masāf al-nahr	θ Erid.
janāḥ ghurāb	γ Corvi
dabarān	α Tauri
dhanab al-jady	δ Capr.
dhanab dajājah	α Cygni
dhanab clulfin	ε Delph.
dhanab al-qayṭus	β Ceti
ra's al-jāthī	α Herc.
ra's ḥawwā'	α Oph.
rijl	β Ori.
rijl yumnā	κ Ori.
sāq sākib al-mā'	δ Aq.
surrat fars	δ Peg.
simāk rāmiḥ	α Boot.
simāk a‘zal	α Virg.
shi‘rā sha'āmī	α Can. Min.
shi‘rā yamānī	α Can. Mai.
ṣadr qayṭus	π Ceti
ṭaraf al-sufīnah	ρ Puppis
‘anāq	ζ Urs. Mai.
‘unq ḥayyah	α Serp.
‘ayn tinnīn	α Drac.
‘ayyūq	α Aur.
ghūl	β Pers.
fard shujā‘	α Hydr.
fam faras	ε Peg.
fam qayṭus	γ Ceti
qā‘idat al-bāṭiyyah	α Crat.
kaff al-jadhmā	α Ceti
mankib faras	β Peg.
minqār dajājah	β Cygni
nasr ṭā'ir	α Aquil.
nayyir fakkah	α Cor. Bor.
wāqi‘	α Lyr.
yad ḥawwā'	δ Oph.
yad yusrā	γ Ori.
yad yumnā	α Ori.

★ STAR LIST PROJECTION FROM NORTH POLE

Name	*Identification*
tālī masāf	θ Erid.
dhanab qayṭus ja	β Ceti
dhanab qayṭus sha	ι Ceti
al-ra's	β Pers.
ra's al-jāthī	α Herc.
ra's al-ḥawwā'	α Oph.
rijl	β Ori.
rijl yumnā	κ Ori.
rukbat ḥawwā'	ζ Ophi.
simāk rāmiḥ	α Boot.
shi‘rā sha'āmī	α Can. Min.
ṣadr qayṭus	π Ceti
ṣarfah	β Leo
ṭaraf sufīnah	ρ Puppis
‘iṣābat al-rāmī	ρ Sag. (?)
‘unq ḥayyah	α Serp.
‘ayn	α Tauri
fard al-shujā‘	α Hydr.
fam faras	ε Peg.
qā‘idat bāṭiyya	α Crat.
qaran al-jady al-thānī	β Cap.
kiffah shā	β Librae
mu'akhkhar sharaṭān	γ Ari.
muqaddam sharaṭān	β Ari.
mankib faras	β Peg.
nasr ṭa'ir	α Aquil.
yad al-ḥawwā'	δ Oph.
yad yumnā	α Ori.

Left: Northern Rete of No. 19
Right: Southern Rete of No. 19

TYMPANS

Tympan	Latitude	Longest Daylight
Ia	19° North	— hours, — ghaṭīs
Ib	22° South	34 ghaṭīs
IIa	32° North	35 ghaṭīs
IIb	40° South	36;6 ghaṭīs
IIIa	27° North	34 ghaṭīs
IIIb	36° South	35 ghaṭīs
IVa	11° North	12;48 hours, — ghaṭīs
IVb	29° North	13;12 hours, 34 ghaṭīs (34 in Nāgarī)

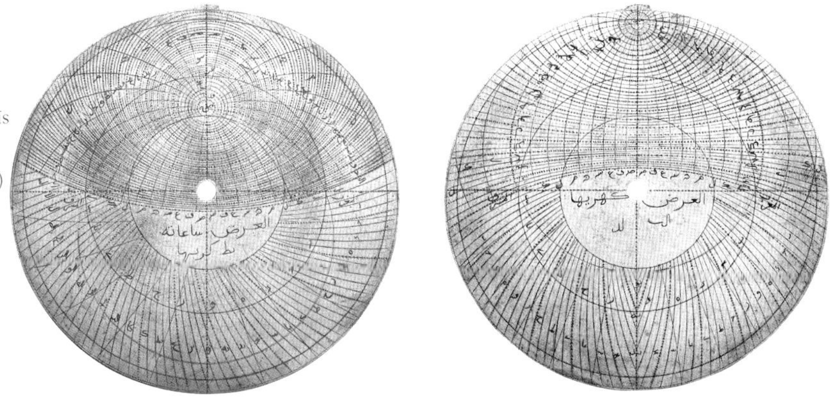

Tympan #1 of No. 19

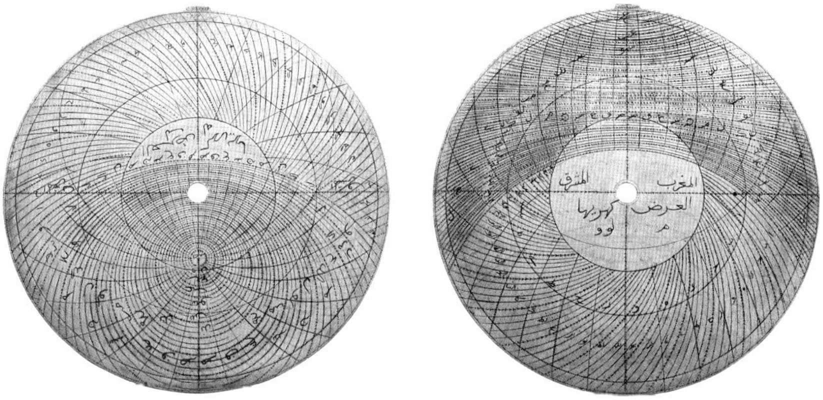

Tympan #2 of No. 19

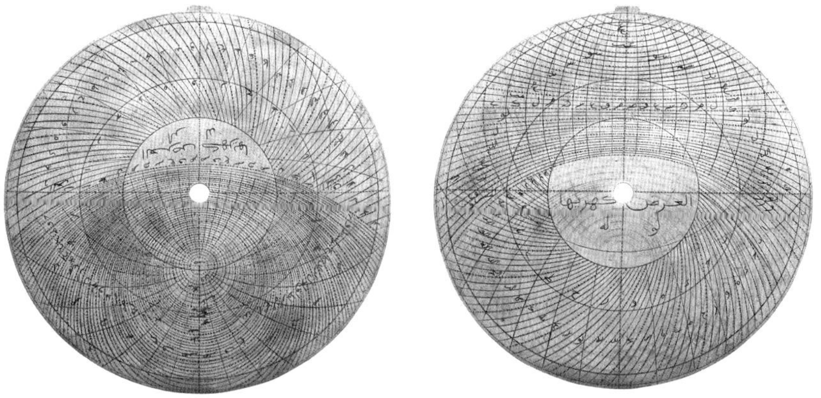

Tympan #3 of No. 19

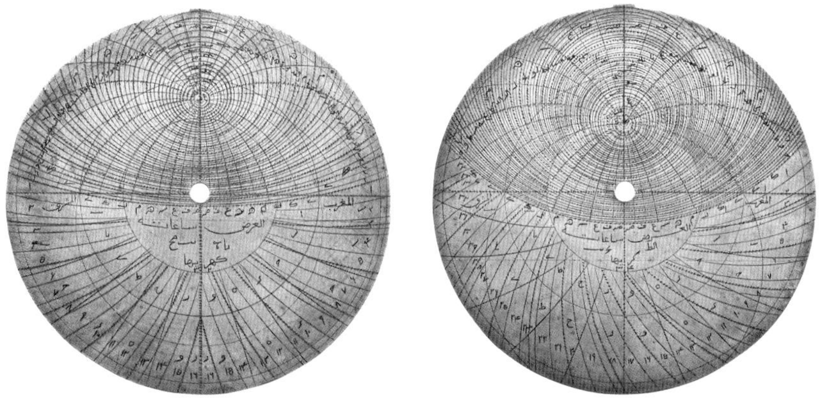

Tympan #4 of No. 19

TYMPANS

Tympan	*Latitude*	*Longest Daylight*
Va top	30° South	
Va bottom	0°	
Vb top	34° South	14;26 hours
Vb bottom	38° South	14;42 hours

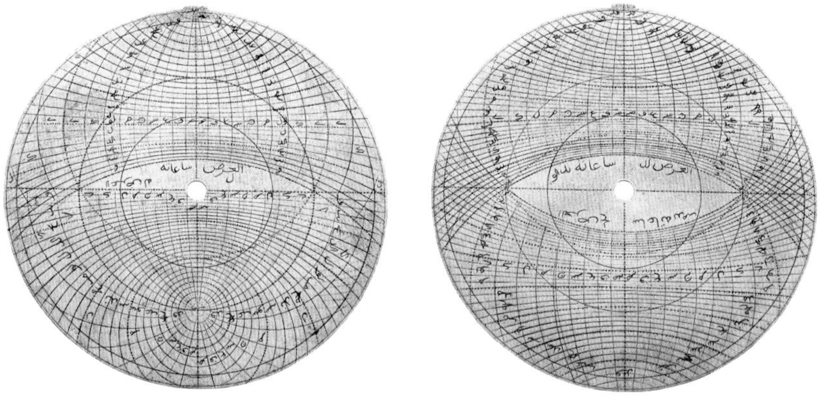

Tympan #5 of No. 19

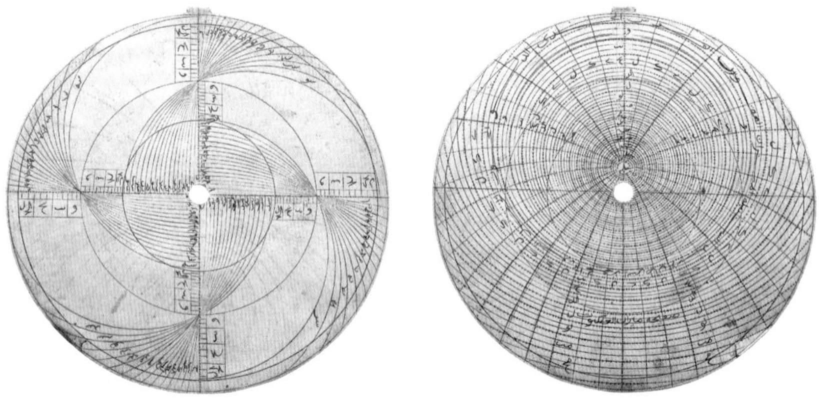

Tympan #6 of No. 19

Tympan #7 of No. 19

OUTER CIRCLE

City	Longitude	Latitude	Identification
From the equator to the first clime			
qanbalah jazīrah	21;0°	3;0°	(Madagascar)
jamkūt	177;0°	5;0°	(Yamakoṭ)
shalā	180;0°	5;0°	(in China)
ghānah	76;0°	11;0°	(coordinates of ʿAdan) (Ghana)
sarāndīb	130;0°	10;0°	(Śrīlaṅkā)
Beginning of the first clime 12;48°			
jublah	75;0°	13;30°	Jubil
janad	75;30°	14;30°	(in Yemen)
burhānbūr	103;0°	20;30°	Burhanpur
junīr	108;0°	19;48°	
tānah	104;0°	19;20°	Thana
juyūl	106;40°	19;0°	(Kennedy-Kennedy p. 167)
bījāpūr	105;30°	17;10°	Bījapur
dalatābād	102;0°	20;30°	Daulatabad
ṭā'if	77;30°	21;20°	Tā'if
Beginning of the second clime 20;31°			
makkah musharrafah	77;10°	21;40°	Mecca
madīnah ṭayyibah	75;10°	25;0°	Medina
hurmūz	92;0°	25;0°	Hormoz
akbarābād	114;0°	26;43°	Agra
ganawj	115;50°	26;35°	Kannauj
ajmīr	111°	26;0°	Ajmer
kwāliyar	114;0°	26;29°	Gwalior
lahknaw	114;53°	26;30°	Lucknow
jawnbūr	119;6°	36;36° (for 26;36°)	Jaunpur
awadah	118;6°	27; 22°	Ayodhya
budāūn	119;59° (for 114;59°)	27;32°	Budaun
banāras	117;20°	26;15°	Varanasi
aḥmadabād	108;40°	23;15°	Ahmedabad
kanbāyat	109;20°	22;20°	Cambay
Beginning of the third clime 27;34°			
kūlkandah	114;39°	28;4°	Golconda
fīrūzābād	87;30°	28;10°	Fīrūzābād
delhī	113;35°	28;39°	Delhi
hānsī	114;25° (for 112;25°)	28;45°	Hansi
bānī pahtaḥ	113;20°	28;52°	Panipat
baran	114;0°	28;48°	Bulandshahr
multān	107;35°	29;40°	Multan
shīrāz	88;0°	29;16°	Shīrāz
baṣraḥ	84;0°	30;0°	Basra
kirmān	91;30°	30;5°	Kermān

MIDDLE CIRCLE

City	Longitude	Latitude	Identification
tahānisar	112;13°	30;10°	Thānīsar
miṣr	63;20°	30;20°	Cairo
sunnām	112;25°	82 (for 30;30°)	Sunam
parshawar	105;55°	31;0°	Pershawar
kūfaḥ	79;30°	31;30°	Kūfah
qayrawān	41;0°	31;40°	Kairawan
bayt al-maqdis	66;30°	31;50°	Jerusalem
lāhawar	109;20°	31;50°	Lahore
yazd	89;0°	32;0°	Yazd
iṣfahān	81;40°	32;25°	Iṣfahān
qandahar	107;40°	33;0°	Kandahar
baghdād	82;0°	33;25°	Baghdād
Beginning of fourth clime 33;43°			
kāshān	86;0°	34;0°	Kashan
herāt	62;10°	34;20°	Herat
sāwah	85;0°	35;0° (for 36; 15°)	Sāveh
kashmīr	108;0°	35;0°	Srinagar
halab	72;0°	35;50°	Aleppo
mashhad	92;30°	36;0°	Mashhad
balkh	101;0°	36;51°	Wazirabad
marāghah	82;0°	37;20°	Marāgheh
Beginning of fifth clime 39°			
samarqand	99;16°	39;37°	Samarkand
bukhārā	97;30°	39;50°	Bukhara
tibat	110;0°	40;0°	(Tibet)

INNER CIRCLE

City	Longitude	Latitude	Identification
khutan	101;0°	42;0°	Khotan
farghāna	102;0°	42;20°	Fargana
īlāg	99; 30°	43;6°	Angren
khujand	97;45° (for 105;35°)	43;30° (for 41;55°)	Leninabād
Beginning of the sixth clime 43;30°			
kāshghar	106;30°	44;0°	Kashgar
alān	43;0° (for 73;0°)	44;0°	Alan
bish bāligh	111;0°	45;50°	Pochongtse
khān bāligh	124;46°	46;0°	Beijing
haraqlah	67;20°	46;30°	Ereğli
Beginning of the seventh clime 47;30°			
araq	85;0°	48;0°	Araq
aklah	108;0°	49;0°	Aklah
qirqir	65;30°	50°	Qirqir (in Crimea)
Outside the seventh clime			
mārī kirmān	64;45°	50;40°	Mari Kirmān
jarāsir fard	--	64;41°	
ṣūdāq	66;0°	51;0°	Sudak

20 Astrolabe

Maker unknown
India (Lahore?)
17th century
16.5 x 10 x 4.1 cm
Mater thickness 0.7 cm
Brass, except that the tympans are silvered
M-39

Face of No. 20

Astrolabe recessed on both sides of the body. Made in India (perhaps Lahore), probably in the seventeenth century. Inscriptions in naskh.

Consists of body, rete, horizon rete, uninscribed diopter with sights, pin, horse, and four tympans which have notches at the bottom fitting about a corresponding tab in the body.

The rims of both the front and the back of the body are graduated for 1°; the numbers, running clockwise, in each quadrant are 5, 10, 15, ...90. In the recesses on both sides are geographical tables in two circles arranged by the climata; the names of the climata are given in Persian. The names of the cities are written between their coordinates. (☽)(●)

On the rete the ecliptic is graduated for every 2°; the numbers, for each zodiacal sign, are 6, 12, 18, ...30. The names of the stars (some pointers are uninscribed) are listed. (★)

The first three tympans, which are rather crudely executed, have altitude circles for every 2° and equal and unequal hour curves, but no azimuths. They indicate their latitudes and longest daylights. (■)

Tympan IVa is a double projection north and south; the inscriptions indicate for latitudes 72° and 35°. IVb is a substitute for the

Reverse of No. 20

normal back. The upper half is graduated for every 2°; the numbers, clockwise from the left to the top and counter-clockwise from the right to the top, are 6, 12, 18, ...90. At the edge of the lower half are indicated the cotangents. In the upper left quadrant are crude cosines for every 6°. In the upper right quadrant are six arcs of 90° concentric with the center of the plate. Below the center is the shadow-square for 7 feet and 12 digits, within which are, on the right, a table of the planetary lords (indicated by their final letters) of the four triplicities and, on the left, a table of the annual excess of revolution for 1 to 9 years and for 10 to 90 years. (■)

The horizon rete has horizons for latitudes 37° and 38°.

PROVENANCE — A.W.M. Mensing, Amsterdam, 1924; Max Adler, Chicago, 1930; AP gift, 1930.

ICA 91

EXHIBITIONS — Adler Planetarium, "Episodes in the Story of Time," April - Sept. 2001.

REFERENCES — Engelmann Catalogue (1924), p. 16; Gunther (1932), pp. 220, 218, and XLVIII; Turner, A. J. (1985), pp. 78-79.

❍ OUTER CIRCLE: SIDE I

City	Longitude	Latitude	Identification
Equator in Clima 1			
jamkūt	180;0°	5;0°	(Yamakoṭi)
ʿadan	76;0°	11;0°	Aden
sarāndīb	130;0°	10;0°	(Śrīlaṅkā)
zabīd	74;20°	14;10°	Zabid
jabal	75;0°	13;30°	Jabal Zuqar Island
janad	75;30°	14;30°	
sawmnāt	106;0°	17;0°	Somnath
zaytūn	154;0°	17;55°	Hang Zhou
bījābūr	105;30°	17;20°	Bijapur
ṭāʾif	77;30°	21;20°	Tāʾif
makkah musharrafah	77;10°	21;40°	Mecca
madīnah ṭayyībah	75;10°	25;0°	Medina
Clima 2			
khaybar	75;20°	74;30°	Khaybar
hurmūz	94;0°	25;0°	Hormoz
akbarābād	115;0°	26;43°	Agra
kwāliyar	114;0°	26;29°	Gwalior
qanawj	115;50°	26;35°	Kannauj
ajmīr	111;5°	26;0°	Ajmer
jiddah	76;0°	21;0°	Jiddah
aḥmadābād	103;40°	23;15°	Ahmedabad
burhānbūr	103;0°	20;30°	Burhanpur
Clima 3			
dawlatābād	111;0°	20;30°	Daulatabad
kanbāyat	109;20°	22;20°	Cambay
qayrawān	41;0°	31;40°	Kairouan
bayt al-maqdis	66;30°	31;50°	Jerusalem
kūfah	79;30°	31;30°	Kūfah
miṣr	63;20°	30;20°	Cairo
madāʾin	72;0°	33;15°	Ctesiphon
dimashq	70;0°	33;15°	Damascus
baghdād	82;0°	33;25°	Baghdād
fīrūzabād	87;30°	28;10°	Fīrūzābād
yazd	89;0°	32;0°	Yazd
iṣfahān	87;40°	32;25°	Eṣfahān
shīrāz	83;0°	29;36°	Shīrāz
kirmān	91;30°	30;5°	Kermān

● INNER CIRCLE: SIDE I

City	Longitude	Latitude	Identification
Clima 3			
qandahār	107;46°	33;0°	Kandahar
multān	107;35°	29;42°	Multan
burushawar	105;15°	35;24°	Peshawar
siyālkūt	109;0°	33;0°	Sialkot
lāhawar	109;20°	31;50°	Lahore
tānīsar	112;33°	30;10°	Thanesar
dihlī	118;35°	28;19°	Dehli
kūlkundah	114;19°	18;20°	Golconda
ʿānah	76;0°	34;0°	ʿĀnah
mawṣil	77;0°	34;30°	Mosul
karaj	85;45°	34;0°	Karaj
Clima 4			
kāshān	86;0°	34;0°	Kāshān
qumm	85;40°	35;15°	Qom
rayy	86;20°	35;45°	Shahr Rey
sāwah	85;0°	34;40°	Sāveh
jurfādaqān	85;30°	34;45°	Golpayegan
hamadān	83;0°	35;0°	Hamadan
raqqah	73;15°	36;0°	Raqqa
mashhad	92;30°	36;0°	Mashhad

Mater of No. 20

OUTER CIRCLE: SIDE II

City	*Longitude*	*Latitude*	*Identification*
Clima 4			
suhraward	83;20°	36;0°	
sabzawār	91;30°	36;5°	Sabzevār
basṭām	89;30°	36;10°	Bastam
nīshābūr	92;30°	36;21°	Neyshābūr
abhar	84;30°	36;45°	Abhar
astarābād	89;35°	36;50°	Gorgān
qazwīn	85;0°	36;15°	Qazvīn
marāghah	82;0°	37;20°	Marāgheh
tabrīz	82;0°	38;0°	Tabrīz
nakhjawān	81;15°	38;40°	Nakhichevan
balkh	101;0°	36;51°	Wazirabad
badakhshān	104;24°	37;10°	Faizabad
marw	97;0°	37;40°	Mary
harāt	94;30°	34;30°	Herat
kābul	104;40°	34;30°	Kabul
kashmīr	108;0°	35;0°	Srinagar
ardabīl	80;30°	38;0°	Ardabīl
nawshahr	78;20°	38;10°	Nowshahr
arjīsh	77;0°	38;30°	Ercis
sarakhs	94;30°	37;30°	Sarakhs
bulūr	108;0°	36;0°	
<a>ndarāb	108;45°	36;0°	Andarab
ṭāliqān	85;45°	36;10°	Taliq-ān
fāryāb	99;0°	36;44°	Doulatabad
jurjān	90;0°	36;50°	Jurjān
bākawiyyah	84;30°	39;30°	Baku
Clima 5			
samarqnad	94;17°	39;37°	Samarkand
bukhārā	97;30°	39;50°	Bukhara
bardaʿah	83;0°	40;30°	Barda
shamākhī	84;30°	40;50°	Shemakha
qūnya	66;30°	41;0°	Konya
khujand	105;35°	41;55°	Leninabad
ṭāj	104;0°	42;0°	Tashkent
khutan	107;0°	42;0°	Khotan
farghānah	102;0°	42;20°	Fargana

INNER CIRCLE: SIDE II

City	*Longitude*	*Latitude*	*Identification*
Clima 5			
bāb al-abwāb	85;0°	43;0°	Derbent
īlāq	99;10°	43;20°	Angren
kāshghar	106;30°	44;0°	Kashgar
Clima 6			
alān	83;0°	44;0°	(Alans)
nabalūnah	42;0°	44;0°	Napoli
shulakh	100;30°	44;0°	Chulak-Kurgan
ṭarāz	99;50°	44;31°	Dzhambul
bīsh bāligh	111;0°	44;50°	Pochongtse
qarāqūrūm	115;0°	46;0°	Karakoram
khān bāligh	124;0°	46;0°	Beijing
haraqlah	67;20°	46;30°	Ereğli
araq	100;0°	48;0°	
Clima 7			
aklah	108;0°	49;0°	
bulghār	90;0°	49;30°	(Bulgars)
aqjā kirmān	62;0°	50;0°	Belgoraddnestrov
qirqir	65;30°	50;0°	
mārī kirmān	65;45°	50;40°	
Outside the Climata			
jinah jurābṣūr (?)	68;40°	62;0°	Hebrides Islands
ṣūdāq	66;0°	51;0°	Sudak

Reverse of No. 20

Rete of No. 20

★ STAR LIST

Name	*Identification*
baṭn ḥūt	γ Andr.
tālī masāf al-nahr	θ Erid.
jāthī	α Herc.
janāḥ al-ghurāb	γ Corv.
al-ḥawwā'	α Oph.
khaḍīb	β Cass.
dabarā<n>	α Tau.
dhanab al-jady	δ Capr.
dhanab al-dajājah	α Cygn
dhanab al-dulfīn	ε Delph.
dhanab al-qayṭīs	ι Ceti
rijl al-jawzā' yusrā	β Ori.
rijl yumnā	κ Ori.
rukbah yusrā	ζ Oph.
sāq sākib al-mā'	δ Aquar.
simāk al-a'zal	α Virg.
sha'āmiyyah	α Can. Min.
shi'rā yamāniyyah	α Can. Mai.
ṣadr al-qayṭus	π Ceti
ṭā'ir	α Aquil.
'anāq	ζ Urs. Mai.
'unq ḥayyah	α Serp.
ghūl	β Pers.
fard al-shujā'	α Hydr.
fakkah	α Cor. Bor.
fam faras	ε Peg.
fam al-qayṭīs	γ Ceti
qā'idat al-bāṭiyyah	α Crat.
minqār dajājah	β Cygn.
mankib faras	β Peg.
wāqi'	α Lyr.
yad yamīn	δ Oph. (?)
yad yamīn	α Ori.

Horizon rete of No. 20

TYMPANS

Tympan	Latitude	Longest Daylight
Ia.	22°	13;22 hours
Ib.	25°	13;25 (read 13;35) hours
IIa.	27°	13;43 hours
IIb.	32°	14;8 hours
IIIa.	30°	13;58 hours
IIIb.	36°	14;32 hours.

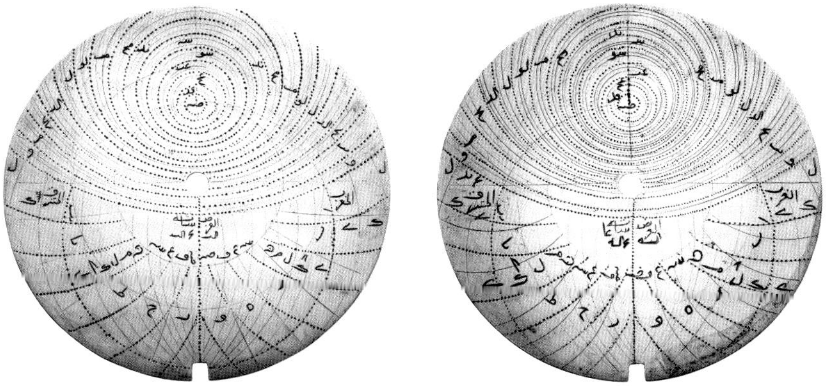

Tympan #1 of No. 20

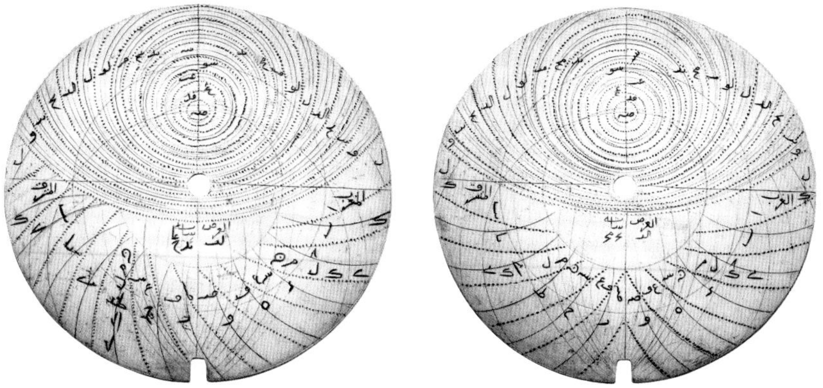

Tympan #2 of No. 20

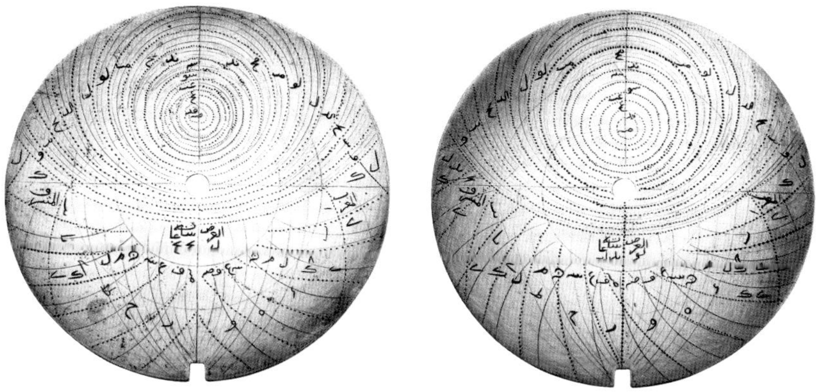

Tympan #3 of No. 20

Tympan #4 of No. 20

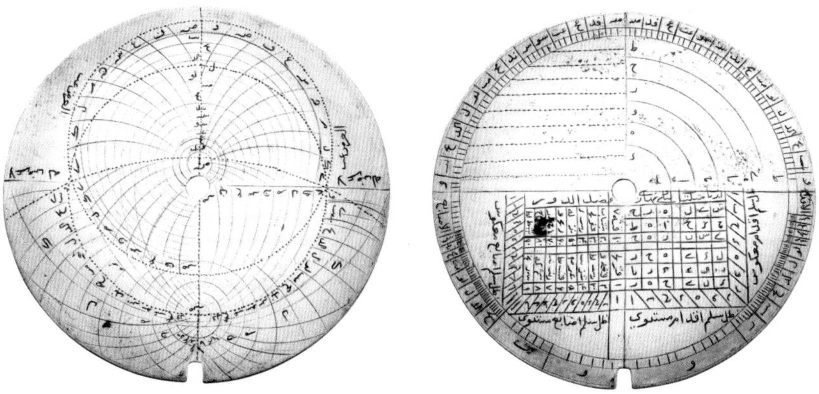

EXCESS OF REVOLUTIONS

Years	Excess	Years	Excess
1	87;53°	10	155;31°
2	175;6°	20	311;2°
3	262;59°	30	106;33°
4	350;12°	40	262;4°
5	77;5°	50	57;35°
6	165;59°	60	213;6°
7	252;52°	70	8;37°
8	340;25°	80	<164>;8°
9	67;18°	90	319;39

21 Astrolabe

al-'Abd Amīn of Mashhad
Iran
1669/70 or 1688/9
"al-'Abd Amīn of Mashhad made it in the year 1080 (or 1100)."
16.8 x 11 x 3.3 cm
Mater thickness 1 cm
Brass
A-40

Face of No. 21

Astrolabe made by al-'Abd Amīn of Mashhad in A.H. 1080 (A.D. 1669/70) or in 1100 (A.D. 1688/9) according to the inscription below the shadow-square on the back: *ṣana'ahu al-'abd amīn al-mashhadī fī sanah '.f.* (or '.*q*.). Inscriptions in naskh.

Consists of body, rete, uninscribed diopter with sights, pin, horse, handle, ring, three tympans, which have tabs at the bottom fitting into a corresponding slot in the body, and a leather pouch.

The rim of the front of the body is graduated for every degree; the numbers, running clockwise in each quadrant, are 5, 10, 15, ...90. In the recess in the front is a geographical table in a circle; it is arranged by increasing longitudes. (1)

The upper half of the back is graduated for every degree; the numbers, running clockwise from the left to the top and counter-clockwise from the right to the top, are 5, 10, 15, ...90. At the edge of the lower half are indicated the cotangents. In the upper left quadrant are 15 concentric 90° arcs and 18 radii. In the upper right quadrant are the noon-altitude curves of the Sun for Iṣfahān and Kāshān, and the azimuths of the qibla for Ṭūs, Yazd, and Qazwīn. Below the center is a shadow-square for 7 feet and 12 digits, within which is a table displaying the elements, the zodiacal signs (indicated by the numbers from 0 to 11), and the diurnal, common, and nocturnal planetary lords (indicated by their final letters) of the four triplicities. Below this are six semicircles indicating the names of the zodiacal signs, the lunar mansions, the planetary lords of the terms (indicated by their final letters), the ends of the terms, the planetary lords of the decans according to the Greek system, and the ends of the decans.

On the rete the ecliptic is graduated for every degree; there are no numbers. The names of the stars (some pointers — e.g. that for a Ori. — are uninscribed) are listed. (★)

Reverse of No. 21

The three tympans have altitude circles for every 3°, azimuths for every 10°, and equal and unequal hour curves. They indicate their latitudes and longest daylights. (■)

IIIb has the horizons for latitudes 18°, 24°, 30°, 36°, 42°, 48°, 54°, 60°, and 66°; for 16°, 22°, 28°, 34°, 40°, 46°, 52°, 58°, and 64°; for 17°, 23°, 29°, 35°, 41°, 47°, 53°, 59°, and 65°; and for 9°, 15°, 21°, 27°, 33°, 39°, 45°, 51°, 57°, and 63°.

PROVENANCE — Ali-Kuli Khan of the House of Persian Art, New York, 1934; AP purchase, 1934.
ICA 2542

Leather pouch of No. 21

① CIRCLE

City	*Longitude*	*Latitude*	*Inclination*	*Identification*
tabrīz	82;0°	38;0°	15;40°	Tabrīz
hamadān	83;0°	35;10°	22;17°	Hamadan
qazwīn	85;0°	36;0°	27;34°	Qazvīn
sāwah	85;0°	35;0°	29;18°	Sāveh
qumm	85;40°	34;45°	31;55°	Qom
kāshān	86;0°	34;0°	34;33°	Kāshān
iṣfahān	86;40°	32;25°	40;29°	Eṣfahān
shīrāz	88;0°	29;36°	53;20°	Shīrāz
simnān	88;0°	36;0°	34;38°	Semnan
yazd	89;0°	32;0°	48;27°	Yazd
astarābād	89;35°	36;50°	38;47°	Gorgān
sabzawār	91;30°	36;0°	44;16°	Sabzevār
ṭūs	92;30°	37;0°	45;9°	Ṭūs
tūn	92;30°	34;30°	50;25°	Ferdows
nīshābūr	92;30°	36;21°	46;26°	Neyshābūr
qayin	93;20°	33;40°	54;13°	Qāyen
harāt	94;20°	34;35°	53;14°	Herat
samarqand	99°	39;37°	52;33°	Samarkand
mūltān	107;35°	29;40°	79;30°	Multan
qandahār	104;40°	33;0°	75;10°	Kandahar
kashmīr	108;0°	35;0°	71;51°	Srinagar
akrah	114;0°	27;13°	91;0°	Agra
makkah mu-ʻaẓẓamah	77;10°	21;40°	0;0°	Mecca

Mater of No. 21

★ STAR LIST

Name	*Identification*
ākhir al-nahr	θ Erid.
a'zal	α Virg.
janāḥ	γ Corv.
jawn	ε Urs. Mai.
ra's	α Oph.
rāmiḥ	α Boot.
rijl	β Ori.
ridf	α Cygn.
sāq	δ Aquar.
sha'āmī	α Can. Min.
ṭā'ir	α Aquil.
'ābarah (?)	in Pegasus.
'unq	α Serp.
'ayn	α Tau.
'ayyūq	α Aur.
fard	α Hydr.
faras	β Peg.
fakkah	α Cor. Bor.
fam	ε Peg.
qayṭus	ι Ceti
kaff	β Cass.
mankib	between Boot. and Urs. Mai.
wāqi'	α Lyr.
yamānī	α Can. Mai.

Rete of No. 21

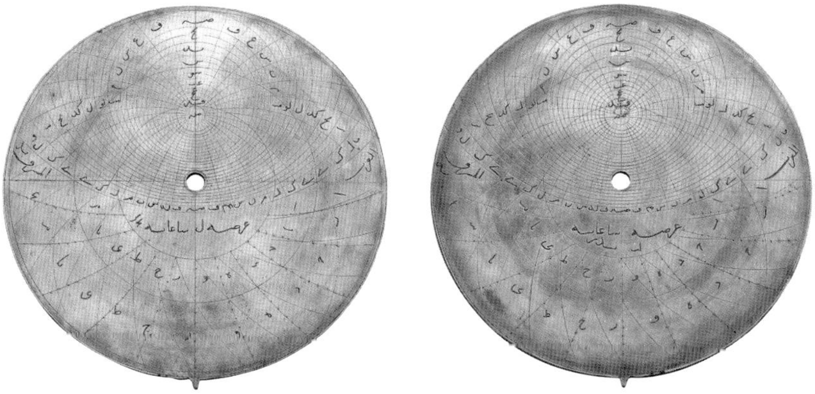

Tympan #1 of No. 21

Tympan #2 of No. 21

■ TYMPANS

Tympan	*Latitude*	*Longest Daylight*
Ia.	30°	13;57 hours
Ib.	32°	14;7 hours
IIa.	34°	14;17 hours
IIb.	36°	14;27 hours
IIIa.	37°	14;34 hours

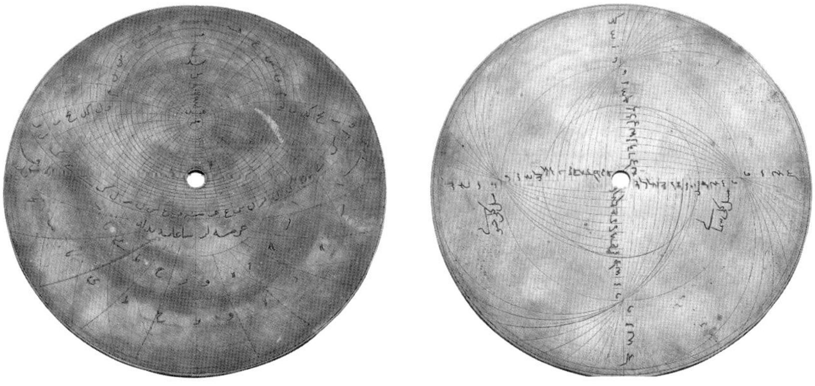

Tympan #3 of No. 21

22 Astrolabe (incomplete)

Khalīl Muḥammad (except for rete)
Iran
1697/8 (rete made in 1830/1)
"Muḥammad Khalīl ibn Ḥasan ʿAlī made it in 1109. Forgive him."
"Determination (of the longitudes) of the stars in 1218;
construction of the rete in 1246" (on the rete).
18.3 x 12 x 3.7 cm
Mater thickness 0.6 cm
Brass
A-82

Face of No. 22

Incomplete astrolabe made by Muḥammad Khalīl (Khalīl Muḥammad) ibn Ḥasan ʿAlī in A.H. 1109 (A.D. 1697/8) according to the inscription below the shadow-square on the back: *ṣanaʿahu ibn ḥasan ʿalī muḥammad khalīl 1109. ʿafā ʿanhu.* The rete was made in A.H. 1246 (A.D. 1830/1) with star positions for A.H. 1218 (A.D.1803/4) according to the inscription in the center: *istikhrāj kuwākib fī 1218. ʿamal ʿankabūt fī 1246.* Inscriptions in naskh.

Consists of body, uninscribed diopter with sights, pin, horse, handle, and ring. The five tympans, which have notches at the bottom fitting about a corresponding tab in the body, seem to be from another instrument, and the rete was made only in 1830/1.

The rim of the front of the body is graduated for every degree; the numbers, running clockwise, are 5, 10, 15, ...360. In the recess in the front is a geographical table in two circles. (☽)(●)

The upper half of the back is graduated for every degree; the numbers, running clockwise from the left to the top and counter-clockwise from the right to the top, are 5, 10, 15, ...90. At the edge of the lower half are indicated the cotangents. In the upper left quadrant are the sines and cosines for each of the 60 parts of the Radius. In the upper right quadrant are the noon-altitude curves of the Sun for latitudes 28°, 30°, 32°, 34°, 36°, 38°, 40°, 42°, 44°, 46°, and 48°; above is the inscription: *dawāʾir inṣāf al-nuhur fī al-ʿurūḍ al-marqūmah ʿalā aṭrāfahā* ("circles of noon in the latitudes written at their ends"). In the same quadrant are the azimuths of the qibla for Ṭūs, Iṣfahān, Baghdād, and Shīrāz; below is the inscription: *khuṭūṭ sumūt al-qiblah fī al-bilād al-marqūmah ʿalā aṭrāfahā bi-irtifāʿ al-gharbī* ("azimuths of the qibla in the cities written at their ends in the western altitude"). Below the center

Reverse of No. 22

is the shadow-square for 7 feet and 12 digits. Below this is a semicircle indicating the planetary lords (denoted by their final letters) of the decans according to the Greek system, the names of the zodiacal signs, the planetary lords (denoted by their final letters) of the terms, and the lengths of the terms.

On the rete the ecliptic is graduated for every 3°. The star pointers are decorative rather than useful; they are shaped like flowers, and the one in the center top is in the shape of scorpion. Many are uninscribed; the names that do occur are listed. (★)

The five tympans have altitude circles 10 every 3°, azimuths for every 10°, and unequal hour curves (IIIb also has equal hour curves). They indicate their latitudes and longest daylights. (■)

Vb has the horizons for latitudes 14°, 22°, 30°, 38°, 46°, 54°, and 62°; for 16°, 24°, 32°, 40°, 48°, 56°, and 64°; for 18°, 26°, 34°, 42°, 50°, 58°, and 66°; and for 20°, 28°, 36°, 44°, 52°, and 60°.

PROVENANCE — Ali-Kuli Khan; AP purchase, before 1945.
ICA 2555

Mater of No. 22

OUTER CIRCLE

City	Longitude	Latitude	Inclination	Distance	Identification
yazd	89;0°	32;0°	48;27°	333	Yazd
iṣfahān	86;40°	32;25°	40;29°	304	Eṣfahān
kāshān	86;0°	34;0°	34;31°	323	Kāshān
qumm	85;40°	34;45°	31;54°	335	Qom
sāwah	85;0°	35;0°	29;16°	333	Sāveh
qazwīn	85;0°	36;0°	27;34°	313	Qazvīn
hamadān	83;0°	35;10°	22;16°	321	Hamadan
rayy	86;20°	35;9°	36;26°	335	Shahr Rey
astarābād	89;35°	36;50°	38;48°	417	Gorgān
simnān	88;0°	36;0°	36;17°	382	Semnān
sabzawār	91;30°	36;0°	44;12°	423	Sabzevār
nīshābūr	92;30°	36;21°	46;25°	440	Neyshābūr
ṭūs	92;30°	37;0°	45;6°	451	Ṭūs
tūn	92;30°	34;30°	50;20°	417	Ferdows
qāyin	93;20°	33;40°	54;1°	414	Qāyen
harāh	94;20°	34;35°	54;8°	440	Herat
qandahār	107;40°	33;0°	74;0°	649	Kandahar
marw	97;0°	37;40°	52;30°	515	Mary
balkh	101;0°	36;41°	60;36°	567	Wazirabad
bust	100;0°	33;0°	68;30°	504	Busht
multān	107;35°	39;40°	80;48°	633	Multan
akrah	114;0°	27;13°	90;58°	752	Agra
kashmīr	108;0°	35;0°	71;51°	667	Srinagar

INNER CIRCLE

City	Longitude	Latitude	Inclination	Distance	Identification
tabrīz	82;0°	38;0°	15;40°	374	Tabrīz
ardabīl	82;30°	38;0°	17;13°	377	Ardabīl
marāghah	82;0°	37;20°	16;17°	378	Marāgheh
sārī	88;0°	37;0°	32;54°	400	Sārī
āmul	88;20°	36;15°	34;37°	393	Amol
liḥsā	83;30°	24;0°	69;30°	133	Aḥsā
baḥrayn	83;0°	25;15°	57;23°	143	Manāmah
baṣrah	84;0°	30;0°	36;5°	229	Basra
baghdād	80;30°	33;25°	12;45°	266	Baghdād
shīrāz	88;0°	29;36°	53;20°	279	Shīrāz
kirmān	92;30°	29;50°	63;0°	370	Kermān

★ STAR LIST

Name	*Identification*
janūbī qayṭus	β Ceti
dhanab al-jady	δ Capr.
simāk aʿzal	α Virg.
simāk rāmiḥ	α Boot.
shiʿrā yamānī	α Can. Mai.
ʿayyūq	α Aur.
fard al-shujāʿ	α Hydr.
kaff al-khaḍīb	β Cass.
nasr ṭā'ir	α Aquil.
nasr wāqiʿ	α Lyr.
nayyir al-fakkah	α Cor. Bor.

Rete of No. 22

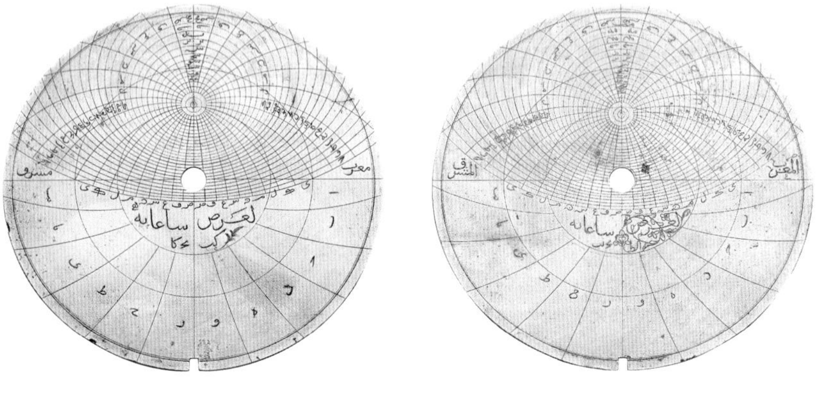

Tympan #1 of No. 22

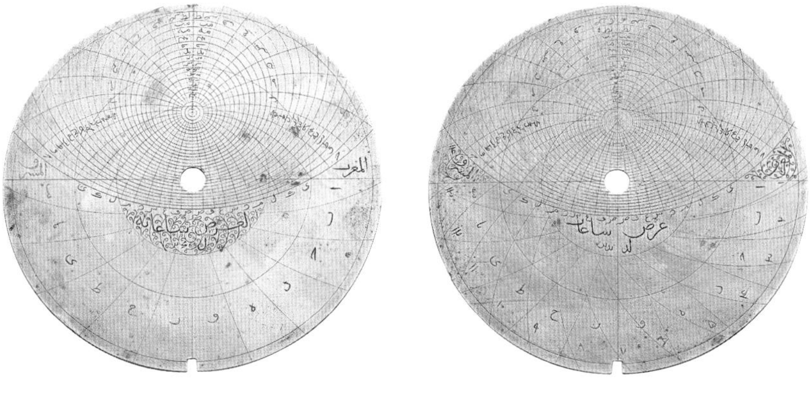

Tympan #2 of No. 22

TYMPANS

Tympan	*Latitude*	*Longest Daylight*
Ia.	22°	13;21 hours
Ib.	29°	13;52 hours
IIa.	30°	13;57 hours
IIb.	34°	14;17 hours
IIIa.	35°	14;22 hours
IIIb.	36°	14;28 hours
IVa.	37;20°	14;35 hours
IVb.	38°	14;39 hours
Va.	40°	14;51 hours

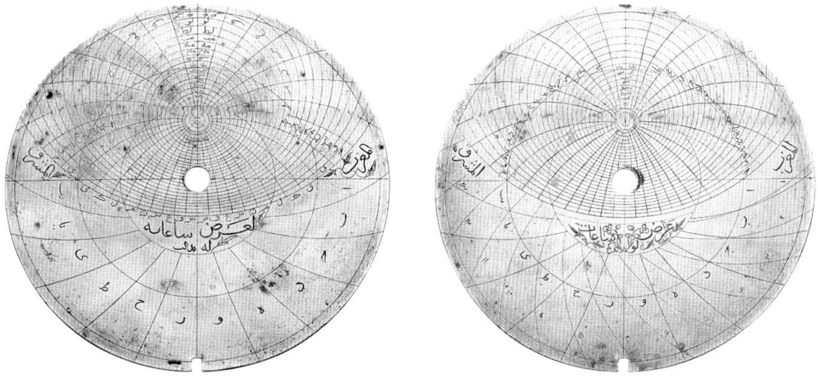

Tympan #3 of No. 22

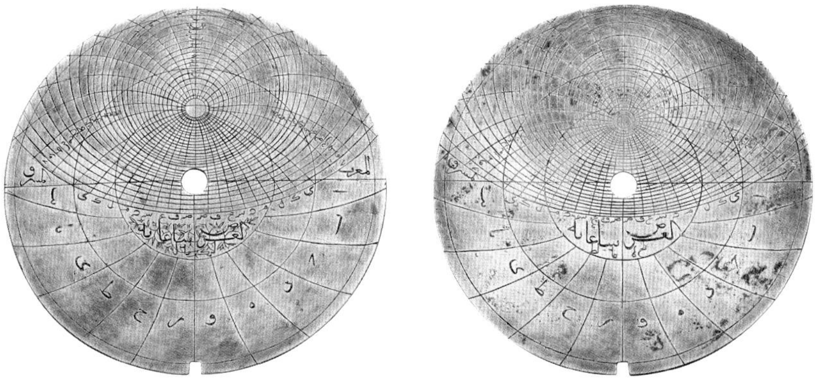

Tympan #4 of No. 22

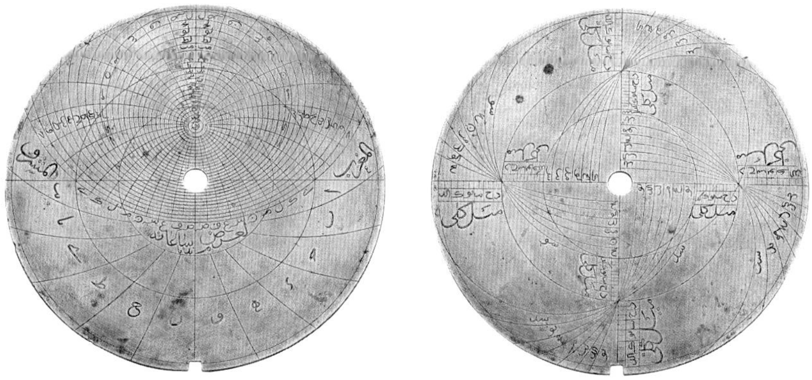

Tympan #5 of No. 22

23 Astrolabe

Khalīl Muḥammad maker; ʿAbd al-Aʾimmah decorator
Iran
ca. 1700
"Muḥammad Khalīl ibn Ḥasan ʿAlī made it."
"ʿAbd al-Aʾimmah ibn ʿAbd al-Ḥusayn decorated it."
28 x 19.5 x 4.7 cm
Mater thickness 0.9 cm
Brass
W-89

Face of No. 23

Astrolabe made by Muḥammad Khalīl (Khalīl Muḥammad) ibn Ḥasan ʿAlī according to the inscription below the shadow-square on the back: *ʿalī muḥammad khalīl ibn ḥasan ṣanaʿahu.* It was decorated by ʿAbd al-Aʾimmah ibn ʿAbd al-Ḥusayn according to the inscription at the bottom of the back: *namaqahu ibn ʿabd al-ḥusayn ʿabd al-aʾimmah.* Inscriptions in naskh.

Consists of body, rete (perhaps from a different instrument since it does not fit perfectly), graduated diopter with scales and sights, pin, washer, horse, handle, ring, and six tympans which have notches at the bottom fitting about a corresponding tab in the body.

The rim of the front of the body is graduated for every degree; the numbers, running clockwise, are 5, 10, 15, ...360. In the recess of the front is a geographical table in two circles. (○)(●)

The upper half of the back is graduated for every degree; the numbers, running clockwise from the left to the top and counter-clockwise from the right to the top, are 5, 10, 15, ...90. At the edge of the lower half are indicated the cotangents. In the upper right quadrant are the sines and cosines for every degree from 1 to 90. In the upper right quadrant are the noon altitude curves of the Sun for latitudes 28°, 30°, 32°, 34°, 36°, 38°, 40°, 42°, 44°, 46°, and 48°; above is the inscription: *dawāʾir inṣāf al-nahur fī al-ʿurūḍ al-marqūmah ʿalā aṭrāfahā.* In the same quadrant are the azimuths of the qibla for Ṭūs, Yazd, Iṣfahān, Hamadān, Baghdād, and Kūfah; below is the inscription: *khuṭūṭ sumūt al-qiblah fī al-bilād al-marqūmah ʿalā aṭrāfahā bi-irtifāʿ al-gharbī.* Below the center is the shadow-square for 7 feet and 12 digits, within which is a table of the elements, zodiacal signs, and planetary lords of the triplicities. Below this are five semicircles

Reverse of No. 23

indicating the planetary lords of the terms, the ends of the terms, the names of the zodiacal signs, the planetary lords (indicated by their final letters) of the decans according to the Greek system, and the names of the lunar mansions.

On the rete the ecliptic is graduated for every 2°; the numbers for each zodiacal sign are 2, 4, 6, ...30. The names of the stars are listed. (★)

The first five tympans have altitude circles for every 2°, azimuths for every 5°, and equal and unequal hour curves. They indicate their latitudes and longest daylights: (■)

VIa gives the cusps of the twelve astrological places for latitude 32° (longest daylight: 14;7 hours). VIb has the horizons for latitudes 10°, 18°, 26°, 34°, 42°, 50°, 58°, and 66°; for 12°, 20°, 28°, 36°, 44°, 52°, and 60°; for 14°, 22°, 30°, 38°, 46°, 54°, and 62°; and for 16°, 24°, 32°, 40°, 48°, 56°, and 64°.

PROVENANCE — Roderick S. Webster, Winnetka, IL, 1983; AP gift, 1983.

EXHIBITIONS — *Islamic Insights: An Introduction to Islamic Art,* Katonah Gallery, Katonah, N.Y., Dec. 27, 1979 – June 30, 1980; *The Heritage of Islam,* Islam Centennial Fourteen, Washington, D.C., Feb. 1, 1982 – Sept. 19, 1983.

REFERENCES — Edwards and Signell (1982), p. 52; Maddison (1976); Newbury (2004a); Saliba (1977b), p. 17; Turner, H. R. (1995), p. 92.

Mater of No. 23

Diopter, pin, and horse of No. 23

❍ OUTER CIRCLE

City	*Longitude*	*Latitude*	*Inclination*	*Direction*	*Identification*
makkah	77;10°	21;40°	0;0°	0;0	Mecca
madīnah	75;20°	25;0°	37;10°	SE	Medina
miṣr	63;20°	30;20°	58;38°	SE	Cairo
ṣana'ā	77;0°	14;30°	1;55°	NE	Ṣan'ā'
liḥsā	83;30°	24;0°	69;30°	SW	Aḥsā
bayt al-maqdis	66;30°	31;50°	45;16°	SE	Jerusalem
dimashq	70;0°	53;15°	30;31°	SE	Damascus
ḥalab	72;10°	35;50°	18;29°	SE	Aleppo
mawṣil	77;0°	34;30°	17;12°	SW	Mosul
nakhjawān	81;15°	18;40°	12;55°	SW	Nakhichevan
marāghah	82;0°	37;20°	16;17°	SW	Marāghah
tabrīz	82;0°	38;0°	15;40°	SW	Tabrīz
ardabīl	82;30°	33;0°	17;13°	SW	Ardabil
surra-manra'ā	79;0°	34;0°	7;56°	SW	Sāmarrā'
kūfah	79;30°	31;30°	12;31°	SW	Kūfah
baghdād	80;0°	53;25°	12;45°	SW	Baghdād
baṣrah	84;0°	30;0°	37;19°	SW	Basra
shūsh	84;30°	31;30°	35;24°	SW	Shūsh
kāzirūn	87;0°	29;15°	51;17°	SW	Kāzerūn
shīrāz	88;0°	29;0°	53;18°	SW	Shīrāz
yazd	89;0°	32;0°	48;18°	SW	Yazd
hamadān	83;0°	35;10°	22;16°	SW	Hamadan
qazwīn	85;0°	36;0°	27;34°	SW	Qazvīn
jurfādaqān	85;30°	34;14°	58;41°	SW	Golpayegan
iṣfahān	86;40°	32;24°	40;58°	SW	Eṣfahān
kāshān	86;0°	34;0°	34;31°	SW	Kāshān
qumm	85;40°	34;45°	31;54°	SW	Qom
ṭāliqān	85;45°	36;10°	29;53°	SW	Taliq-an
astarābād	89;35°	36;50°	58;48°	SW	Gorgān
simnān	88;0°	36;0°	36;17°	SW	Semnān
dāmghān	88;55°	36;20°	53;0°	SW	Dāmghān

● INNER CIRCLE

City	*Longitude*	*Latitude*	*Inclination*	*Direction*	*Identification*
basṭām	89;30°	36;10°	58;53°	SW	Bastam
sabzawār	91;30°	36;0°	44;13°	SW	Sabzevār
nīshābūr	92;30°	36;21°	46;25°	SW	Neyshābūr
ṭūs	92;30°	37;0°	45;6°	SW	Ṭūs
tūn	92;30°	34;30°	50;20°	SW	Ferdows
zūzan	93;30°	35;20°	51;13°	SW	Zūzan
marw	97;0°	37;40°	52;30°	SW	Mary
harāh	94;20°	34;35°	33;54°	SW	Herat
qāyin	93;20°	33;40°	54;20°	SW	Qāyen
ṭabas kīlak	92;0°	33;0°	53;2°	SW	Ṭabas
samarqand	99;36°	39;37°	52;53°	SW	Samarkand
bukhārā	97;30°	35;50	49;37°	SW	Bukhara
badakhshān	104;24°	37;10°	64;9°	SW	Faizabad
qandahār	104;30°	-----	74;59°	SW	Kandahar
multān	107;35°	29;40°	79;30°	SW	Multan

★ STAR LIST

Name	*Identification*
janāḥ al-ghurāb	γ Corv.
dhanab jady	δ Capr.
dhanab al-‘uqāb	ζ Aquil
dhanab qayṭus	β Ceti
ra’s al-ḥawwā’	α Oph.
rijl al-musalasalah	γ Andr.
rijl al-yusrā	β Ori.
rijl al-yumnā	κ Ori.
ridf	α Cygn.
sāq ayman sākib al-mā’	δ Aquar.
simāk a‘zal	α Virg.
simāk rāmiḥ	α Boot.
shi‘rā sha ’āmī	α Can. Min.
ṣadr qayṭus	π Ceti
ṭaraf al-sufīnah	ρ Puppis
‘anāq	ζ Urs. Mai.
‘unq al-ḥayyah	α Serp
‘ayn al-thawr	α Tau.
‘ayyūq	α Aur.
fard al-shujā‘	α Hydr.
fam al-faras	ε Peg.
fam qayṭus	γ Ceti
qā’id	η Urs. Mai.
qā‘idat al-bāṭiyyah	α Crat.
qalb al-asad	α Leo
qalb al-‘aqrab	α Scorp.
kaff al-khaḍīb	β Cass.
kiffah janūbī	α Libr.
matn al-faras	α Peg.
masāfat al-nahr	θ Erid.
minqār dajājah	β Cygn.
mankib al-faras	β Peg.
nasr ṭā’ir	α Aquil.
nasr wāqi‘	α Lyr.
nayyir fakkah	α Cor. Bor.
yad yusrā	γ Ori.
yad al-yumnā	α Ori.
yamāniyyah	α Can. Mai.

Rete of No. 23

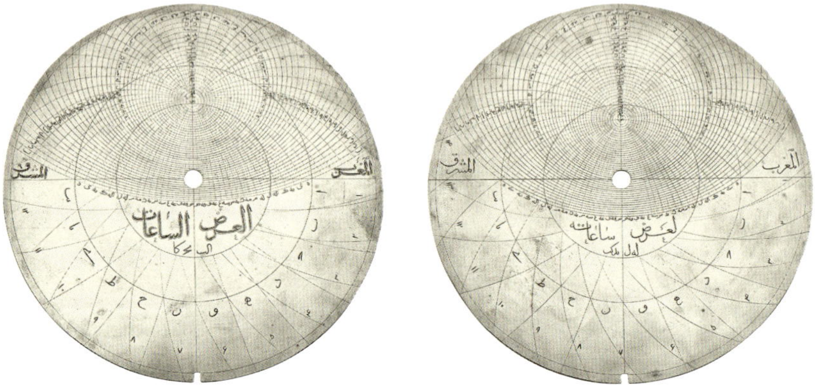

Tympan #1 of No. 23

Tympan #2 of No. 23

Tympan	*Latitude*	*Longest Daylight*
Ia.	22°	13;21 hours
Ib.	35;30°	14;22 hours
IIa.	29°	13;52 hours
IIb.	38°	14;39 hours
IIIa.	30°	13;56 hours
IIIb.	36°	14;28 hours
IVa.	32°	14;7 hours
IVb.	37°	14;38 hours
Va.	34°	14;17 hours
Vb.	40°	14;51 hours

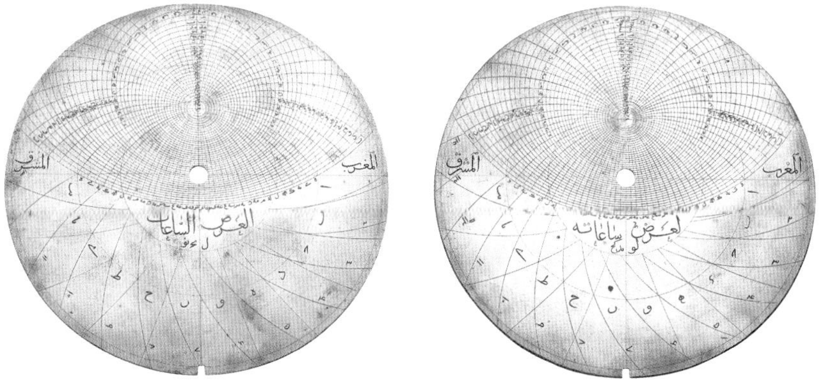

Tympan #3 of No. 23

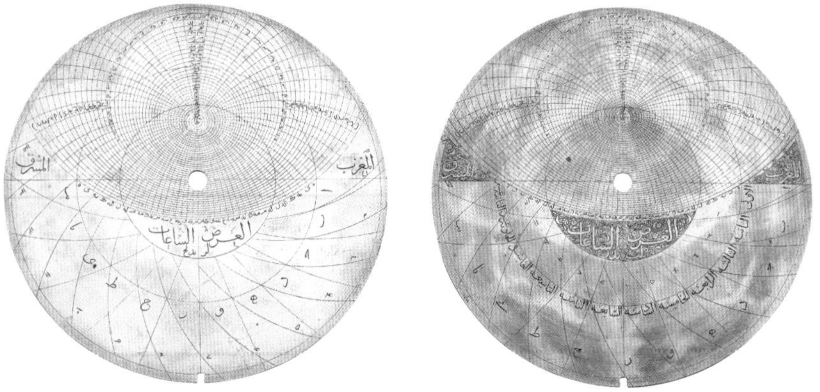

Tympan #4 of No. 23

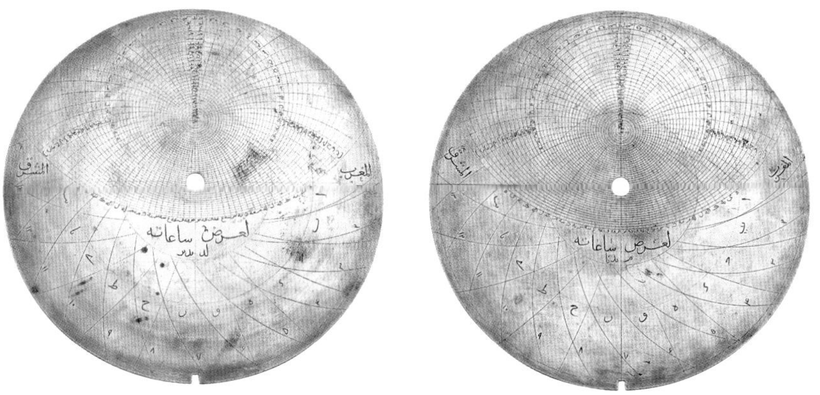

Tympan #5 of No. 23

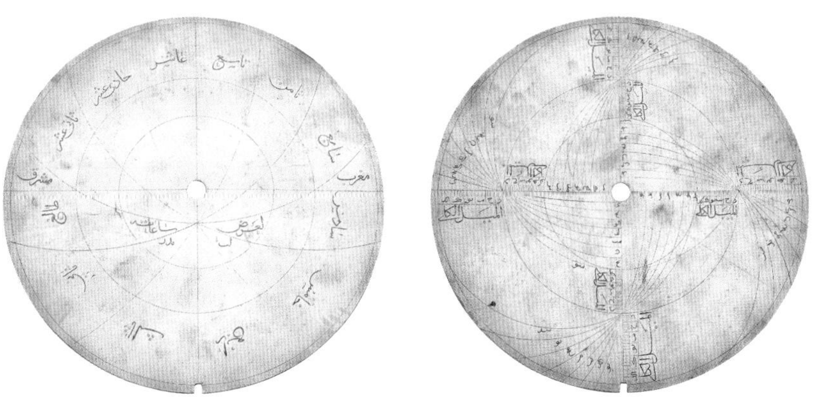

Tympan #6 of No. 23

24 Astrolabe

Probably ʿAbd al-Aʾimmah
Iran
ca. 1700
18.3 x 11.6 x 3.7 cm
Mater thickness 0.8 cm
Brass
A-75

Face of No. 24

Astrolabe in Iranian style of the late seventeenth or early eighteenth century. Throne in front left blank to receive inscription. The script and the decoration are identical with those on the instruments of ʿAbd al-Aʾimmah; *cf.* especially No. 23. Inscriptions in naskh.

Consists of body, rete, inscribed diopter with scales and sights, pin, horse, handle, ring, and five tympans which have notches at the bottom fitting about a corresponding tab in the body.

The rim of the front of the body is graduated for every degree; the numbers, running clockwise, are 5, 10, 20, 5, 10 ...100, 5, 10, ...200, 5, 10, ...300, 5, 10, ...360. In the recess in the front is a geographical table in two circles. (○)(●)

The upper half of the back is graduated for every degree; the numbers, running clockwise from the left to the top and counter-clockwise from the right to the top, are 5, 10, 15, ...90. At the edge of the lower half are indicated the cotangents. In the upper left quadrant are the sines and cosines for each of the 60 parts of the Radius. In the upper right quadrant are the noon-altitude curves of the Sun for latitudes 28°, 30°, 32°, 34°, 36°, 38°, 40°, and 42°; above is the inscription: *dawāʾir inṣāf al-nuhur fī al-ʿurūd al-marqūmah ʿalā aṭrāfahā.* In the same quadrant are the azimuths of the qibla for Ṭūs, Yazd, Iṣfahān, Baghdād, and Shīrāz; below is the inscription: *khuṭūṭ sumūt al-qiblah fī al-bilād al-marqūmah ʿalā aṭrāfahā bi-irtifāʿ al-gharbī.* Below the center is the shadow-square for 7 feet and 12 digits, within which is the table of the elements, zodiacal signs, and planetary lords of the four triplicities. Below this are five semicircles indicating the planetary lords of the terms, the ends of the terms, the names of the zodiacal signs, the planetary lords (denoted by their final letters) of the decans according

Reverse of No. 21

to the Greek system, and the names of the lunar mansions.

On the rete the ecliptic is graduated for every 3°; the numbers for each zodiacal sign are 3, 6, 9, ... 30. The names of the stars are listed. (★)

The five tympans have altitude circles for every 3°, azimuths for every 10°, and unequal hour curves (equal hour curves also on IIb, IVa, and Vb). They indicate their latitudes and longest daylights. (■)

Ia has the horizons for latitudes 9°, 21°, 33°, 45°, and 57°; for 12°, 24°, 36°, 48°, and 60°; for 15°, 27°, 39°, 51°, and 63°; and for 18°, 30° 42°, 54°, and 66°.

PROVENANCE — E.S. David, New York, 1939; AP purchase, 1940.
ICA 2548
REFERENCES — Newbury (2004a).

Mater of No. 24

Diopter, horse and pin of No. 24

OUTER CIRCLE

City	Longitude	Latitude	Inclination	Identification
makkah	77;10°	21;40°	0;0°	Mecca
madīnah	75;20°	25;0°	37;10°	Medina
miṣr	63;20°	30;20°	58;38°	Cairo
ṣana'ā	77;0°	14;30°	1;55°	Ṣan'ā'
liḥsā	83;30°	24;0°	69;30°	Aḥsā
ḥalab	72;10°	35;50°	18;29°	Aleppo
mawṣil	77;0°	34;30°	17;12°	Mosul
nakhjawān	81;15°	18;40°	12;15°	Nakhichevan
marāghah	82;0°	37;20°	16;17°	Marāghah
tabrīz	82;0°	38;0°	15;40°	Tabrīz
ardabīl	82;30°	38;0°	17;13°	Ardabīl
baghdād	80;0°	33;25°	12;45°	Baghdād
baṣrah	84;0°	30;0°	37;19°	Basra
shūshtar	84;30°	31;30°	35;24°	Shustar
shīrāz	88;0°	29;0°	53;18°	Shīrāz
yazd	89;0°	32;0°	48;18°	Yazd
iṣfahān	86;40°	32;24°	40;48°	Eṣfahān
kāshān	86;0°	34;0°	34;31°	Kāshān
qumm	85;40°	34;45°	31;54°	Qom
qazwīn	85;0°	36;0°	27;34°	Qazvīn
hamadān	83;0°	35;10°	22;16°	Hamadan
astarābād	89;35°	36;50°	58;48°	Gorgān
simnān	88;0°	36;0°	36;17°	Semnān

INNER CIRCLE

City	Longitude	Latitude	Inclination	Identification
basṭām	89;30°	36;10°	58;53°	Bastam
sabzawār	91;30°	36;0°	44;13°	Sabzevār
nīshābūr	92;30°	36;21°	46;25°	Neyshābūr
ṭūs	92;30°	37;0°	45;6°	Ṭūs
tūn	92;30°	34;30°	50;20°	Ferdows
marw	97;0°	37;40°	52;30°	Mary
kābul	104;40°	34;26°	69;35°	Kabul
badakhshān	104;24°	37;10°	64;9°	Faizabad
bukhārā	97;30°	39;50	49;18°	Bukhara
samarqand	99;56°	39;37°	54;54°	Samarkand
qandahār	104;40°	18;0°	74;31°	Kandahar
ṭāliqān	85;45°	36;10°	29;13°	Taliq-an
dāmghān	88;55°	36;20°	18;0°	Dāmghān
jurfādaqān	85;30°	34;14°	18;41°	Golpayegan
kāzirūn	87;0°	29;15°	51;17°	Kāzerūn

Rete of No. 24

★ STAR LIST

Name	*Identification*
janāḥ al-ghurāb	γ Corv.
dhanab al-ʻuqāb	ζ Aquil
dhanab qayṭus	ι Ceti
ra's al-ḥawwā'	α Oph.
rijl al-yusrā	β Ori.
ridf	α Cygn.
sāq ayman	δ Aquar.
simāk aʻzal	α Virg.
simāk rāmiḥ	α Boot.
sha'āmī	α Can. Min.
shiʻrā yamānī	α Can. Mia
ṣadr qayṭus	π Ceti
ẓahr al-dubb	α Urs. Mai.
ʻunq al-ḥayyah	α Serp
ʻayn al thawr	α Tau.
ʻayyūq	α Aur.
fard al-shujāʻ	α Hydr.
fam al-faras	ε Peg.
fam qayṭus	γ Ceti
qāʻidah	α Crat.
qalb al-asad	α Leo
qalb al-ʻaqrab	α Scorp.
kaff al-khaḍīb	β Cass.
masāfat al-nahr	θ Erid.
mankib al-faras	β Peg.
nasr ṭā'ir	α Aquil.
nasr wāqiʻ	α Lyr.
nayyir fakkah	α Cor. Bor.
yad yumnā	α Ori.

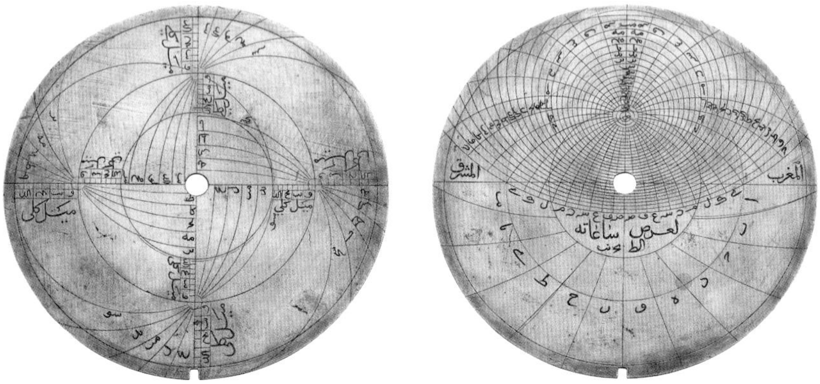

Tympan #1 of No. 24

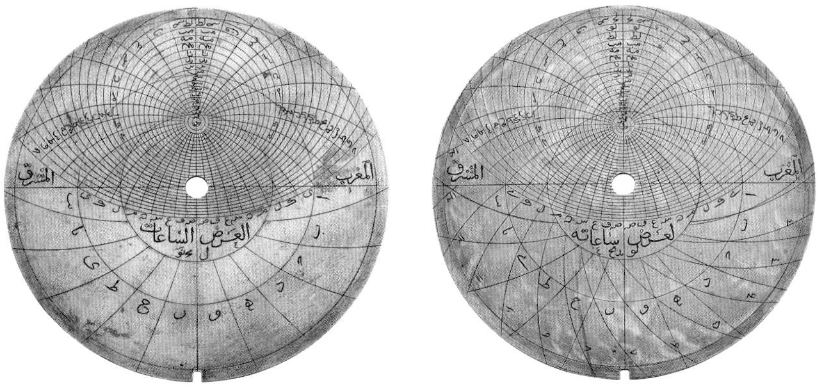

Tympan #2 of No. 24

TYMPANS

Tympan	*Latitude*	*Longest Daylight*
Ib.	29°	13;52 hours
IIa.	30°	13;56 hours
IIb.	36°	14;28 hours
IIIa.	31°	14;1 hours
IIIb.	40°	14;51 hours
IVa.	32°	14;7 hours
IVb.	38°	14;39 hours
Va.	34°	14;16 hours
Vb.	37°	14;38 hours

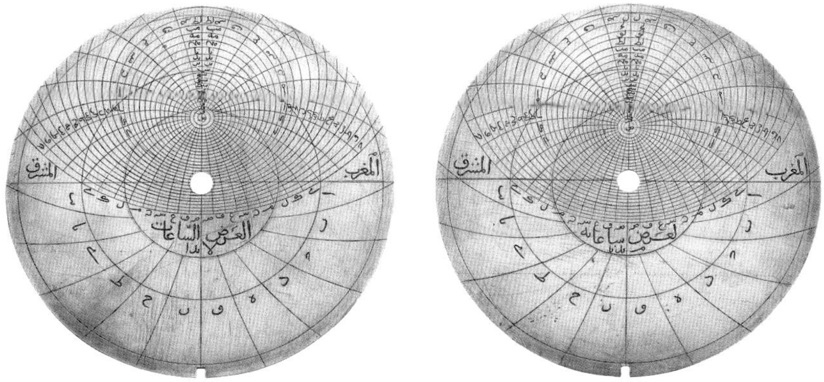

Tympan #3 of No. 24

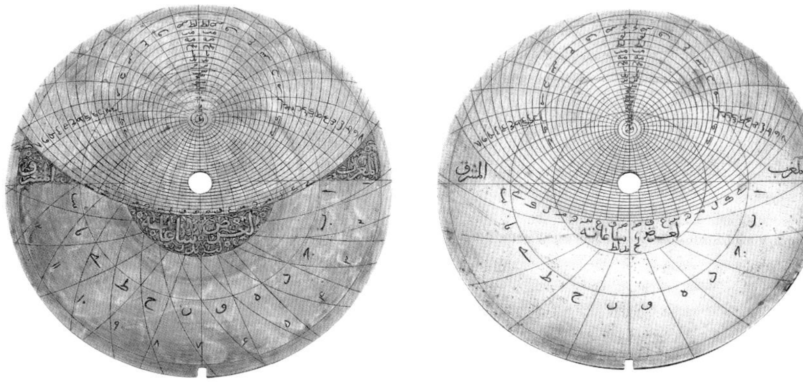

Tympan #4 of No. 24

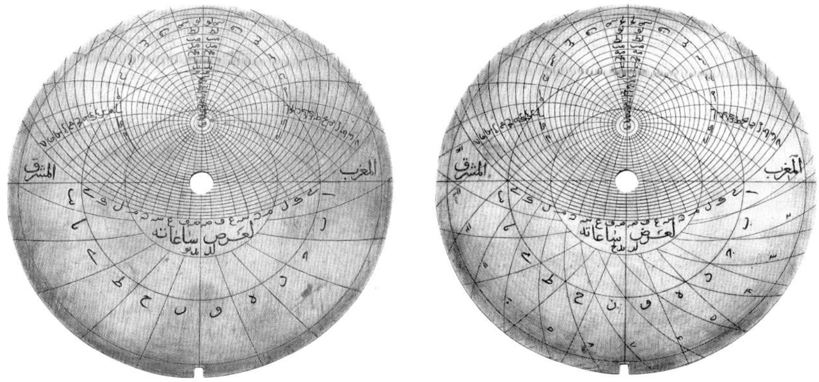

Tympan #5 of No. 24

25 Astrolabe

Muḥammad Ibrāhīm al-Ḥasanī
Iran
1699/1700
"Muḥammad Ibrāhīm al-Ḥasanī made it in the year 1111."
"Muḥammad Ibrāhīm decorated it."
10.7 x 7.2 x 2.5 cm
Mater thickness 0.5 cm
Brass
A-37

Face of No. 25

Astrolabe made by Muḥammad Ibrāhīm al-Ḥasanī in A.H. 1111 (A.D. 1699/1700) according to the inscription within and below the shadow-square on the back: *ṣanaʿahu muḥammad ibrāhīm al-ḥasanī sanah 1111.* He also decorated it, as is indicated by an inscription at the bottom of the back: *namaqahu muḥammad ibrāhīm.* The style and the names of the cities included in the geographical table show that it was made in Iran. Inscriptions in naskh.

Consists of body, rete, graduated diopter with scales and sights, pin, horse, handle, ring, and four tympans which have notches at the bottom fitting about a corresponding tab in the body.

The rim of the front of the body is graduated for every degree; the numbers, running clockwise, are 5, 10, 5, 20, 5, ... 100, 5, 10, ...200, 5, 10, ...300, 5, 10, ...60. In the recess in the front is a geographical table in a circle. (①)

The upper half of the back is graduated for every degree; the numbers, running clockwise from the left to the top and counter-clockwise from the right to the top, are 5, 10, 5, 20, 5, ...90. At the edge of the bottom half are indicated the cotangents. In the upper left quadrant are the cosines for every 2 of the 60 parts of the Radius. In the upper right quadrant are the noon-altitude curves of the Sun for latitudes 32°, 34°, 36°, 38°, and 40°; above is the inscription: *dawāʾir inṣāf al-nuhur fī al-ʿurūḍ al-marqūmah ʿalā aṭrāfahā.* In the same quadrant are the azimuths of the qibla for Ṭūs, Iṣfahān, and Baghdād; below is the inscription: *khuṭūṭ sumūt al-qiblah fī al-bilād al-marqūmah ʿalā aṭrāfahā bi-irtifāʿ al-gharbī.* Below the center is the shadow-square for 7 feet and 12 digits. Below this are five semicircles indicating the planetary lords of the terms (denoted by their final letters), the ends of the terms, the names of

Reverse of No. 25

the zodiacal signs, the planetary lords of the decans (denoted by their final letters) according to the Greek system, and the lunar mansions.

On the rete the ecliptic is graduated for every 6°; the numbers in each zodiacal sign are 6, 12, 18, ... 30. The names of the stars are listed. (★)

The four tympans have altitude circles for every 6°, azimuths for every 10°, and equal and unequal hour curves. They indicate their latitudes and longest daylights: (■)

Ia has the horizons for latitudes 18°, 30°, 42°, 54°, and 66°; for 21°, 33°, 45°, and 57°; for 24°, 36°, 48°, and 60°; and for 27°, 39°, 51°, and 63°.

PROVENANCE — China, date unknown; Leo Metzenburg, Chicago, date unknown; Max Adler, date unknown; AP purchase, date unknown.

ICA 2541

① CIRCLE

City	Longitude	Latitude	Inclination	Direction	Identification
iṣfahān	86;40°	32;25°	40;29°	SW	Eṣfahān
kāshān	86;0°	34;0°	34;31°	SW	Kāshān
qumm	85;40°	34;45°	31;54°	SW	Qom
qazwīn	85;0°	36;0°	27;34°	SW	Qazvīn
yazd	89;0°	32;0°	48;38°	SW	Yazd
mashhad	92;30°	37;0°	44;6°	SW	Mashhad
harānh	94;20°	34;35°	55;20°	SW	Herat
qandahār	107;40°	33;0°	75;0°	SW	Kandahar
marw	97;0°	37;40°	52;30°	SW	Mary
sabzawār	91;30°	36°	44;12°	SW	Sabzevār
nīshābūr	92;30°	36;21°	46;25°	SW	Neyshābūr
simnān	88;0°	36;0°	36;17°	SW	Semnān
baghdād	80;0°	33;25°	12;45°	SW	Baghdād
baṣrah	84;0°	30;0°	37;19°	SW	Basra
tabrīz	82;0°	37;30°	16;17°	SW	Tabrız
shirwān	91;30°	36;0°	44;13°	SW	Shīrvān
shīrāz	88;0°	29;36°	53;18°	SW	Shīrāz
kirmān	92;30°	29;50°	62;41°	SW	Kermān
hamadān	83;0°	34;10°	22;16°	SW	Hamadan

★ STAR LIST

Name	Identification
janāḥ	γ Corv.
al-khaḍīb	β Cass.
dhanab qayṭus	β Ceti
ra's al-ḥawwā'	α Oph.
rāmiḥ	α Boot.
rijl al-yusrā	β Ori.
ridf	α Cygn.
sāq	δ Aquar.
simāk a'zal	α Virg.
sha'āmī	α Can. Min.
shi'rā yamānī	α Can. Mai
ṣadr	π Ceti
'unq	α Serp
'ayn al-thawr	α Tau.
'ayyūq	α Aur.
fard al-shujā'	α Hydr.
fam al-faras	ε Peg.
qā'idah	α Crat.
qalb al-'aqrab	α Scorp.
misfah	θ Erid.
nasr ta'ir	α Aquil.
nasr wāqi'	α Lyr.
nayyir fakkah	α Cor. Bor.
yad yumnā	α Ori.

Mater of No. 25

Rete of No. 25

TYMPANS

Tympan	*Latitude*	*Longest Daylight*
Ib.	24°	13;30 hours
IIa.	27°	13;42 hours
IIb.	36°	14;18 hours
IIIa.	30°	13;36 hours
IIIb.	33°	14;12 hours
IVa.	32°	14;7 hours
IVb.	38°	14;39 hours

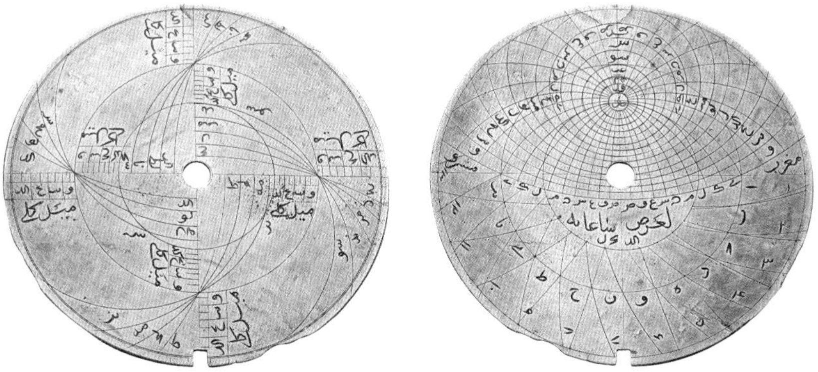

Tympan #1 of No. 25

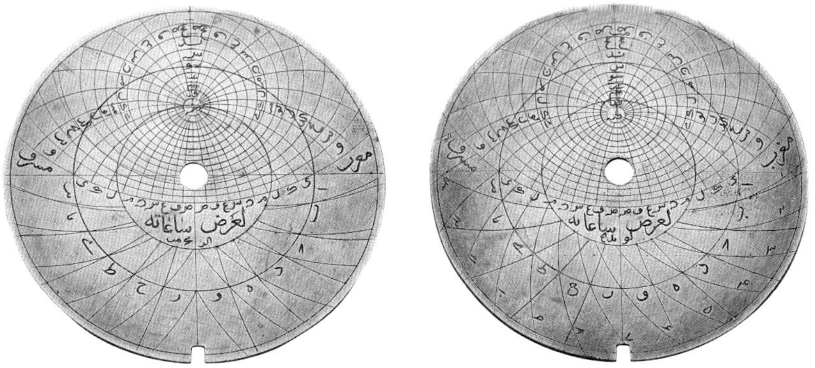

Tympan #2 of No. 25

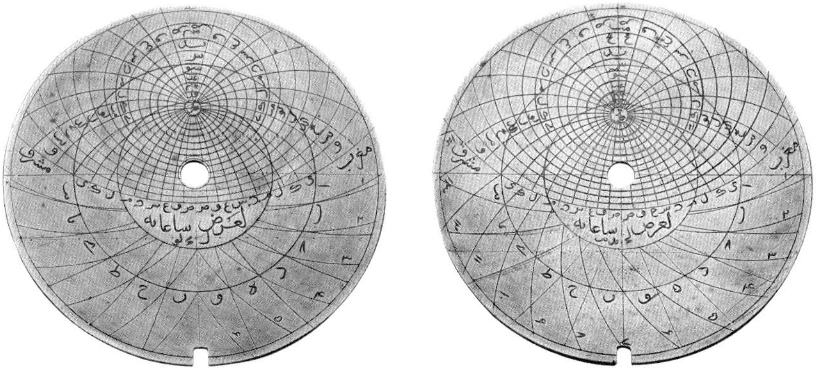

Tympan #3 of No. 25

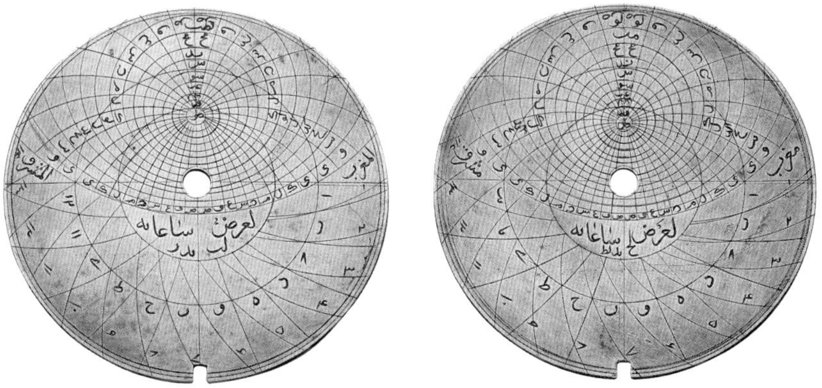

Tympan #4 of No. 25

Diopter of No. 25

26 Astrolabe

'Abd al-A'immah
Iran
1715
"'Abd al-A'immah made it in 1127."
"According to the order of the servants of the minister (?), high in power and position, the exalted, the infallible, the imperial Mahdī Qulī Khān Zanganah it was completely finished."
21 x 13.5 x 1.1 cm
Mater thickness 0.9 cm
Brass
A-90

Face of No. 26

Astrolabe made by 'Abd al-A'immah in A.H. 1127 (A.D. 1715) according to the inscription below the shadow-square on the back: *'abd al-a'immah ṣana'ahu 1127.* There is a lengthy Persian inscription on the front of the throne indicating that it was made for Qulī Khān: *ḥasb al-farmūdah bandagān dastyār (?) sumūw al-qadr wa al-makān 'ālījāh muqarrar al-khāqānī mahdī qulī khān zankanah bā tamām rasīd.* At the bottom of the back is a Persian inscription appearing in several other astrolabes as well: *gharḍ naqshīst kaz mā bāz mānd* ("the desire is that an image remain behind"). Inscriptions in naskh. The body is the only original part of this instrument.

The original part consists of the body with handle and ring. The rete and four tympans (which have notches at the bottom fitting about a corresponding tab in the body) are later fakes. The diopter, pin , and horse are missing; the instrument is held together by a modern nut screw.

The rim of the front of the body is graduated for every degree; the numbers, running clockwise in each quadrant, are 5, 10, 5, 20, 5, ...90. In the recess in the front is a geographical table in two circles. (❍)(●)

The upper half of the back is graduated for every degree; the numbers, running clockwise from the left to the top and counterclockwise from the right to the top, are 5, 10, 15, ...90. At the edge of the lower half are indicated the cotangents. In the upper left quadrant are the sines and cosines for all of the 60 parts in the Radius. In the upper right quadrant are the noon altitude curves of the sun for latitudes 28°, 30°, 32°, 34°, 36°, 38°, and 40°; above is the inscription: *dawā'ir inṣāf al-nuhur fī al- 'urūḍ al-marqūmah 'alā aṭrāfahā.* In the same quadrant are the azimuths of the qibla for Ṭūs, Iṣfahān, Baghdād, and Shīrāz; below is the inscription: *khuṭūṭ sumūt al-qiblah fī al-bilād al-marqūmah 'alā*

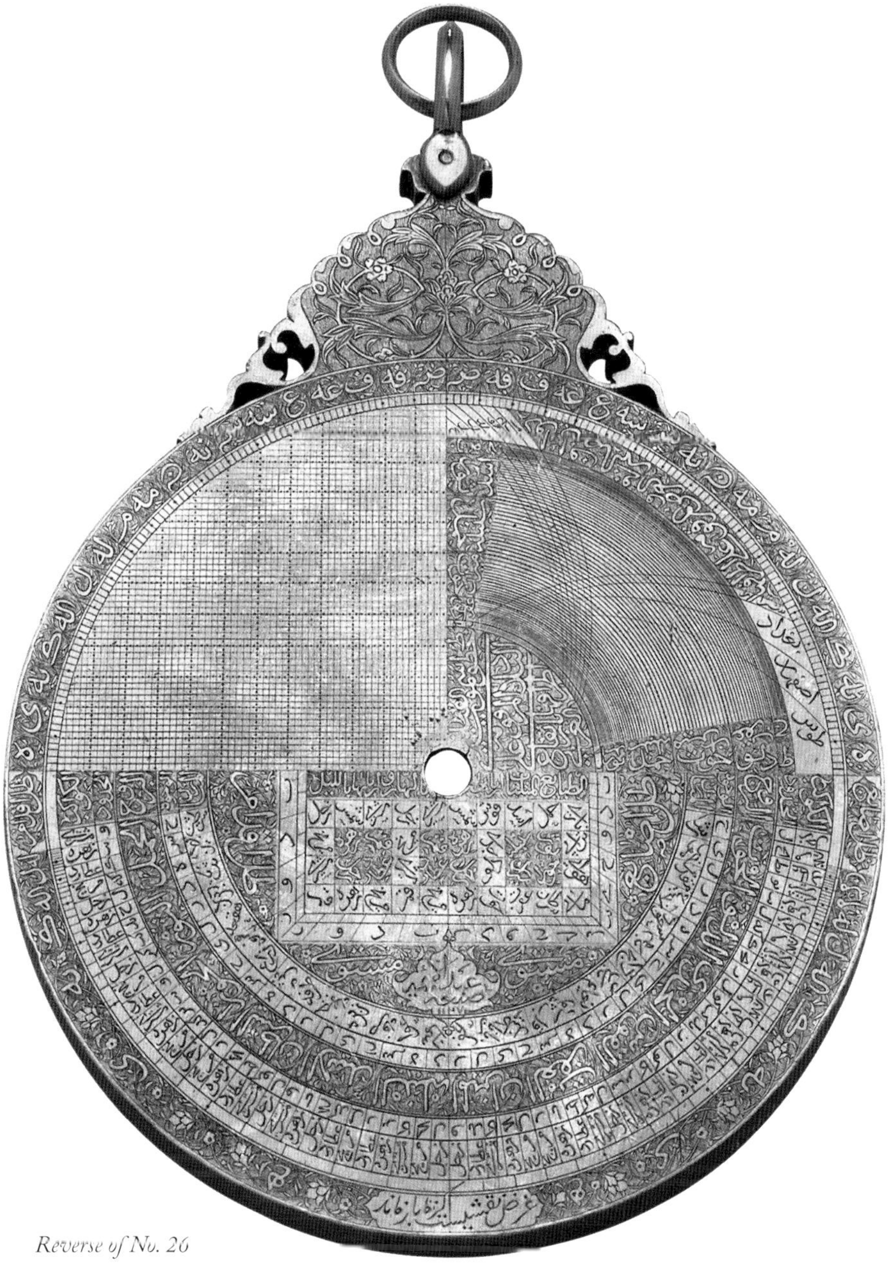

Reverse of No. 26

aṭrāfahā bi-irtifāʿ al-gharbī. Below the center is the shadow-square for 7 feet and 12 digits, within which is a table of the elements, zodiacal signs, and planetary lords of the four triplicities. Below this are five semicircles indicating the planetary lords of the terms, the ends of the terms, the names of the zodiacal signs, the planetary lords (denoted by their final letters) of the decans according to the Greek system, and the names of the lunar mansions.

The fake rete is symmetrical and devoid of star names. The four fake tympans are crude and useless.

PROVENANCE — E.S. David, New York, 1943; AP purchase, 1943.
ICA 2562
REFERENCES — Gingerich, King, and Saliba (1972), p. 189; Gingerich (1982).

OUTER CIRCLE

City	*Longitude*	*Latitude*	*Inclination*	*Direction*	*Identification*
makkah	77;10°	21;40°	0;0°	0;0°	Mecca
madīnah	75;20°	25;0°	27;10°	SE	Medina
bayt al-maqdis	66;30°	31;50°	45;16°	SE	Jerusalem
dimashq	70;0°	33;15°	30;31°	SE	Damascus
ḥalab	72;10°	35;50°	18;29°	SE	Aleppo
miṣr	63;20°	30;20°	58;38°	SE	Cairo
baghdād	82;0°	33;25°	12;45°	SW	Baghdād
baṣrah	84;0°	30;0°	37;19°	SW	Basra
surra-manra'ā	79;0°	34;0°	7;56°	SW	Sāmarrā'
kūfah	79;30°	31;30°	12;31°	SW	Kūfah
ahwāz	85;0°	31;0°	40;30°	SW	Ahvāz
yazd	89;0°	32;0°	48;18°	SW	Yazd
hamadān	83;0°	35;10°	22;16°	SW	Hamadan
qazwīn	85;0°	36;0°	27;34°	SW	Qazvīn
jurfādaqān	85;30°	34;14°	38;0°	SW	Golpayegan
iṣfahān	86;40°	32;25°	40;29°	SW	Eṣfahān
kāshān	86;0°	34;0°	34;31°	SW	Kāshān
rayy	86;20°	35;0	36;26°	SW	Shahr Rey
ṭāliqān	85;45°	36;10°	29;33°	SW	Taliq-an
astarābād	89;35°	36;50°	38;48°	SW	Gorgān
simnān	88;0°	36;0°	36;17°	SW	Semnān
dāmghān	88;55°	37;20°	38;15°	SW	Dāmghān
basṭām	89;30°	36;10°	39;13°	SW	Bastam
nīshābūr	92;30°	36;21°	46;25°	SW	Neyshābūr
ṭūs	92;30°	37;0°	45;6°	SW	Ṭūs
tūrshīz	92;0°	35;0°	48;11	SW	Kashmar
tūn	92;30°	34;30°	50;20°	SW	Ferdows
qāyin	93;20°	33;40°	54;1°	SW	Qāyen
marw	97;0°	37;40°	52;30°	SW	Mary
harāh	94;20°	34;30°	54;8°	SW	Herat
qandahār	107;40°	33;0°	75;0°	SW	Kandahar

INNER CIRCLE

City	*Longitude*	*Latitude*	*Inclination*	*Direction*	*Identification*
tabrīz	82;0°	38;0°	15;40°	SW	Tabrīz
ardabīl	82;30°	38;0°	17;13°	SW	Ardabīl
marāghah	82;0°	37;20°	16;17°	SW	Marāghah
nakhjawān	81;15°	38;40°	12;49°	SW	Nakhichevan
shīrwān	84;30°	40;50°	20;9°	SW	Shīrvān
ayrawān	89;15°	38;30°	36;20°	SW	Yerevan
kirmānshāhān	83;0°	34;30°	23;18°	SW	Kermānshāh
nihāwand	83;45°	34;20°	24;29°	SW	Nahāvand
shūsh	84;30°	31;30°	35;24°	SW	Shūsh
kāzirūn	87;0°	29;0°	51;43°	SW	Kāzerūn
shīrāz	88;0°	29;36°	53;18°	SW	Shīrāz
kābul	104;40°	34;26°	69;35°	SW	Kabul
lahāwar	109;20°	31;50°	78;26°	SW	Lahore
multān	107;35°	29;40°	80;48°	SW	Multan
kirmān	92;30°	29;50°	62;11°	SW	Kermān

Mater of No. 26

Detail of throne of No. 26

27 Astrolabe

'Abd al-A'immah
Iran.
ca. 1720
"'Abd al-A'immah made it".
15 x 9.2 x 3.2 cm
Mater thickness 0.7 cm
Brass
A-74

Face of No. 27

Astrolabe made by 'Abd al-A'immah according to the inscription below the shadow-square on the back: *'abd al-a'immah ṣana'ahu.* Inscriptions in naskh.

Consists of body, rete, graduated and inscribed diopter with scales and sights, pin, horse, handle, ring, cord, and four typmpans which have notches at the bottom fitting about a corresponding tab in the body.

The rim of the front of the body is graduated for every degree; the numbers, running clockwise, are 5, 10, 5, 20, 5, ...100, 5, 10, ...200, 5, 10, ...300, 5, 10, ...60. In the recess in the front is a geographical table in two circles. (☽)(●)

The upper half of the back is graduated for every degree; the numbers, running clockwise from the left to the top and counter-clockwise from the right to the top, are 5, 10, 15, ...90. At the edge of the lower half are indicated the cotangents. In the upper left quadrant are the sines and cosines for all 60 parts of the Radius. In the upper right quadrant are the noon altitude curves of the Sun for latitudes 28°, 30°, 32°, 34°, 36°, 38°, and 40°. In the same quadrant are the azimuths of the qibla for Ṭūs, Iṣfahān, Baghdād, and Shīrāz; below is the inscription: *khuṭūṭ sumūt al-qiblah fī al-buldān al-marqūmah 'alā aṭrāfahā bi-irtifā' al-gharbī.* Below the center is the shadow-square for 7 feet and 12 digits, within which is a table of the elements, zodiacal signs, and planetary lords of the four triplicities identical with that in No. 23. Below this are five semicircles also identical with those in No. 23.

On the rete the ecliptic is graduated for every 6°; the numbers in each zodiacal sign are 6, 12, 18, ...30. The names of the stars are: (★)

The four tympans have altitude circles for every 3°, azimuths for every 10°, and unequal hour curves (IIb, III and IV have equal hour curves as well). They indicate their latitudes and longest daylights: (■)

★ STAR LIST

Name	*Identification*
janāḥ al-ghurāb	γ Corv.
dhanab qayṭus	ι Ceti
ra's al-ḥawwā'	α Oph.
rāmiḥ	α Boot.
rijl al-yusrā	β Ori.
ridf	α Cygn.
sāq ayman	δ Aquar.
simāk a‘zal	α Virg.
sha'āmī	α Can. Min.
shi‘rā yamānī	α Can. Mai
ṣadr qayṭus	π Ceti
ẓahr al-dubb	α Urs. Mai
‘unq al-ḥayyah	α Serp
‘ayn al-thawr	α Tau.
‘ayyūq	α Aur.
ghūl	β Pers.
fard al-shujā‘	α Hydr.
fam al-faras	ε Peg.
fam qayṭus	γ Ceti
qā‘idah	α Crat.
qalb al-‘aqrab	α Scorp.
kaff al-khaḍīb	β Cass.
masāfah	θ Erid.
mankib al-faras	β Peg.
nasr tā'ir	α Aquil.
nasr wāqi‘	α Lyr.
nayyir fakkah	α Cor. Bor.
yad yumnā	α Ori.

Reverse of No. 27

Ia has the horizons for latitudes 14°, 22°, 30°, 40°, 50°, 58°, and 66°; for 16°, 24°, 34°, 42°, 52°, and 60°; for 18°, 26°, 16° (read 36°), 44°, 54°, and 62°, and for 20°, 28°, 38°, 48°, 56°, and 64°.

PROVENANCE — E.S. David, New York; AP purchase, 1940.
ICA 2547
REFERENCES — Gingerich, King, and Saliba (1972), pp. 189, 192; Newbury (2004a); Stephenson, Bolt, and Friedman (2000), p. 28.

Rete of No. 27

OUTER CIRCLE

City	Longitude	Latitude	Inclination	Identification
makkah	77;10°	21;40°	0;0°	Mecca
madīnah	75;20°	25;0°	37;10°	Medina
baghdād	80;0°	33;25°	12;45°	Baghdād
sāmarrah	79;0°	34;0°	7;56°	Sāmarrā'
baṣrah	84;0°	30;0°	37;19°	Basra
shūshah	84;30°	31;30°	35;24°	Shūsh
kāzirūn	87;0°	29;15°	51;7°	Kāzerūn
shīrāz	88;0°	29;36°	53;18°	Shīrāz
yazd	89;0°	32;0°	48;28°	Yazd
hamadān	83;0°	35;10°	22;16°	Hamadan
qazwīn	85;0°	36;0°	27;34°	Qazvīn
jurfādaqān	84;30°	34;15°	18;41°	Golpayegan
kirmānshāh	93;0°	34;30°	23;9°	Kermānshāh
iṣfahān	86;40°	32;25°	40;29°	Eṣfahān
kāshān	86;0°	34;0°	34;31°	Kāshān
qumm	85;40°	34;45°	31;54°	Qom
rayy	86;20°	35;0°	37;26°	Shahr Rey
astarābād	89;35°	36;50°	38;48°	Gorgān
simnān	88;0°	36;0°	56;17°	Semnān
dāmghān	88;55°	36;20°	38;5°	Dāmghān
basṭām	89;30°	36;10°	38;58°	Bastam
sabzawār	91;30°	36;5°	44;12°	Sabzevār
nīshābūr	92;30°	36;21°	46;25°	Neyshābūr

Mater of No. 27

INNER CIRCLE

City	Longitude	Latitude	Inclination	Identification
mashhad	92;30°	37;0°	45;6°	Mashhad
marw	97;0°	37;40°	52;30°	Mary
harāh	94;20°	34;30°	54;0°	Herat
qandahār	107;40°	33;0°	75;0°	Kandahar
tabrīz	82;0°	38;15°	15;40°	Tabrīz
ardabīl	82;30°	38;0°	17;13°	Ardabīl
shīrwān	84;30°	40;50°	20;9°	Shīrvān
ayrawān	89;45°	38;30°	36;20°	Yerevan
lāhījān	84;0°	37;45°	22;30°	Lāhījān
kirmān	92;33°	29;50°	62;51°	Kermān
sīstān	97;0°	32;30°	63;18°	Sistan

Tympan #1 of No. 27

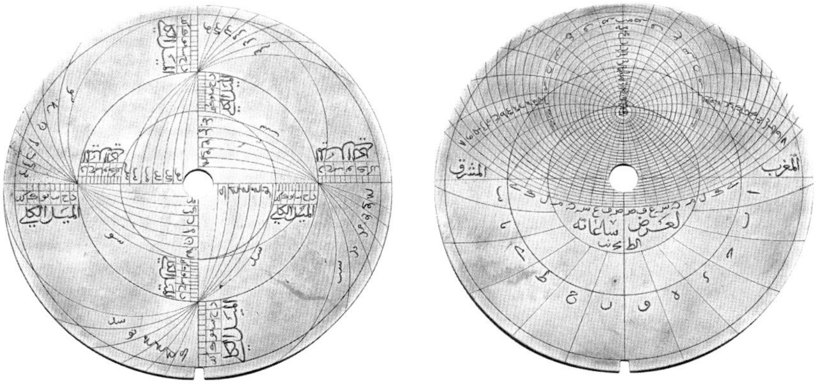

■ TYMPANS

Tympan	*Latitude*	*Longest Daylight*
Ib.	29°	13;52 hours
IIa.	30°	13;56 hours
IIb.	36°	14;28 hours
IIIa.	32°	14;7 hours
IIIb.	34°	14;16 hours
IVa.	37°	14;38 hours (read 14;33 hours)
IVb.	38°	14;39 hours

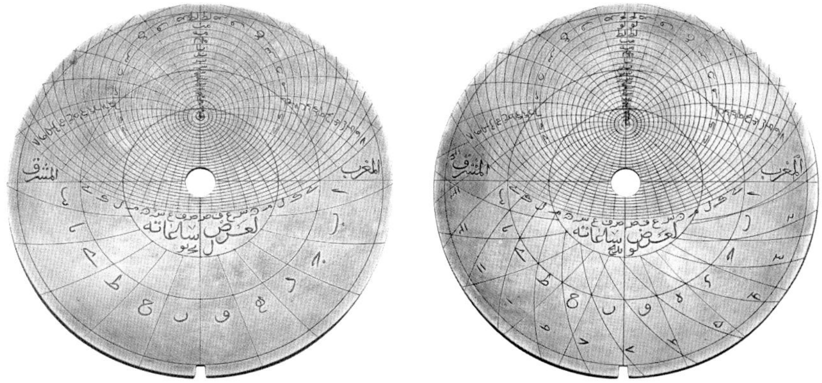

Tympan #2 of No. 27

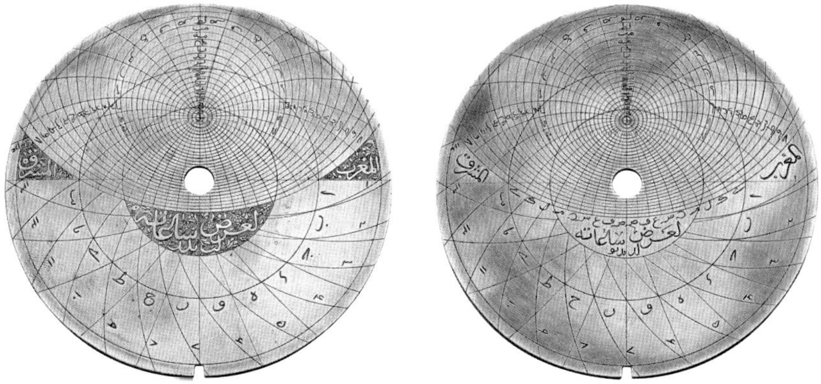

Tympan #3 of No. 27

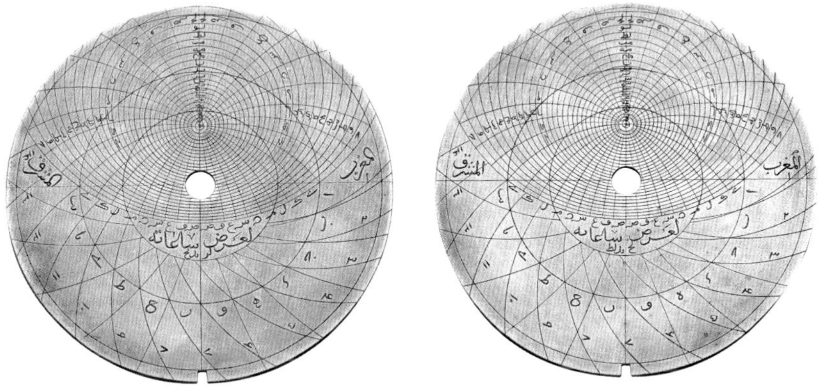

Tympan #4 of No. 27

28 Astrolabe

'Abd al-A'immah
Iran
ca. 1720
"Abd al-A'immah made it".
15.5 x 9.3 x 3.2 cm
Mater thickness 0.7 cm
Brass
A-88

Face of No. 28

Astrolabe made by 'Abd al-A'immah according to the inscription below the shadow-square on the back: *'abd al-a'immah ṣana'ahu.* There is the same inscription at the bottom of the back as is found on instrument No. 26. Inscriptions in naskh.

Consists of body, rete, diopter with scales and sights, pin, horse, handle, ring, chord, and four tympans.

The rim of the front of the body is graduated for every degree; the numbers, running clockwise, are 5, 10, 5, 20, 5, ...100, 5, 10, 5, ...200, 5, 10, 5, ...300, 5, 10, 5, ...60. In the recess of the front is a geographical table in two columns. (☽)(●)

The upper half of the back is graduated for every degree; the numbers, running clockwise from the left to the top and counter-clockwise from the right to the top, are 5, 10, 15, ...90. At the edge of the lower half are indicated the co-tangents. In the upper left quadrant are the sines and cosines for each of the 60 parts of the Radius. In the upper right quadrant are the noon-altitude curves of the Sun for latitudes 28°, 30°, 32°, 34°, 36°, 38°, and 40°; above is the inscription: *dawā'ir inṣāf al-nuhur fī al-'urūḍ al-marqūmah 'alā aṭrāfahā.* In the same quadrant are the azimuths of the qibla for Ṭūs, Iṣfahān, Baghdād, and Shīrāz; below is the inscription: *khuṭūṭ sumūt al-qiblah fī al-buldān al-marqūmah 'alā aṭrāfahā bi-irtifā' al-gharbī.* Below the center is the shadow-square for 7 feet and 12 digits, within which is a table of the elements, the zodiacal signs, and the planetary lords of the four triplicities. Below this are five semicircles listing the planetary lords of the terms, the ends of the terms, the names of the zodiacal signs, the planetary lords (indicated by their final letters) of the decans according to the Greek system, and the names of the lunar mansions.

Reverse of No. 28

On the rete the ecliptic is graduated for every 6°; the numbers, in each zodiacal sign, are 6, 12, 18, ...30. The names of the stars are listed. (★)

The four tympans have altitude circles for every 6°, azimuths for every 10°, and unequal hour curves (Ib, IIb, and IIIa and b also have equal hour curves). They indicate their latitudes and longest daylights. (■)

Ia has horizons for latitudes 15°, 27°, 39°, 51°, and 63°; for 18°, 30°, 42°, 54°, and 66°; for 21°, 33°, 45°, and 57°; and for 24°, 36°, 48°, and 60°.

PROVENANCE — E.S. David, New York, 1943; AP purchase, 1943.
ICA 2560
REFERENCES — Gingerich, King, and Saliba (1972), pp. 189-90; Newbury (2004a); Stephenson, Bolt, and Friedman (2000), p. 64.

★ STAR LIST

Name	*Identification*
janāḥ al-ghurāb	γ Corv.
dhanab qayṭus	ι Ceti
ra's al-ḥawwā'	α Oph.
rāmiḥ	α Boot.
rijl al-yusrā	β Ori.
ridf	α Cygn.
sāq ayman	δ Aquar.
simāk a'zal	α Virg.
sha'āmī	α Can. Min.
shi'rā yamānī	α Can. Mai
ṣadr qayṭus	π Ceti
ẓahr al-dubb	α Urs. Mai
'unq al-ḥayyah	α Serp
'ayn al-thawr	α Tau.
'ayyūq	α Aur.
ghūl	β Pers.
fard al-shujā'	α Hydr.
fam al-faras	ε Peg.
fam qayṭus	γ Ceti
qā'idah	α Crat.
qalb al-asad	α Leo
qalb al-'aqrab	α Scorp.
kaff al-khaḍib	β Cass.
masāfah	θ Erid.
mankib al-faras	β Peg.
nasr tā'ir	α Aquil.
nasr wāqi'	α Lyr.
nayyir fakkah	α Cor. Bor.
yad yumnā	α Ori.

Mater of No. 28

Rete of No. 28

OUTER CIRCLE

City	*Longitude*	*Latitude*	*Inclination*	*Identification*
makkah	77;10°	21;40°	0;0°	Mecca
madīnah	75;20°	25;0°	27;10°	Medina
baghdād	82;0°	33;25°	12;45°	Baghdād
baṣrah	84;0°	30;0°	37;59°	Basrah
shīrāz	88;0°	29;36°	53;18°	Shīrāz
shūshah	84;30°	31;30°	35;24°	Shūsh
jurfādaqān	85;30°	34;15°	18;0°	Golpayegan
iṣfahān	86;40°	32;25°	40;29°	Eṣfahān
kāshān	86;0°	34;0°	34;31°	Kāshān
qumm	85;40°	34;45°	31;54°	Qom
rayy	86;20°	35;0°	37;26°	Shahr Rey
qazwīn	85;0°	36;0°	27;34°	Qazvīn
sāwah	85;0°	35;0°	29;16°	Sāveh
hamadān	83;0°	35;10°	22;16°	Hamadan
simnān	88;0°	36;0°	36;17°	Semnān
dāmghān	88;55°	36;20°	38;15°	Dāmghān
basṭām	89;30°	36;10°	39;13°	Bastam
sabzawār	91;30°	36;5°	44;12°	Sabzevār
nīshābūr	92;30°	36;21°	46;25°	Neyshābūr
mashhad	92;30°	37;0°	45;6°	Mashhad
harāt	94;20°	34;30°	54;8°	Herat
marw	97;0°	37;40°	52;30°	Mary
qandahār	107;40°	33;0°	75;0°	Kandahar

INNER CIRCLE

City	*Longitude*	*Latitude*	*Inclination*	*Identification*
tabrīz	82;0°	38;0°	15;40°	Tabrīz
ardabīl	82;30°	38;0°	17;13°	Ardabīl
ayrawān	89;15°	38;30°	36;20°	Yerevan
shīrwān	84;30°	40;50°	20;9°	Shīrvān
marāghah	82;0°	37;20°	16;17°	Marāgheh
kanjah	83;0°	41;20°	15;49°	Kirovabad
barda‘	83;0°	40;30°	16;27°	Barda
lāhījān	84;0°	36;10°	29;20°	Lāhījān
astarābād	89;35°	36;50°	38;48°	Gorgān
āmul	87;20°	36;15°	35;0°	Amol
sārī	88;0°	37;0°	32;45°	Sārī

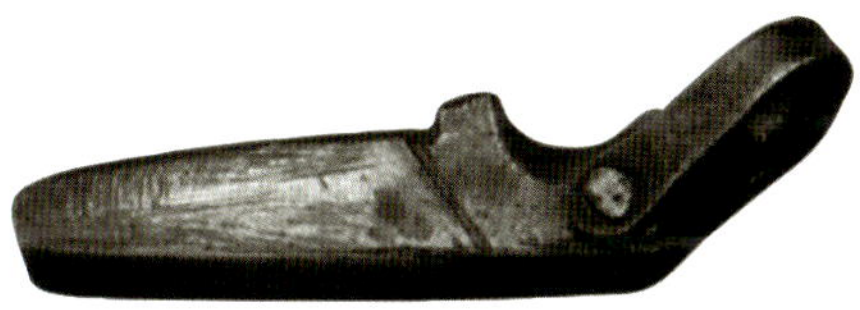

Pin, horse and diopter of No. 28

TYMPANS

Tympan	Latitude	Longest Daylight
Ib.	29°	13;52 hours
IIa.	30°	13;56 hours
IIb.	36°	14;28 hours
IIIa.	32°	14;7 hours
IIIb.	34°	14;16 hours
IVa.	37°	14;38 hours (read 14;33 hours)
IVb.	38°	14;39 hours

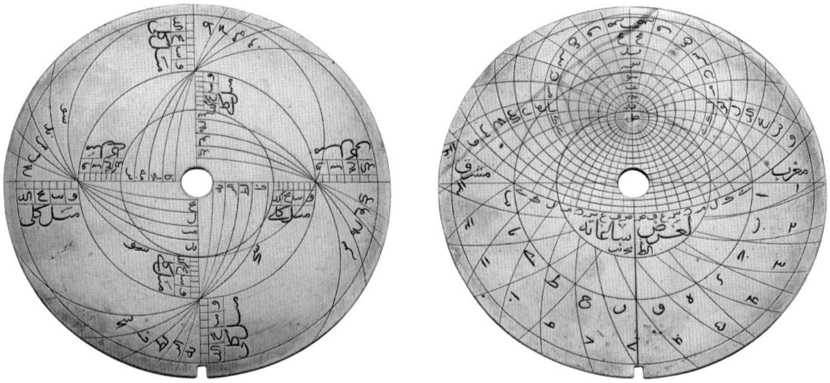

Tympan #1 of No. 28

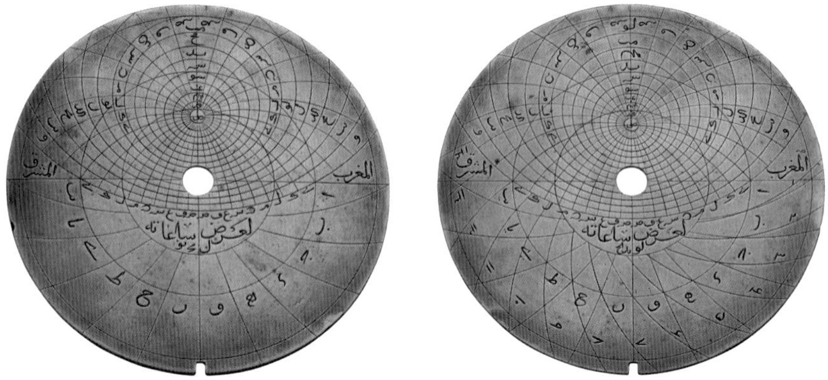

Tympan #2 of No. 28

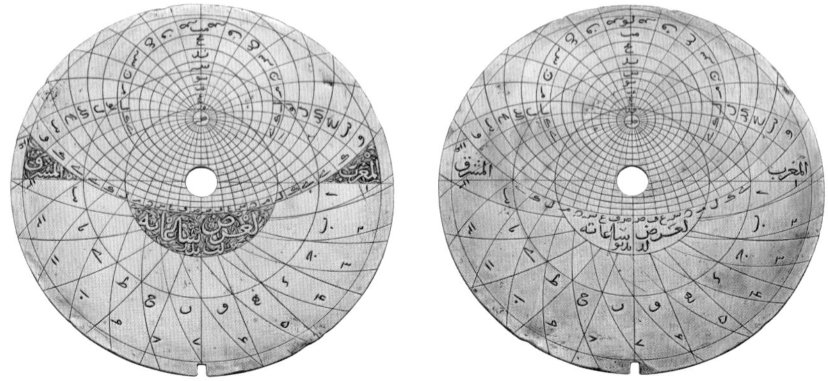

Tympan #3 of No. 28

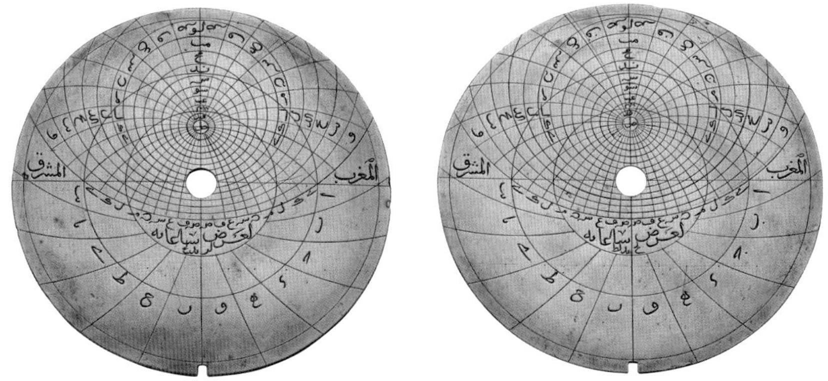

Tympan #4 of No. 28

29 Astrolabe

'Abd al-'Alī
Iran
ca. 1710
"'Abd al-'Alī made it."
13.7 x 9 x 2 cm
Mater thickness 0.7 cm
Brass
A-69

Face of No. 29

Incomplete astrolabe made by 'Abd al-'Alī according to the inscription under the shadow-square on the back: *'abd al-'alī ṣana'ahu.* 'Abd al-'Alī is known to have made instruments in Iran between 1701 and 1714. For a fake instrument ascribed to 'Abd al-'Alī see No. 40. Quotation from the *"Qur'an"* on the front of the throne. Inscriptions in naskh.

Consists of body, graduated diopter with scales and sights, pin, handle, and ring. The five tympans, which have notches at the bottom fitting about a corresponding tab in the body, are later additions; the rete is a fake; and a bent bar has replaced the horse.

The rim of the front of the body is graduated for every degree; the numbers, running clockwise, are 5, 10, 5, 20, 5, ...100, 5, 10, ...200, 5, 10, ...300, 5, 10, ...60. In the recess in the front is a geographical table in a circle. (1)

The upper half of the back is graduated for every degree; the numbers, running clockwise from the left to the top and counter-clockwise from the right to the top, are 5, 10, 15, ...90. At the edge of the lower half are indicated the cotangents. In the upper left quadrant are the cosines for each of the 60 parts of the Radius. In the upper right quadrant are the noon-altitude curves of the Sun for latitudes 28°, 30°, 32°, 34°, and 36°; above is the inscription: *dawā'ir inṣāf al-nuhur fī al-'urūḍ al-marqūmah 'alā aṭrāfahā.* In the same quadrant are the azimuths of the qibla for Shīrāz, Ṭūs, Yazd, Iṣfahān, and Qazwīn; below is the inscription: *khuṭūṭ sumūt al-qiblah fī al-bilād al-marqūmah 'alā aṭrāfahā bi-irtifā' al-gharbī.* Below the center is the shadow-square for 7 feet and 12 digits, within which is the table of the elements, zodiacal signs, and planetary lords of the four triplicities. Below this are five semicircles indicating the planetary lords of the terms, the ends of the terms, the names of the zodiacal

Reverse of No. 29

signs, the planetary lords (denoted by their final letters) of the decans according to the Greek system, and the names of the lunar mansions.

The rete is purely decorative; the only names on it are those of the zodiacal signs.

The five tympans, which are engraved by someone other than ʿAbd al-ʿAlī, have altitude circles for every 6°, azimuths for every 10°, and equal and unequal hour curves. They indicate their latitudes and longest daylights. (■)

Ia has horizons for latitudes 15°, 27°, 39°, 51°, and 63°; for 18°, 30°, 42°, 54°, and 66°; for 21°, 33°, 45°, and 57°; and for 24°, 36°, 48°, and 60°.

PROVENANCE — E.S. David, New York, 1940; AP purchase, 1940.
ICA 2543
REFERENCES - Newbury (2004a); Rose (1968).

Tympan #1 of No. 29

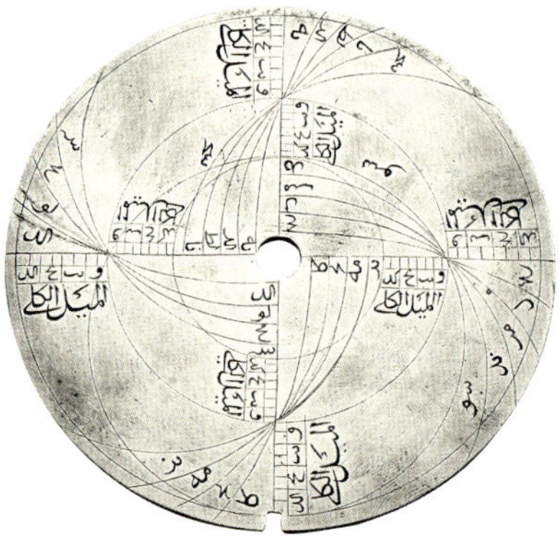

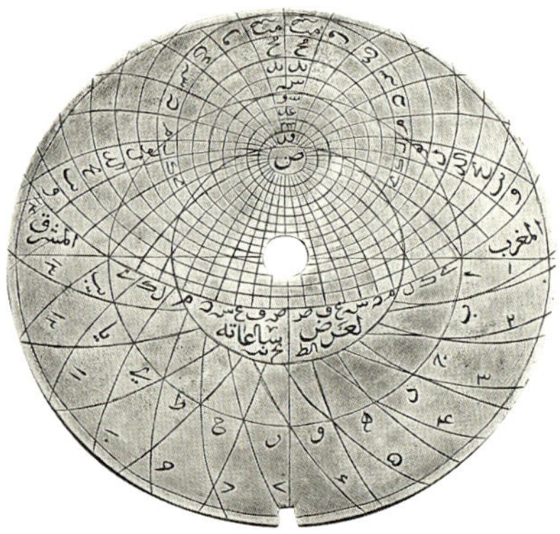

TYMPANS

Tympan	*Latitude*	*Longest Daylight*
Ib.	29°	13;52 hours
IIa.	29°	13;52 hours
IIb.	37°	14;48 hours
IIIa.	32°	14;7 hours
IIIb.	38°	14;39 hours
IVa.	32°	14;7 hours
IVb.	<34°>	14;16 hours
Va.	34°	14;16 hours
Vb.	37°	14;38 hours (read 14;33 hours)

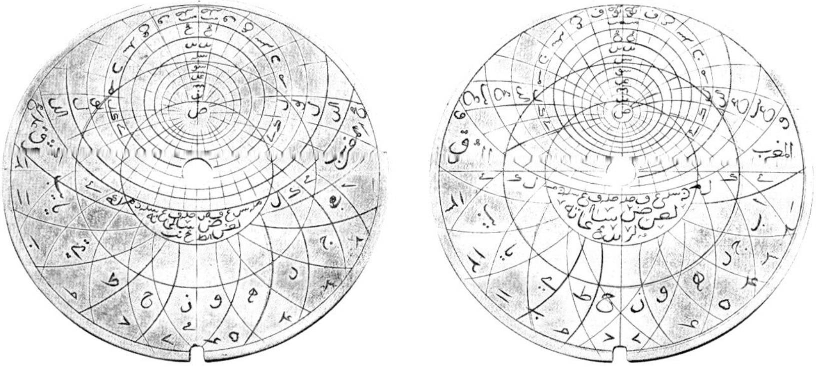

Tympan #2 of No. 29

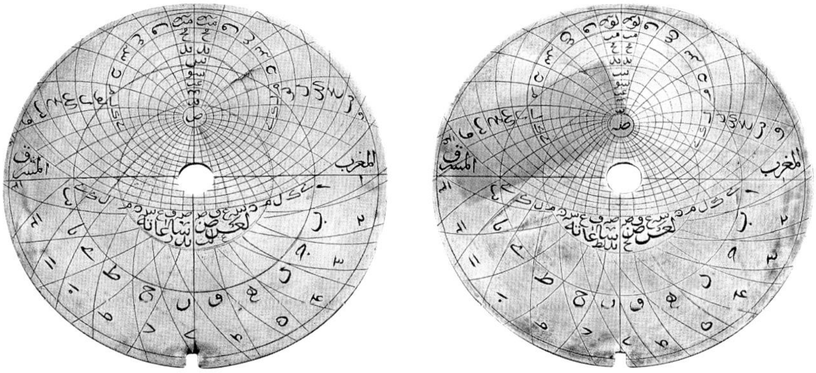

Tympan #3 of No. 29

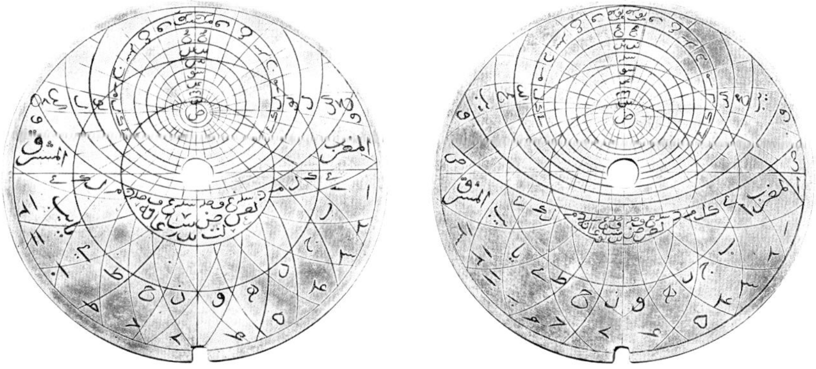

Tympan #4 of No. 29

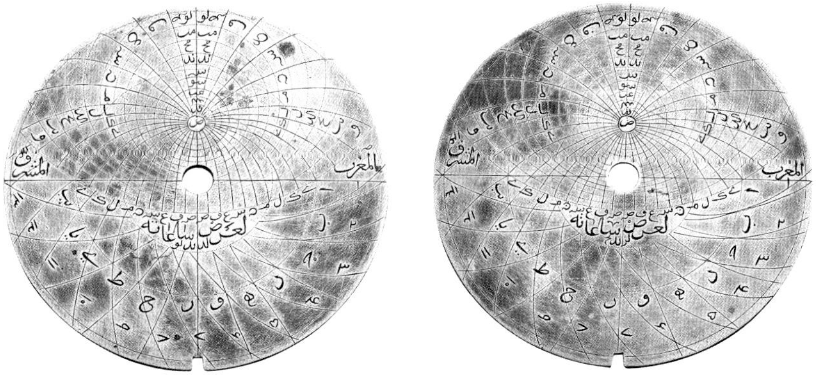

Tympan #5 of No. 29

Mater of No. 29

Detail of throne of No. 29

① CIRCLE

City	Longitude	Latitude	Inclination	Direction	Dist.	Identification
qandahār	107;40°	33;0°	75;0°	SW	656	Kandahar
harāh	94;20°	34;30°	54;0°	SW	439	Herat
tūn	92;30°	34;30°	50;20°	SW	413	Ferdows
mashhad	92;30°	37;0°	45;6°	SW	451	Mashhad
shirwān	91;30°	36;5°	44;12°	SW	422	Shīrvān
basṭām	89;30°	36;10°	39;53°	SW	400	Bastam
simnān	88;0°	36;0°	36;17°	SW	381	Semnān
astarābād	89;35°	36;50°	38;48°	SW	413	Gorgān
ṭāliqān	85;45°	36;10°	29;33°	SW	362	Taliq-an
rayy	86;20°	35;0°	37;26°	SW	356	Shahr Rey
qumm	85;40°	34;45°	31;55°	SW	339	Qom
kāshān	84;0°	34;0°	34;31°	SW	324	Kāshān
iṣfahān	86;40°	32;25°	40;29°	SW	303	Eṣfahān
jurfādaqān	85;30°	34;15°	32;21°	SW	323	Golpayegan
qazwīn	85;0°	36;0°	27;34°	SW	352	Qazvīn
hamadān	83;0°	35;10°	22;17°	SW	230	Hamadan
yazd	89;0°	32;0°	48;29°	SW	331	Yazd
shīrāz	88;0°	29;36°	53;23°	SW	279	Shīrāz
kāzirūn	87;0°	29;15°	51;43°	SW	259	Kāzerūn
ahwāz	85;0°	31;0°	38;46°	SW	259	Ahvāz
shūsh	84;0°	31;30°	35;24°	SW	262	Shūsh
baṣrah	85;0°	30;0°	37;59°	SW	230	Basra
wāsiṭ	81;30°	32;20°	20;54°	SW	251	Wasit
baghdād	80;0°	33;25°	12;46°	SW	266	Baghdād
surra-manra'ā	79;0°	34;0°	7;56°	SW	276	Sāmarrā'
ardabīl	82;30°	38;0°	17;13°	SW	377	Ardabīl
tabrīz	82;0°	38;0°	15;40°	SW	367	Tabrīz
marāghah	82;0°	37;20°	16;17°	SW	360	Marāghah
ḥalab	72;10°	35;50°	18;33°	SE	329	Aleppo
liḥsā	83;30°	24;0°	69;28°	SW	139	Aḥsā
ṣana'ā	77;0°	14;30°	1;55°	NE	92	Ṣan'ā'
miṣr	63;20°	30;20°	58;38°	SE	335	Cairo
bayt al-maqdis	66;30°	31;50°	45;43°	SE	309	Jerusalem
madīnah	75;20°	25;10°	26;3°	SE	86	Medina
makkah	77;10°	21;40°	0;0°	0;0°	0	Mecca

Diopter and pin of throne of No. 29

30 Astrolabe

'Ali ibn Ṣādiq
Iran
18th. century
"Work of 'Ali ibn Ṣādiq"
14.7 x 9 x 2.8 cm
Mater thickness 0.5 cm
Brass
W-151

Face of No. 30

Astrolabe made by 'Ali ibn Ṣādiq according to the inscription at the bottom of the back: *'amal 'ali ibn ṣādiq.* Made most probably in Iran in the eighteenth century as it shares many features with the 'Abd al-A'immah instruments. Inscriptions in naskh.

Consists of body, rete, graduated diopter with seals and sights, pin, horse, handle, ring and four tympans, which have notches at the bottom fitting about a corresponding tab in the body.

The rim of the front of the body is graduated for every degree; the numbers, running clockwise, are 5, 10, 5, 20, 5, ...100, 5, 10, ...200, 5, 10, ...300, 5, 10, ...60. In the recess in the front is a geographical table in two circles. (○)(●)

The upper half of the back is graduated for every degree; the numbers, running clockwise from the left to the top and counter-clockwise from the right to the top, are 5, 10, 15, ...90. At the edge of the lower half are indicated the cotangents. In the upper left quadrant are the sines and cosines for every one of the 60 parts of the Radius. In the upper right are the noon altitude curves of the Sun for latitudes 28°, 30°, 32°, 34°, 36°, 38°, and 40°; above is the inscription: *dawā'ir inṣāf al-nuhur fī al-'urūḍ al-marqūmah 'alā aṭrāfahā.* In the same quadrant are the azimuths of the qibla for Ṭūs, Iṣfahān, Baghdād, and Shīrāz; below is the inscription: *khuṭūṭ sumūt al-qiblah fī al-bilād al-marqūmah 'alā aṭrāfahā bi-irtifā' al-gharbī.* Below the center is the shadow-square for 7 feet and 12 digits, within which is the table of the elements, the zodiacal signs, and planetary lords of the four triplicities. Below this are five semicircles indicating the planetary lords of the terms, the ends of the terms, the names of the zodiacal signs, the planetary lords (denoted by their final letters) of the decans according to the Greek system, and the names of the lunar mansions.

Reverse of No. 30

On the rete the ecliptic is graduated for every 6°; the numbers, for each zodiacal sign, are 6, 12, 18, ...30. The names of the stars are listed. (★)

The four tympans have altitude circles for every 2°, azimuths for every 10°, and equal and unequal hour curves. They indicate their latitudes and longest daylights. (■)

Ia has the horizons for latitudes 15°, 27°, 39°, 51°, and 63°; for 18°, 30°, 42°, 54°, and 66°; for 21°, 33°, 45°, and 57°; and for 24°, 36°, 48°, and 60°.

PROVENANCE — Sotheby's, Geneva, 1985; Ahuan Islamic Art, 1985; Roderick S. Webster, Winnetka, IL, 1985; AP gift, 1985.

REFERENCES — Sotheby's (1985), lot 236; Stephenson, Bolt, and Friedman (2000), pp. 11, 65.

★ STAR LIST

Name	*Identification*
janāḥ al-ghurāb	γ Corv.
dhanab al-ʻuqāb	ζ Aquil.
dhanab qayṭus	ι Ceti
ra's al- ḥawwā'	α Oph.
rāmiḥ	α Boot.
rijl al-yusrā	β Ori.
ridf	α Cygn.
sāq ayman	δ Aquar.
simāk a'zal	α Virg.
sha'āmī	α Can. Min.
shi'rā yamānī	α Can. Mai
'unq al-ḥayyah	α Serp
'ayyūq	α Aur.
ṣadr qayṭus	π Ceti
ẓahr al-dubb	α Urs. Mai
fard al-shujā'	α Hydr.
fam al-faras	ε Peg.
fam qayṭus	γ Ceti
qā'idah	α Crat.
qalb al-asad	α Leo
qalb al-'aqrab	α Scorp.
kaff al-khaḍīb	β Cass.
masāfah	θ Erid.
mankib al-faras	β Peg.
nasr tā'ir	α Aquil.
nasr wāqi'	α Lyr.
nayyir fakkah	α Cor. Bor.
yad yumnā	α Ori.

Mater of No. 30

Rete of No. 30

OUTER CIRCLE

City	*Longitude*	*Latitude*	*Inclination*	*Identification*
makkah	77;10°	21;40°	0;0°	Mecca
madīnah	75;20°	25;0°	37;10°	Medina
baghdād	80;0°	33;25°	12;55°	Baghdād
baṣrah	84;0°	30;0°	36;19°	Basra
shīrāz	88;0°	29;36°	53;18°	Shīrāz
shūshtar	84;30°	31;30°	35;24°	Shustar
jurfādaqān	85;30°	34;30°	33;41°	Golpayegan
iṣfahān	86;40°	32;25°	40;28°	Eṣfahān
kāshān	86;0°	34;0°	34;31°	Kāshān
qumm	85;40°	34;45°	35;15°	Qom
rayy	86;20°	35;0°	36;26°	Shahr Rey
qazwīn	85;0°	36;0°	26;34°	Qazvīn
sāwah	85;0°	35;0°	29;36°	Sāveh
hamadān	83;0°	35;10°	22;16°	Hamadan
simnān	88;0°	36;0°	36;17°	Semnān
dāmghān	88;55°	36;20°	38;0°	Dāmghān
basṭām	89;30°	36;10°	39;53°	Bastam
shīrwān	91;30°	36;0°	44;12°	Shīrvān
nīshābūr	92;30°	36;21°	46;25°	Neyshābūr
mashhad	92;30°	37;0°	45;6°	Mashhad
harāt	94;30°	34;30°	54;8°	Herat
marw	96;0°	37;40°	57;30°	Mary
qandahār	106;40°	33;0°	75;0°	Kandahar

INNER CIRCLE

City	*Longitude*	*Latitude*	*Inclination*	*Identification*
tabrīz	82;0°	38;0°	15;40°	Tabrīz
ardabīl	82;30°	33;0°	17;13°	Ardabīl
ayrawān	89;15°	38;30°	16;20°	Yerevan
yazd	89;0°	32;0°	48;28°	Yazd
marāghah	82;0°	37;20°	16;17°	Marāghah
kanjah	83;0°	41;20°	15;49°	Kirovabad
kāzirān	87;0°	29;15°	51;17°	Kāzerūn
kirmān	92;30°	29;50°	65;11°	Kermān
astarābād	89;35°	36;50°	58;48°	Gorgān
āmul	88;20°	36;15°	34;37°	Amol
tūrshīz	92;0°	35;0°	48;51°	Kāshmar

Diopter, horse and pin of No. 30

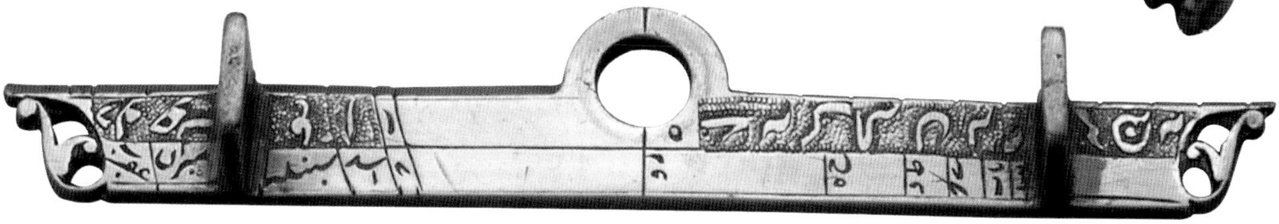

TYMPANS

Tympan	*Latitude*	*Longest Daylight*
Ib.	29°	13;52 hours
IIa.	30°	13;56 hours
IIb.	37°	14;28 hours
IIIa.	32°	14;7 hours
IIIb.	34°	14;16 hours
IVa.	37°	14;28 hours
IVb.	38°	14;19 (!) hours

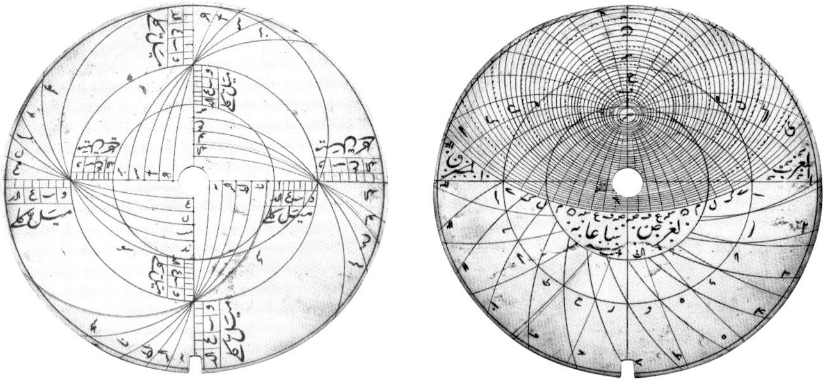

Tympan #1 of No. 30

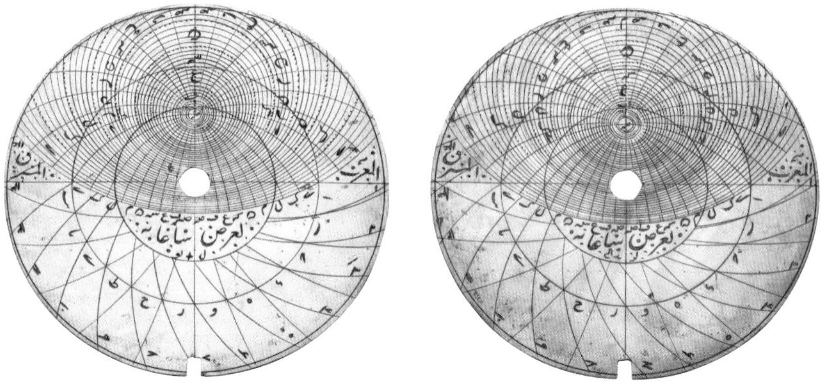

Tympan #2 of No. 30

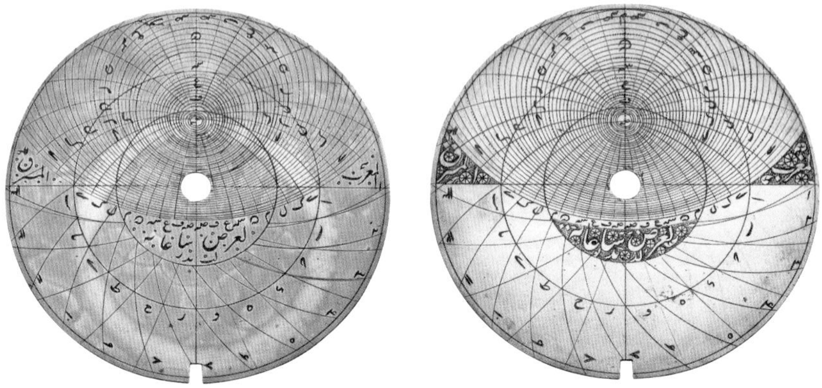

Tympan #3 of No. 30

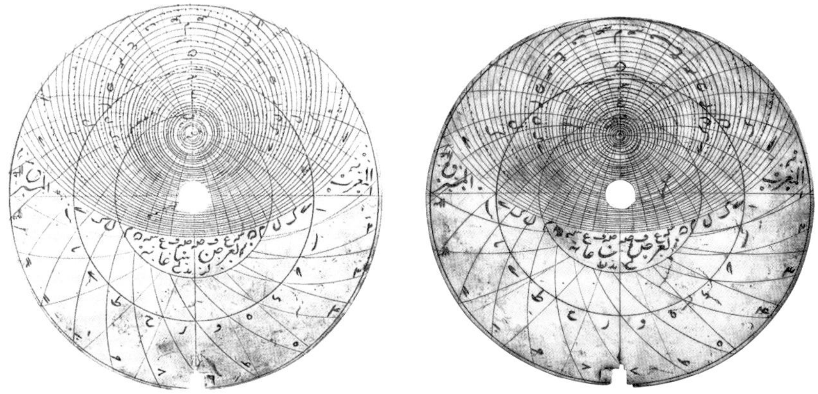

Tympan #4 of No. 30

31 Astrolabe

Muḥammad Mu'min
Iran
18th century
"Muḥammad Mu'min made it."
19 x 12.5 x 3.2 cm
Mater thickness 1 cm
Brass
A-146

Face of No. 31

Incomplete astrolabe made by Muḥammad Mu'min according to the inscription below the shadow-square on the back: *muḥammad mu'min ṣana'ahu.* Made in Iran in the eighteenth century; the decoration and script are similar to those on instruments of 'Abd al-A'immah. Pious inscription on the front of the throne. Inscriptions in naskh.

Consists of body, rete, inscribed and graduated diopter with scales and sights, pin, horse, handle, ring, and six tympans, which have notches at the bottom fitting about a corresponding tab in the body.

The rim of the front of the body is graduated for every degree; the numbers, running clockwise, are 5, 10, 5, 20, 5, ...100, 5, 10, ...200, 5, 10, ...300, 5, 10, ...60. In the recess in the front is a skeleton for a geographical table that was never filled in.

The upper half of the back is graduated for every degree; the numbers, running clockwise from the left to the top and counter-clockwise from the right to the top, are 5, 10, 15,...90. At the edge of the lower half are indicated the cotangents. In the upper left quadrant are the sines and cosines for every 5 of the 60 parts of the Radius; those for each part presumably would have been engraved later. In the upper right quadrant it was intended to engrave some noon-altitude curves of the sun and azimuths of the qibla; what has been engraved is the skeleton and, above is the inscription: *dawā'ir inṣāf al-nahār fī al-bilād al-marqūmah 'alā aṭrāfahā.* Below the center is the shadow-square for 7 feet and 12 digits, within which is the table of the elements, the zodiacal signs, (denoted by the numbers from 0 to 11), and the planetary lords (denoted by their final letters) of the four triplicities. Below this are five semicircles indicating the planetary lords of the terms, the ends of the terms, the names of the zodiacal signs, the

Reverse of No. 31

planetary lords (denoted by their final letters) of the decans according to the Greek system, and the names of the lunar mansions. For a similar back, see No. 32.

On the rete the ecliptic is graduated for every 6°; the numbers, for each zodiacal sign, are 6, 12, 18, ...30. The names of the stars (some pointers have nonsensical inscriptions) are listed. (★)

One of the six tympans is incomplete; the other five have altitude circles for every 3°, azimuths for every 10°, and equal and unequal hour curves (except that Ia omits the equal hour curves). They indicate their latitudes and longest daylights. (■)

Vb has the horizons for latitudes 27°, 39°, 51°, and 63°; for 24°, 36°, 48°, 60°, and 72°; for 21°, 33°, 45°, 57°, and 69°; and for 18°, 30°, 42°, 54°, and 66°.

PROVENANCE — D. Black, 1938; Percy Webster, London, 1938; AP Purchase, 1939.
ICA 2564
EXHIBITIONS — *Making Time Marking Memory,* Miami University Art Museum, Oxford, OH, Feb. 21 – May 1, 2001.
REFERENCES — Newbury (2004a).

★ STAR LIST

Name	*Identification*
janāḥ al-ghurāb	γ Corv.
dhanab al-ʻuqāb	ζ Aquil.
dhanab qayṭus	ι Ceti
ra's al- ḥawwā'	α Oph.
rijl al-yusrā	β Ori.
ridf	α Cygn.
sāq ayman	δ Aquar.
simāk a'zal	α Virg.
simāk rāmiḥ	α Boot.
sha'āmī	α Can. Min.
shi'rā yamānī	α Can. Mai.
ṣadr qayṭus	π Ceti
ẓahr al-dubb	α Urs. Mai.
'unq al-ḥayyah	α Serp.
'ayn al-thawr	α Tau.
'ayyūq	α Aur.
fard al-shujā'	α Hydr.
fam al-faras	ε Peg.
qā'idah	α Crat.
qalb al-asad	α Leo
qalb al-'aqrab	α Scorp.
kaff al-khaḍīb	β Cass.
masāfat al-nahr	θ Erid.
mankib al-faras	β Peg.
nasr ṭā'ir	α Aquil.
nasr wāqi'	α Lyr.
nayyir fakkah	α Cor. Bor.
yad yumnā	α Ori.

Rete of No. 31

Diopter, horse and pin of No. 31

Mater of No. 31

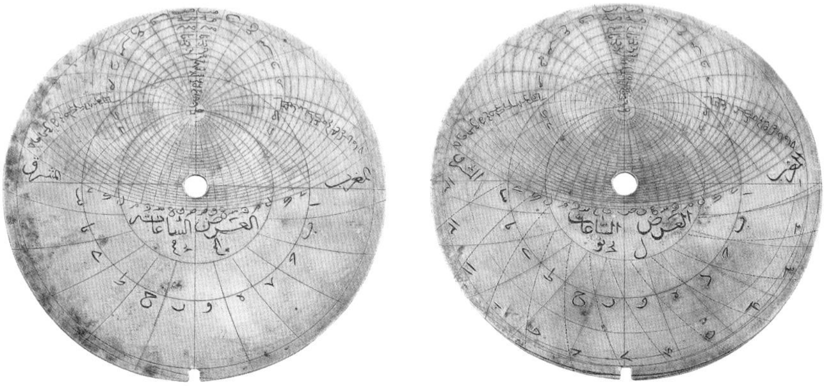

Tympan #1 of No. 31

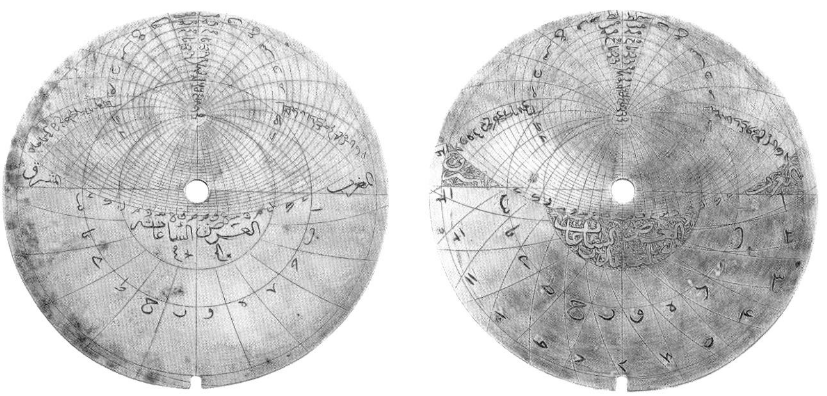

Tympan #2 of No. 31

TYMPANS

Tympan	*Latitude*	*Longest Daylight*
Ia.	28°	13;48 hours
Ib.	30°	13;56 hours
IIa.	31°	14;8 hours
IIb.	32°	14;10 hours
IIIa.	33°	14;12 hours
IIIb.	34°	14;16 hours
IVa.	35°	14;22 hours
IVb.	37°	14;38 hours
Va.	39°	14;46 hours

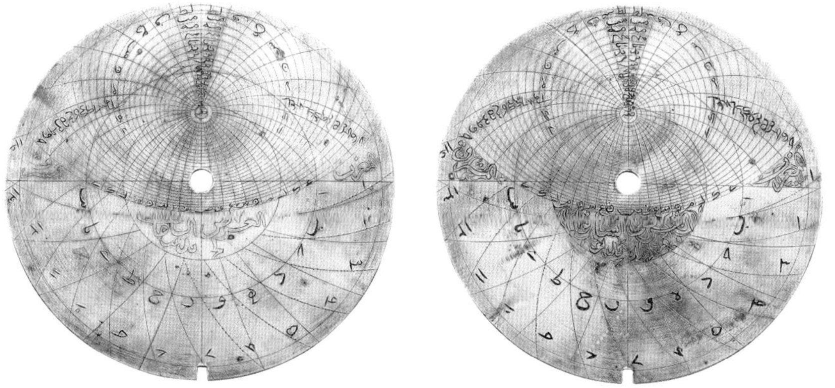

Tympan #3 of No. 31

Tympan #4 of No. 31

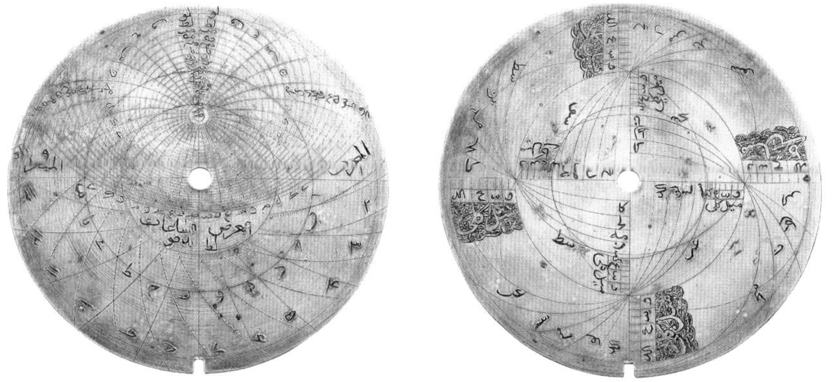

Tympan #5 of No. 31

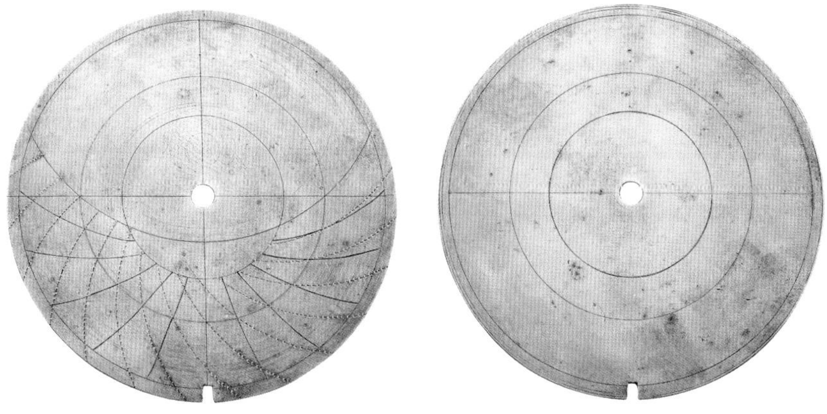

Tympan #6 of No. 31

32 Astrolabe

Maker unknown
Iran
18th. century
22 x 13.4 x 3.6 cm
Mater thickness 1.1 cm
Brass, but tympans silvered
M-41

Face of No. 32

Astrolabe in Iranian style of the eighteenth century. Cartouche for name of maker below shadow-square on back uninscribed. Pious inscriptions on the front and back of the throne. Inscriptions in naskh.

Consists of body, rete, uninscribed diopter with sights, pin, pointer, horse, handle, ring, and four tympans, which have notches at the bottom fitting about a corresponding tab in the body.

The rim of the front of the body is graduated for every degree; the numbers, running clockwise, are in each quadrant 5, 10, 15,...90. In the recess in the front is a geographical table in two circles (◯)(●)

The upper half of the back is graduated for every degree; the numbers, running clockwise from the left to the top and counter-clockwise from the right to the top, are 5, 10, 15,...90. At the edge of the lower half are indicated the co-tangents. In the upper left quadrant are the sines and cosines for every 5°; and the hour curves for 1 to 6 hours, labeled: *khuṭūṭ sāʿāt muʿawwaj<ah>* ("lines of unequal hours"). In the upper right quadrant are a number of arcs concentric with the center of the back; above is the inscription for the non-existent noon altitude curves of the Sun: *dawāʾir inṣāf al-nuhur fī al-ʿurūḍ al-marqūmah ʿalā aṭrāfahā.* Below the center is the shadow-square for 7 feet and 12 digits, within which is the table of the elements, the zodiacal signs (identified by the numbers from 0 to 11), and the planetary lords of the four triplicities. Below this are five semicircles indicating the planetary lords of the terms, the ends of the terms, the names of the zodiacal signs, the planetary lords (denoted by their final letters) of the decans according to the Greek system, and the names of the lunar mansions. For a similar back see No. 31.

On the rete the ecliptic is graduated for every 3°; the numbers, for each zodiacal sign, are 3, 6, 9,...30.

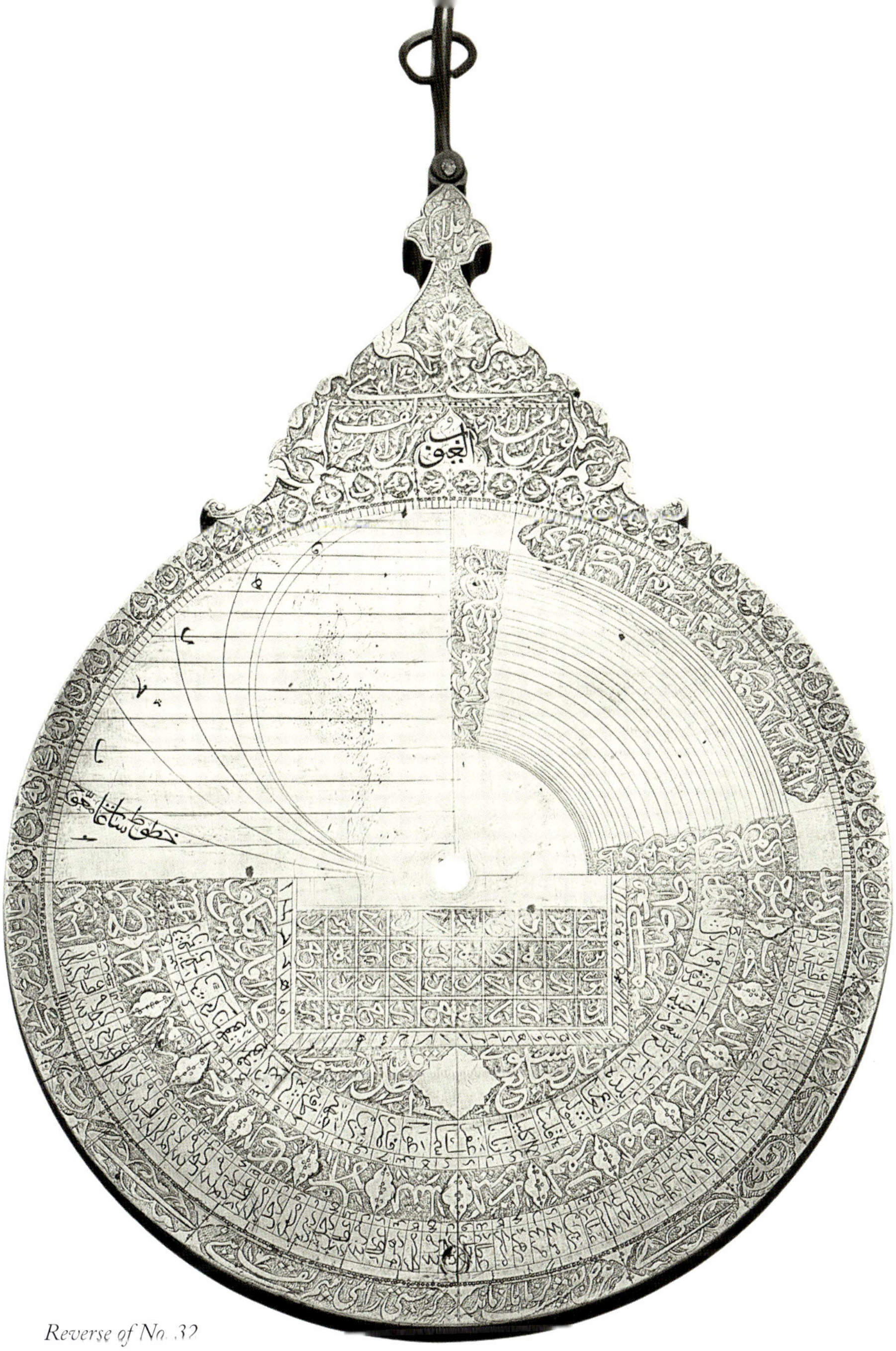

Reverse of No. 32

The names of the stars (some pointers contain only flower designs) are listed. (★)

The four tympans have altitude circles for every 3°, azimuths for every 10°, and equal and unequal hour curves. They indicate their latitudes and longest daylights. (■)

Ia has the horizons for latitudes 12°, 24°, 36°, 48°, and 60°; for <15°>, 27°, 39°, 51°, and 63°; for 18°, 30°, 42°, 54°, and 66°; and for 21°, 33°, 45°, and 57°.

PROVENANCE — A.W.M. Mensing, Amsterdam, 1924; Max Adler, Chicago, 1930; AP gift, 1930.

ICA 1797

EXHIBITIONS —Michigan State University, East Lansing MI., April 27 – Nov. 30, 1977.

REFERENCES — Engelmann Catalogue (1924), p. 16; Fox (1932), pp. 335-37; Fox (1933), pp. 35-36; Gunther (1932), p. 167; Newbury (2004a).

Mater of No. 32

OUTER CIRCLE

City	Longitude	Latitude	Inclination	Identification
makkah	77;6°	21;40°	0;0°	Mecca
madīnah	75;20°	25;0°	37;10°	Medina
miṣr	63;20°	30;20°	58;38°	Cairo
ṣana'ā	77;0°	14;30°	1;55°	Ṣan'ā'
liḥsā	83;30°	24;0°	69;30°	Aḥsā
bayt magdis	66;30°	31;50°	45;16°	Jerusalem
dimashq	70;0°	33;15°	30;31	Damascus
ḥalab	72;10°	35;50°	18;29°	Aleppo
mawṣil	77;0°	34;30°	17;2°	Mosul
nakhjawān	81;15°	38;40°	12;15°	Nakhichevan
marāghah	82;0°	37;20°	16;17°	Marāgheh
tabrīz	82;0°	38;0°	15;40°	Tabrīz
ardabīl	82;30°	38;0°	17;13°	Ardabīl
surramanra'ā	79;31°	34;0°	7;56°	Sāmarrā'
kūfah	79;30°	31;30°	12;31°	Kūfah
baghdād	80;0°	33;25°	82;45°	Baghdād
baṣrah	84;0°	30;0°	37;19°	Basra
shūshah	84;0°	31;30°	36;24°	Shūsh
yazd	89;0°	32;0°	28;28°	Yazd
hamadān	83;0°	35;10°	22;16°	Hamadan
qazwīn	85;0°	36;0°	25;34°	Qazvīn
jurfādaqān	85;30°	34;14°	28;41°	Golpayegan
iṣfahān	86;40°	32;25°	40;28°	Eṣfahān
kāshān	86;0°	34;0°	34;31°	Kāshān
qumm	85;40°	34;45°	34;31°	Qom
ṭālighān	85;45°	36;10°	29;23°	Taliq-an
astarābād	89;35°	36;50°	28;48°	Gorgān
simnān	88;0°	36;0°	37;17°	Semnān
damghān	88;55°	36;20°	28;0°	Dāmghān
kāzirān	87;0°	29;45°	51;17°	Kāzerūn
shīrāz	88;0°	29;0°	53;18°	Shīrāz

INNER CIRCLE

City	Longitude	Latitude	Inclination	Identification
basṭām	89;2°	36;10°	28;58°	Bastam
sharāwār	91;30°	36;0°	41;13°	Sabzevār
nīshābūr	92;30°	36;0°	46;23°	Neyshābūr
ṭūs	92;30°	37;0°	45;6°	Ṭūs
tūn	92;30°	34;30°	52;50°	Ferdows
zūzan	93;30°	35;20°	51;23°	Zūzan
marw	97;0°	37;40°	52;30°	Mary
kābul	104;40°	34;26°	69;30°	Kabul
kirdah	104;0°	27;13°	91;0°	Agra
badakhshān	104;24°	37;10°	64;19°	Faizabad
bukhārā	97;30°	39;50	45;48°	Bukhara
samarqand	99;36°	39;36°	54;54°	Samarkand
qandahār	104;40°	38;0°	74;31°	Kandahar
multān	107;35°	29;40°	80;48°	Multan
qāyin	93;20°	33;40°	55;1°	Qāyen

★ STAR LIST

Name	Identification
janāḥ al-ghurāb	γ Corv.
al-dajājah	α Cygn
dhanab al-qayṭus janūbī	β Ceti
ra's al- ḥawwā'	α Oph.
rijl al-yusrā	β Ori.
sāq ayman	δ Aquar.
simāk a'zal	α Virg.
simāk rāmiḥ	α Boot.
shi'rā sha'āmiyyah	α Can. Min.
shi'rā yamānī	α Can. Mai.
'unq al-ḥayyah	α Serp
'ayn al-thawr	α Tau.
fard al-shujā'	α Hydr.
fakkah	α Cor. Bor.
fam al-faras	ε Peg.
qalb al-asad	α Leo
kaff al-khaḍīb	β Cass.
matn faras	α Peg.
mankib al-faras	β Peg.
nasr ṭā'ir	α Aquil.
nasr wāqi'	α Lyr.
nayyir qā'id	η Urs. Mai.

Rete of No. 32

Diopter, horse, pin and pointer of No. 32

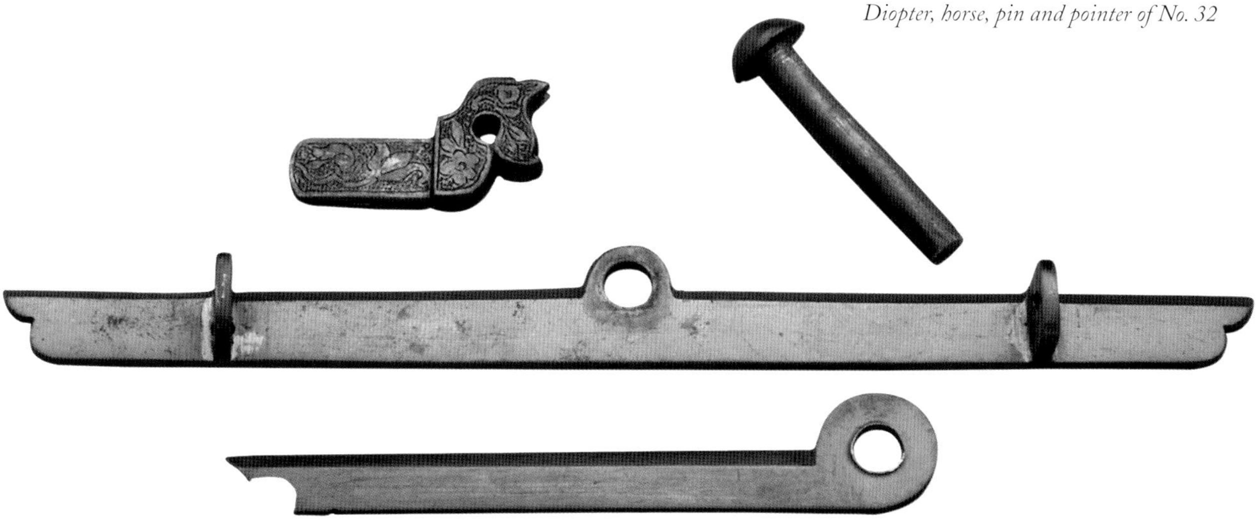

■ TYMPANS

Tympan	*Latitude*	*Longest Daylight*
Ib.	32°	14;6 hours
IIa.	33°	14;12 hours
IIb.	35°	14;22 hours
IIIa.	34°	14;17 hours
IIIb.	36°	14;28 hours
IVa.	37°	14 (later corrected to 14;34) hours
IVb.	38°	14;43 hours

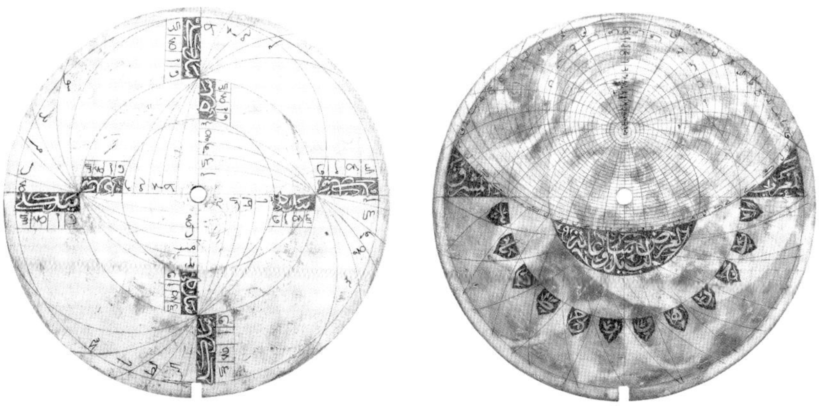

Tympan #1 of No. 32

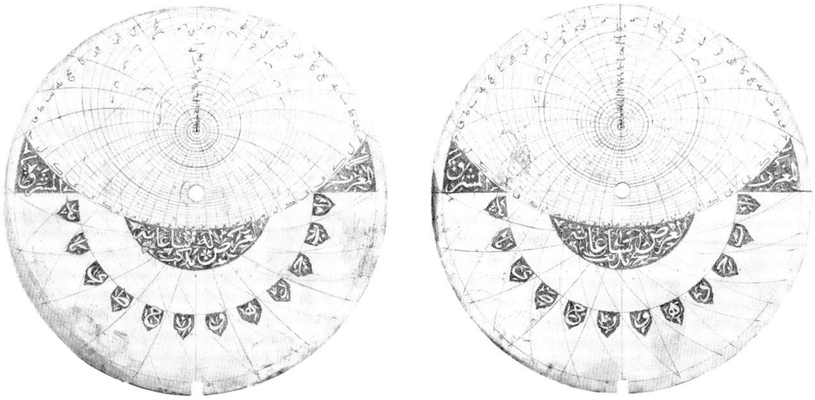

Tympan #2 of No. 32

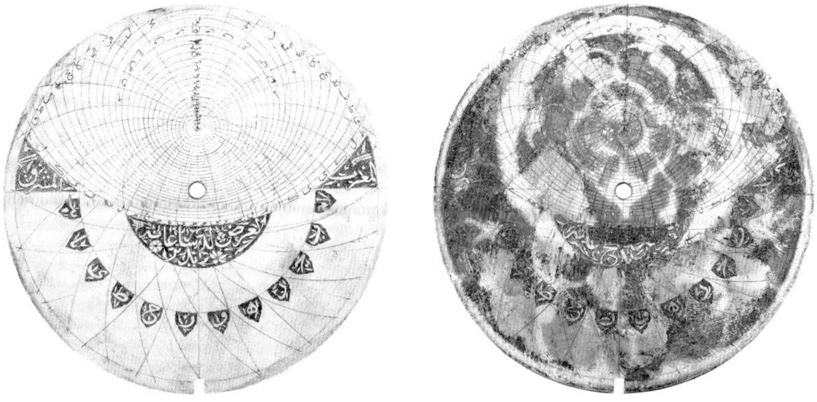

Tympan #3 of No. 32

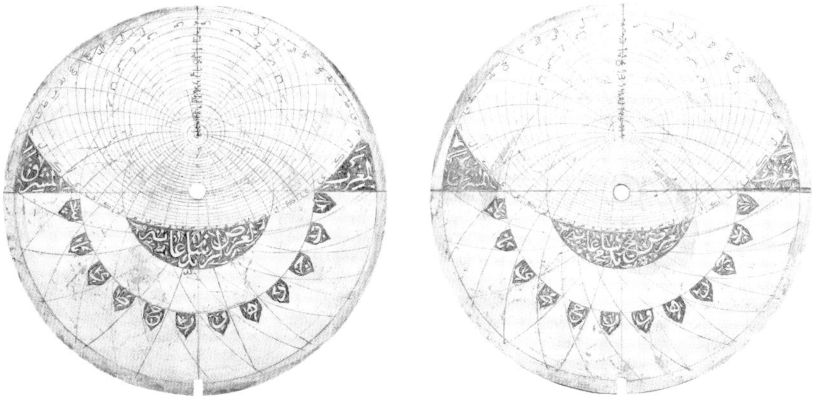

Tympan #4of No. 32

33 Astrolabe

Maker unknown
India
18th. century
22.4 x 14.1 x 5.1 cm
Mater thickness 0.9 cm
Brass
N-68

Face of No. 33

Astrolabe similar in several ways to No. 34; it was probably made in India, as indicated by the use of ghaṭīs, in the late eighteenth century. Inscriptions in naskh.

Consists of body, rete, graduated diopter with sights, (a bolt has been inserted through the upper pair of sights), pin, horse, two washers, handle, ring, and seven tympans (six genuine and one uncertain) which have notches at the bottom fitting about a corresponding tab in the body.

The rim of the front of the body is graduated for every half degree; the numbers, running clockwise, are 5, 10, 15, ...360. In the recess in the front is a geographical table in three circles. (●)(●)(❍)

The upper half of the back is graduated for every degree; the numbers, running clockwise from the left to the top and counter-clockwise from the right to the top, are 5, 10, 15, ...90. At the edge of the lower half are indicated the cotangents. In the upper left quadrant are the sines and cosines for each one of the 60 parts in the Radius. In the upper right quadrant are 30 arcs concentric with the center of the instrument, five (labeled 6, 12, 18, 24, and 30) for each pair of zodiacal signs, and the noon altitude curves of the Sun for the latitudes 19°, 27°, 32°, 36°, 40°, 44°, 49°, 56°, and 60°; below is the inscription: *khuṭūt nisf al-nahār bi-ʿurūd mawḍuʿah* ("lines of noon in local latitudes"). Below the center to the left is the shadow-square for 12 digits, within which is a table of the excess of revolution (*faḍl al-dawr*) for 1 to 9 and 10 to 90 years (20 to 90 labeled 11 to 18). (■)

Below the center to the right is the shadow-square for 7 feet, within which is the table of the elements and the planetary lords (denoted by their final letters) of the four trip-licities. Below this are six semicircles indicating the names of the lunar

■ EXCESS OF REVOLUTIONS

Years	Excess	Years	Excess
1.	87;33°	10.	155;31°
2.	175;6°	20.	311;2°
3.	262;39°	30.	106;33°
4.	350;12°	40.	262;4°
5.	77;45°	50.	57;35°
6.	165;19°	60.	213;6°
7.	252;52°	70.	0,8;37°
8.	340;25°	80.	164;8°
9.	67;58°	90.	319;39°

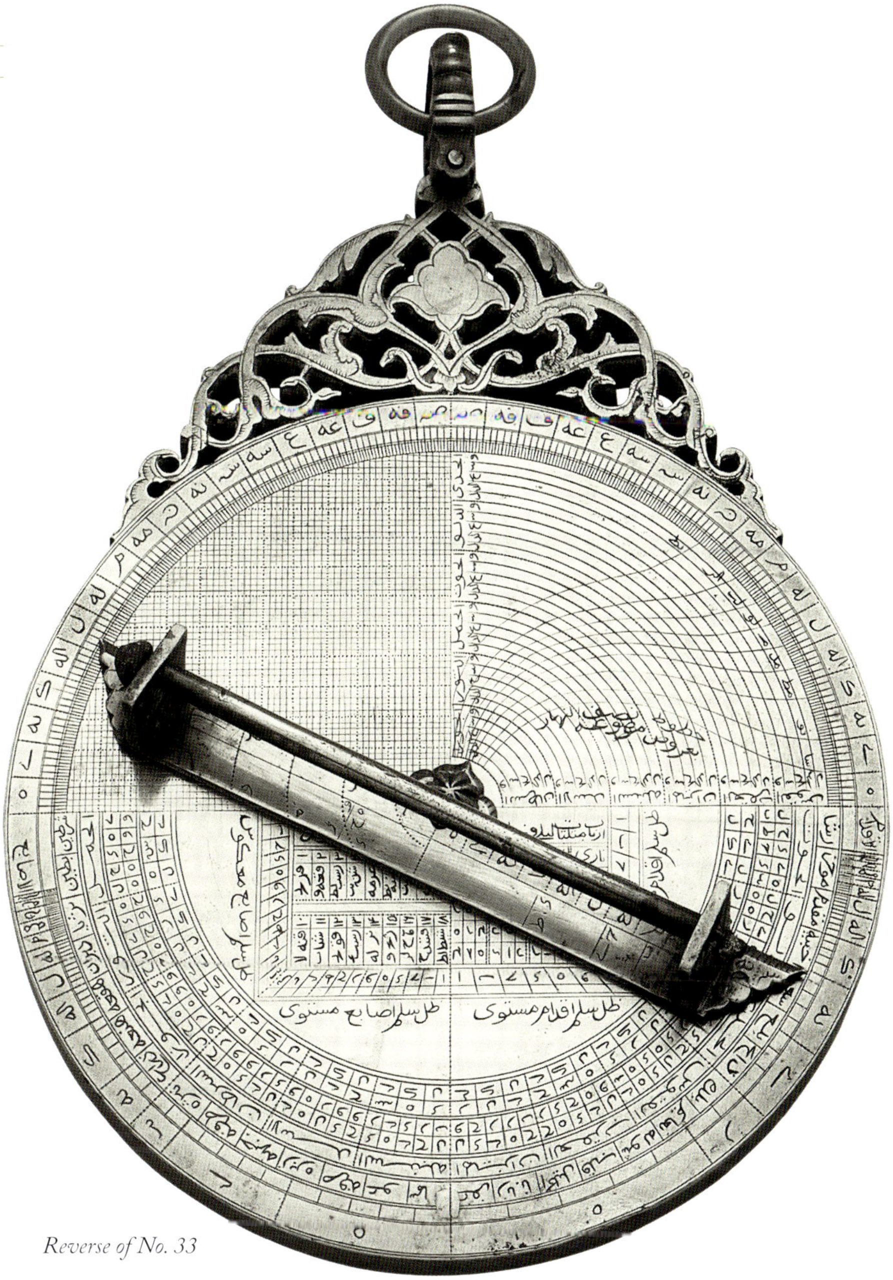

Reverse of No. 33

mansions, the names of the zodiacal signs, the planetary lords (denoted by their final letters) of the terms, the ends of the terms, the planetary lords (denoted by their final letters) of the decans according to the Greek system, and the ends of the decans.

On the rete the ecliptic is graduated for every 6°; the numbers, for each zodiacal sign, are 6, 12, 18, ...30. The names of the stars are listed. (★)

The first six tympans have altitude circles for every 1°, azimuths for every 10°, and equal ghaṭikā curves. They indicate their latitudes and their longest daylights in hours and in kurrīs (ghaṭīs). (■)

VIa has horizons for latitudes 6°, 10°, 18°, 26°, 34°, 42°, 50°, 58°, and 66°; for 8°, 16°, 24°, 32°, 40°, 48°, 56°, and 64°; for 9°, 17°, 25°, 33°, 41°, 49°, 57°, 61°, and 65°; and for 11°, 19°, 27°, 35°, 43°, 51°, 59°, 63°, and 67°.

VIb has a projection labeled: *ṣafīḥat mīzān al-ʿankabūt.* The seventh tympan has unusual projections that I do not understand.

PROVENANCE — Sotheby's & Co., London, 1968; Kenneth Nebenzahl, Glencoe, IL, 1985; AP gift, 1986.
REFERENCES — Newbury (2004a); Newbury (2006); Sotheby's (1968), lot 21.

INNER CIRCLE

City	Longitude	Latitude	Identification
az khaṭṭ istiwā' ("from the equator")			
qanbalah jazīrah	21;0°	3;0°	(Madagascar)
jamkūt	177;0°	5;0°	(Yamakoṭi)
shalā	180;0°	5;0°	(In China)
ghānah	39;0°	10;0°	(Ghana)
'adan	76;0°	11;0°	Aden
sarāndīb	130;0°	10;0°	(Śrīlankā)
al-iqlīm al-awwal ("the first clime")			
bajjah az barbar	65;0°	14;0°	(in Somalia)
ṣan'ā'	77;0°	14;30°	Ṣan'ā'
jublah	75;0°	13;30°	Jubil
janad	76;30°	14;30°	(in Yemen)
sawmnāt	107;0°	22;0°	Somnath
mu'jam	74;15°	17;15°	
jurash	77;50°	17;15°	(in Yemen)
zaytūn	154;0°	17;15°	Hang Zhov
bījāpūr	105;30°	17;20°	Bijapur
ṣuḥār	84;15°	19;20°	Sohar
ṭā'if	77;30°	21;20°	Tā'if
al-iqlīm al-thānī ("the second clime")			
makkah mu'aẓẓam	77;10°	21;40°	Mecca
madīnah ṭayyibah	75;20°	25;0°	Medina
jiddah	76;0°	21;0°	Jiddah
khaybar	75;20°	24;20°	Khaybar
māhūr	116;0°	24;40°	Mathura
nahlawarah	108;20°	28;30°	Patan

Mater of No. 33

MIDDLE CIRCLE

City	*Longitude*	*Latitude*	*Identification*
tatimma iqlīm al-thānī ("completion of the second clime")			
akbarābād	115;0°	26;46°	Agra
banāras	117;20°	26;15°	Varanasi
kwāliyar	114;0°	26;29°	Gwalior
qanawj	114;50°	26;35°	Kannauj
kurah	118;10°	26;49°	Manikpur
jawnbūr	119;6°	26;36°	Jaunpur
lahknaw	114;53°	26;30°	Lucknow
budāūn	114;59°	27;32°	Budaun
awadah	118;6°	27;22°	Ayodhya
ajmīr	111;5°	26;0°	Ajmer
kūpāmaw	116;33°	26;45°	Gopamaw
burhānbūr	113;0°	20;30°	Burhanpur
dawlatābād	111;0°	20;30°	Daulatabad
aḥmadābād	108;40°	23;15°	Ahmedabad
kanbāyat	109;20°	22;20°	Cambay
al-iqlīm al-thālath ("the third clime")			
qayrawān	41;0°	31;40°	Kairouan
mahdiyyah	42;0°	32;31°	Mahdia
bayt al-maqdis	66;30°	31;50°	Jerusalem
dimashq	70;0°	33;15°	Damascus
baghdād	82;0°	33;25°	Baghdād
kūfah	79;30°	31;30°	Kūfah
baṣrah	84;0°	30;0°	Basra
yazd	89;0°	32;0°	Yazd
madā'in	80;20°	33;15°	Ctesiphon
iṣfahān	87;40°	32;25°	Eṣfahān
shīrāz	88;0°	29;30°	Shīrāz
fīrūzābād	87;30°	28;10°	Fīrūzābād
kirmān	91;30°	30;5°	Kermān
hurmūz	92;0°	25;0°	Hormoz
siyālkūt	109;0°	33;0°	Sialkot
lāhawar	109;20°	31;50°	Lahore
thānīsar	112;33°	30;10°	Thanesar
sunnām	112;25°	30;30°	Sunam
bānī palital	113;20°	28;52°	Panipat
hānsī	112;25°	29;45°	Hansi
dihlī	113;35°	28;39°	Dehli
kawil jalālī	114;19°	28;4°	Aligarh
kūlkandah	114;9°	28;4°	Golconda

OUTER CIRCLE

City	*Longitude*	*Latitude*	*Identification*
al-iqlīm al-rābi' ("the fourth clime")			
karaj	84;45°	34;0°	Karaj
tūn	92;30°	34;30°	Ferdows
harāh	94;20°	34;30°	Herat
kābul	104;0°	34;30°	Kabul
qumm	85;40°	34;45°	Qom
rayy	86;20°	35;0°	Shahr Rey
sāwah	85;0°	36;15°	Sāveh
kashmīr	108;0°	35;0°	Srinagar
hamadān	83;0°	36;10°	Hamadan
ḥalab	72;10°	35;50°	Aleppo
mashhad	92;30°	36;0°	Mashhad
iskandariyyah	61;54°	30;58°	Alexandria
balkh	101;0°	36;51°	Wazirabad
nīshābūr	92;30°	36;21°	Neyshābūr
astarābād	89;35°	36;50°	Gorgān
qazwīn	85;0°	36;55°	Qazvīn
badakhshān	104;24°	36;10°	Faizabad
marāghah	82;0°	37;20°	Marāgheh
marw	97;0°	37;40°	Mary
tabrīz	82;0°	38;0°	Tabrīz
nakhjawān	81;15°	38;40°	Nakhichevan
al-iqlīm al-khāmis ("the fifth clime")			
samarqand	99;16°	39;37°	Samarkand
baykand	97;30°	39;45°	Baykand
bukhārā	97;30°	39;50°	Bukhara
tibat	110;0°	40;0°	(Tibet)
barda'ah	83;0°	40;30°	Barda
shamākhī	84;30°	40;50°	Shemakha
qūnyah	66;30°	41;0°	Konya
khujand	105;35°	41;55°	Leninabād
ḥajāj	84;0°	42;0°	
khutan	107;0°	42;0°	Khotan
farghānah	102;0°	42;20°	Fargana
īlāq	99;10°	43;20°	Angren
al-iqlīm al-sadis ("the sixth clime")			
kāshghar	106;30°	44;0°	kashgar
binidiqiyyah	42;0°	44;0°	Veince
ṭarāz	99;50°	44;31°	Dzhambul
bīsh bāligh	111;0°	45;50°	Pochongtse
khān bāligh	124;0°	46;0°	Beijing
haraqlah	67;20°	46;30°	Ereğli
al-iqlīm al-sābi' ("the seventh clime")			
araq	105;0°	48;0°	
aklah	108;0°	49;0°	
bulghār	90;0°	49;30°	(Bulgars)
aqjaḥ	62;0°	50;0°	Belgortoddnestrov
qirqir	65;30°	50;0°	
khārij al-iqlīm al-sābi' (" outside the seventh clime")			
mārī	65;45°	50;40°	
rūdajard jazā'ir		62;0°	(Hebrides)
ṣūdāq	66;0°	51;0°	Sudak

★ STAR LIST

Name	*Identification*
a'zal	α Virg.
baṭn al-ḥūt	β Andr.
tālī masāfat al-nahr	θ Erid.
janāḥ al-ghurāb	γ Corv.
al-jāthī	α Herc.
dhanab al-jady	δ Capr.
dhanab dulfīn	ε Delph.
dhanab al-qayṭus	ι Ceti
dhanab al-qayṭīs	β Ceti
ra's al-ḥawwā'	α Oph.
ra's al-ghūl	β Pers.
rijl al-jawzā' yusrā	β Ori.
rijl al-jawzā' yumnā	κ Ori.
rijl musalsalah	γ Andr.
rukbat al-ḥawwā' yusrā	ζ Oph.
sāq sākib al-mā'	δ Aquar.
surrat faras	δ Peg.
simāk rāmiḥ	α Boot.
shi'rā sha'āmiyyah	α Can. Min.
shi'rā yamāniyyah	α Can. Mai.
ṣadr al- qayṭus	π Ceti
ẓahr dubb al-akbar	α Urs. Mai.
'anāq	ζ Urs. Mai.
'unq al-ḥayyah	α Serp.
'ayn al-tinnīn	α Drac.
'ayn al-thawr	α Tau.
'ayyūq	α Aur.
fard al-shujā'	α Hydr.
fakkah	α Cor. Bor.
fam faras	ε Peg.
fam qayṭus	γ Ceti
qā'idat al-bāṭiyyah	α Crat.
kaff al-jadhmā'	α Ceti
kaff al-khaḍīb	β Cass.
minkḥār (?) mutaqaddim	ρ Serp (?)
minqār dajājah	β Cygn.
mankib faras	β Peg.
nasr ṭā'ir	α Aquil.
nasr wāqi'	α Lyr.
yad al-jawzā' al-yusrā	α Ori.
yad al-jawzā' yumnā	α Ori.

Rete of No. 33

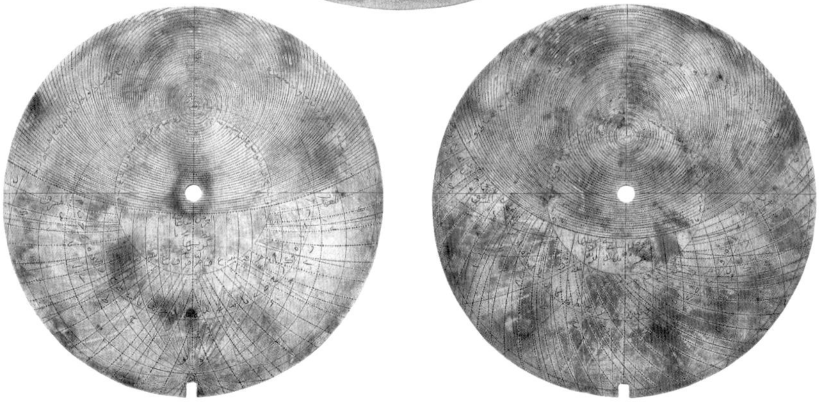

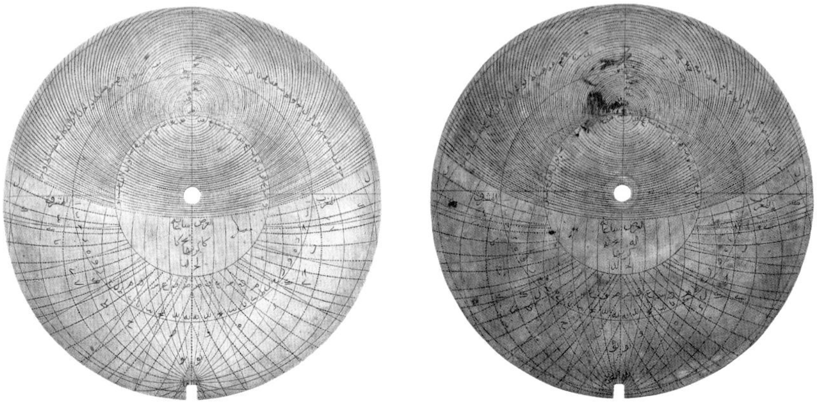

Tympan #1 of No. 33

Tympan #2 of No. 33

Tympan #3 of No. 33

TYMPANS

Tympan	Latitude	Longest Daylight	
Ia	19°	13;11 hours	32;58 ghaṭīs
Ib	40°	14;54 hours	37;15 ghaṭīs
IIa	21°	13;21 hours	33;22 ghaṭīs
IIb	25°	13;25 hours	33;34 ghaṭīs
IIIa	23;30°	13;24 hours	33;30 ghaṭīs
IIIb	38°	14;50 hours	37;11 ghaṭīs
IVa	27°	13;46 hours	34;25 ghaṭīs
IVb	32°	14;8 hours	35;8 ghaṭīs
Va	30°	13;58 hours	34;55 ghaṭīs
Vb	35°	14;21 hours	35;51;30 ghaṭīs

Tympan #4 of No. 33

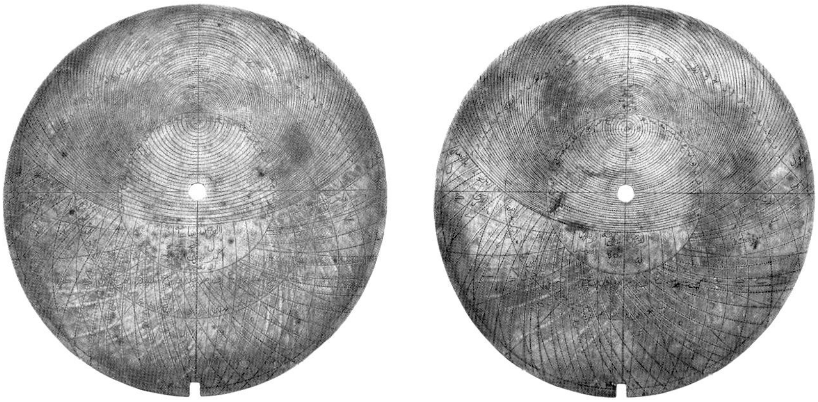

Tympan #5 of No. 33

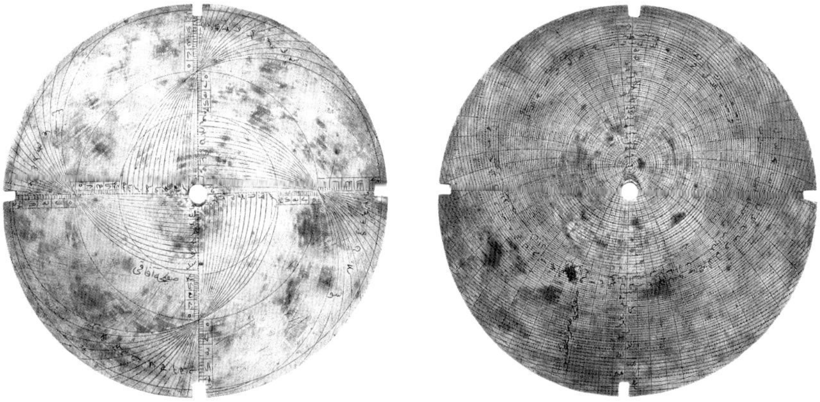

Tympan #6 of No. 33

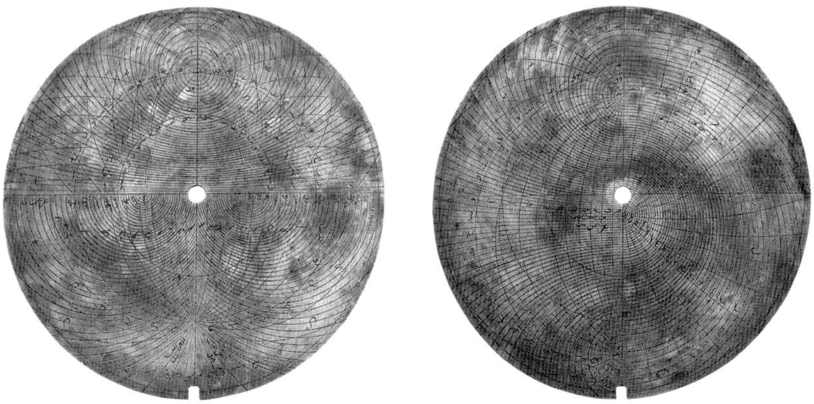

Tympan #7 of No. 33

34 Astrolabe, Imperfect

Muḥammad Sharīf of Kirmān
India
ca. 1780
"From the manufacture of Muḥammad Sharīf ibn Muḥammad."
17.2 x 10.2 x 4.1 cm
Mater thickness 1.1 cm
Brass
DPW-52

Face of No. 34

Incomplete astrolabe made by Muḥammad Sharīf ibn Muḥammad according to the inscription in the center of the recess in the front: *min ṣanaʿat muḥammad sharīf ibn muḥammad.* As was noted by W. Hartner, this maker is probably identical with Muḥammad Muḥsin ibn Muḥammad ʿAlī al-Sharīf al-Kirmānī. Inscriptions in naskh.

Consists of body, diopter with scales and sights, pin, horse, handle, ring, and six tympans, which have notches at the bottom; the corresponding tab in the body has fallen out. The rete is fake.

The rim of the front of the body is graduated for every degree; the numbers, running clockwise, in each quadrant are 6, 12, 18,...90. In the recess in the front is a geographical table in three circles. (❍)(●)(●)

The upper half of the back is graduated for every degree; the numbers, running clockwise from the left to the top and counterclockwise from the right to the top, are 6, 12, 18,...90. At the edge of the lower half are indicated the cotangents. In the upper left quadrant are cosines and for every 2 of the 60 parts of the Radius. In the upper right quadrant are the curves of the noon altitudes of the Sun for latitudes 27° and 32°. Below the center is the shadow-square for 7 feet and 12 digits, within which is a semicircle indicating the elements and the planetary lords (denoted by their final letters) of the four triplicities. Below this are two semicircles giving the names of the lunar mansions and of the zodiacal signs.

On the rete the ecliptic is graduated for every 6°; the numbers for each zodiacal sign are 6, 12, 18, ...30. The names of the stars (many pointers are uninscribed, many stars are misplaced, and some stars are given in two different places) are listed. (★)

Reverse of No. 34

The six tympans have altitude circles for every 6° (Va omits them), azimuths for every 6° (Ia and IIa), for every 10° (IIIb, IVa, Va, and Vb), or not at all (Ib, IIb, IIIa, IVb, and VIa), and equal and unequal hour curves. They indicate their latitudes and longest daylights. (■)

VIb has horizons for latitudes 6°, 14°, 22°, 30°, 38°, 46°, 54°, 62°, and 72°; for 8°, 16°, 24°, 32°, 40°, 48°, 56°, and 64°; for 18° (read 10°), 18°, 24° (read 26°), 34°, 42°, 50°, 58°, and 66°; and for 12°, 20°, 28°, 36°, 44°, 52°, 60°, and 68°.

PROVENANCE — Mercator, Paris, 1921; Richard Wheatland, Sr., 1921; David P. Wheatland, Topsfield, MA, 1943; AP gift, 1988.

EXHIBITIONS — *Making Time Marking Memory*, Miami University Art Museum, Oxford, OH, Feb. 21 – May 1, 2001.

Mater of No. 34

Diopter, horse and pin of No. 34

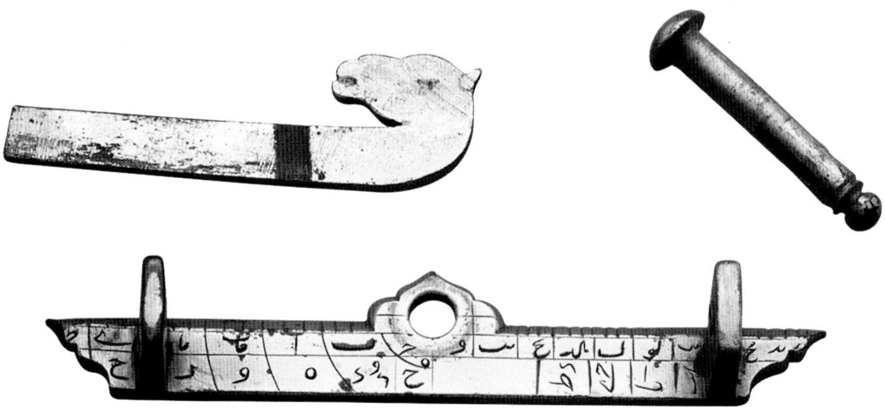

OUTER CIRCLE

City	*Longitude*	*Latitude*	*Identification*
sarandīp	130;0°	10;0°	(Srīlankā)
kanbāyat	109;10°	22;20°	Cambay
kūlkundah	114;19°	18;10°	Golconda
bījāpūr	105;30°	16;20°	Bijapur
burhānbūr	108;0°	20;30°	Burhanpur
jawnbūr	119;6°	26;36°	Jaunpur
ujayn	110;50°	23;30°	Ujjain
sarūj	72;40°	34;50°	Suruch
jiddah	76;0°	21;0°	Jiddah
jamkūt	177;0°	5;0°	(Yamakoti)
purushawar	105;14°	31;0°	Peshawar
nawshahr	78;20°	28;10°	Nowshahr
siyālkūt	109;0°	38;0°	Sialkot
kawil jalālī	114;19°	28;10°	Aligarh
nānbīr (?)	112;23°	30;10°	
sanbhal	115;0°	28;28°	Sambhal
sulṭānkūt	115;0°	23;30°	
budā'un	114;19°	27;32°	Budaun
sunnam	112;25°	30;30°	Sunam
bānī batah	113;20°	28;52°	Panipat
hānsī	112;25°	29;45°	Hansi
kurah wa mānikbūr	113;10°	29;49°	Manikpur
(illegible)	114;33°	26;30°	Lucknow
kūpāmaw	116;43°	26;45°	Gopāmaū
awadah	113;6°	27;22°	Ayodhya
ajmīr	111;5°	26;0°	Ajmer
kwāliyar	114;0°	26;29°	Gwalior
banāras	117;20°	26;52°	Varanasi
qanawj	115;50°	26;35°	Kannauj
akrah	115;0°	27;43°	Agra
ḥaṣr dihlī	113;35°	28;19°	Dehli

MIDDLE CIRCLE

City	*Longitude*	*Latitude*	*Identification*
lāhawar	109;20°	31;50°	Lahore
multān	107;35°	29;40°	Multan
kashmir	108;0°	35;0°	Srinagar
tibat	110;0°	40;0°	(Tibet)
qandahār	107;40°	33;0°	Kandahar
daybul	102;30°	25;10°	Tatta
kābul	104;40°	34;50°	Kabul
ghaznah	104;20°	33;35°	Ghazni
kirmān	91;30°	30;5°	Kermān
hurmūz	92;0°	25;0°	Hormoz
khutan	107;0°	42;0°	Khotan
kāshghar	106;30°	44;0°	Kashgar
khujand	100;35°	41;55°	Leninabad
samarqand	99;36°	39;32°	Samarkand
bukhārā	94;30°	39;50°	Bukhara
badakhshān	94;24°	34;10°	Faizabad
balkh	91;0°	36;41°	Wazirabad
harāt	94;20°	34;30°	Herat
mashhad	92;30°	32;0°	Mashhad
sabzawār	91;30°	36;0°	Sabzevār
basṭām	89;30°	36;10°	Bastam
kāshān	86;0°	34;0°	Kāshān
iṣfahān	87;40°	22;26°	Eṣfahān

INNER CIRCLE

City	*Longitude*	*Latitude*	*Identification*
qazīn	85;0°	36;0°	Qazvīn
yazd	89;0°	32;0°	Yazd
shīrāz	83;0°	29;36°	Shīrāz
fīrūzābād	84;30°	28;10°	Firuzābād
baṣrah	84;0°	30;11°	Basra
baghdād	82;0°	33;25°	Baghdād
kūfah	79;10°	31;30°	Kūfah
tabrīz	82;0°	38;0°	Tabrīz
maraghah	82;0°	37;20°	Marāgheh
ḥalab	72;10°	35;50°	Aleppo
dimashq	70;0°	33;15°	Damascus
bayt al-maqdis	66;30°	31;50°	Jerusalem
makkah musharrafah	77;10°	21;40°	Mecca
madīnah ṭayyibah	75;20°	25;0°	Medina
miṣr	63;20°	30;20°	Cairo

★ STAR LIST

Name	*Identification*
baṭn qayṭus	ζ Ceti
tālī masāfat al-nahr	θ Erid.
jāthī	α Herc.
janāḥ al-ghurāb	γ Corv.
al-ḥawwā’	α Oph.
khaḍīb (see kaff)	β Cass.
dhanab al-jady	δ Capr.
dhanab al-qayṭus ja	β Ceti
dhanab al-qayṭus shamā	ι Ceti
rāmiḥ	α Boot.
rijl yusrā	β Ori.
rijl yamānī	κ Ori.
sāq sākib al-mā’	δ Aquar.
surrat	δ Peg.
simāk a‘zal	α Virg.
shi‘rā sha‘āmi	α Can. Min.
shi‘rā yamānī	α Can. Mai.
ṣadr al- qayṭus	π Ceti
ṭarf al-safīnah	ρ Puppis
ẓahr dubb al-akbar	α Urs. Mai.
al-‘anāq	ζ Urs. Mai.
‘ayyūq (in two places)	α Aur.
ghūl	β Pers.
faras (see mankib)	β Peg.
fakk	α Cor. Bor.
qā‘idat al-bāṭiyyah	α Crat.
qalb al-asad	α Leo
kaff al-khaḍīb (see khaḍīb)	β Cass.
minqār	β Cygn.
mankib (see faras)	β Peg.
mu’akhkhar shujā‘	σ Hydr.
nasr ṭā’ir	α Aquil.
<wā>qi‘	α Lyr.
yad yusrā	γ Ori.
yad yumnā	α Ori.

Rete of No. 34

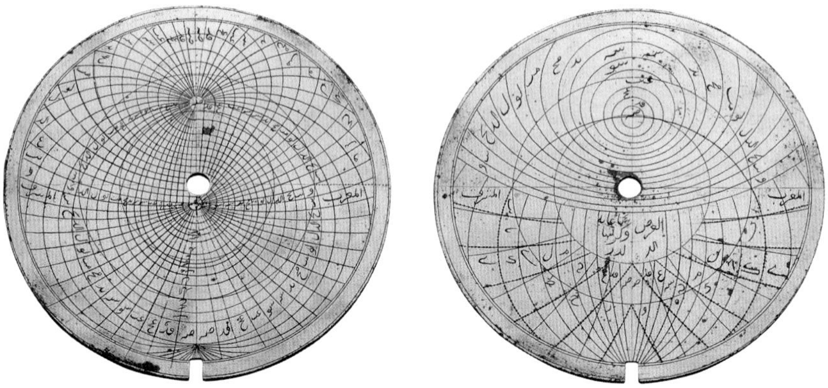

Tympan #1 of No. 34

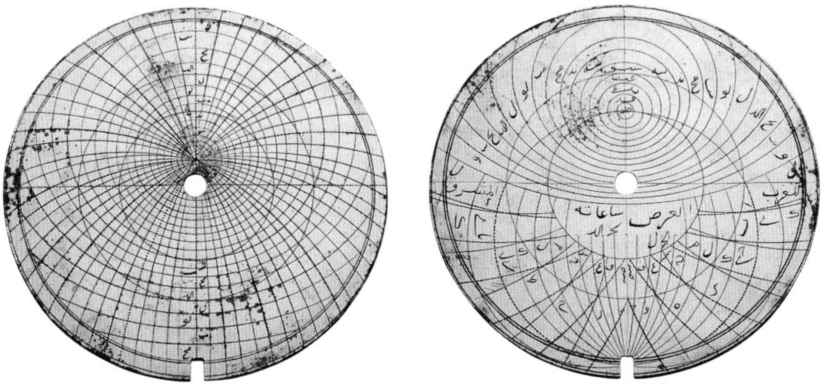

Tympan #2 of No. 34

TYMPANS

Tympan	*Latitude*	*Longest Daylight*
Ia.	72° and 18° (a double projection)	
Ib.	27	34;7 kurrī (ghaṭīs)
IIa.	36° (altitude circles and azimuths in both hemispheres)	
IIb.	33°	13;24 hours
IIIa.	32°	14;8 hours
IIIb.	18°	13;9 hours
IVa.	36°	14;32 hours
IVb.	40°	14;54 hours
Va.	not completed	
Vb.	21°	13;19 hours
VIa.	25°	13;25 hours.

Tympan #3 of No. 34

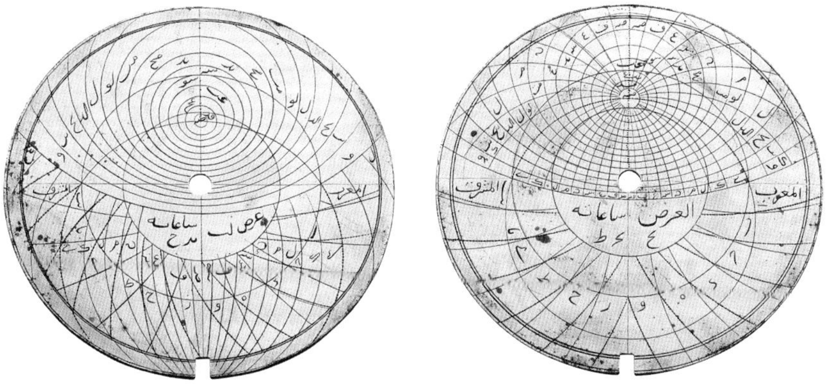

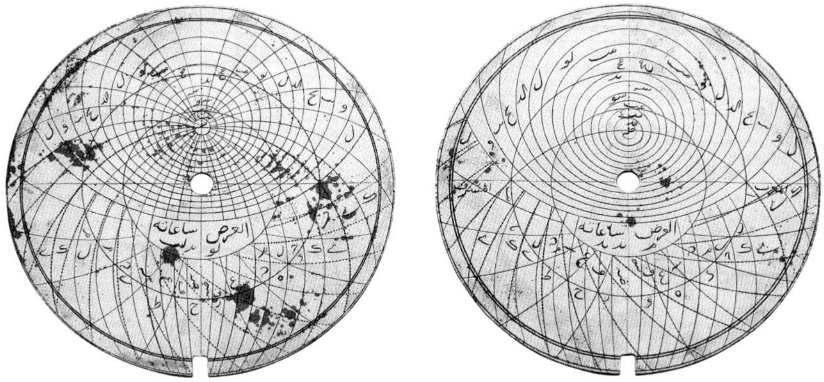

Tympan #4 of No. 34

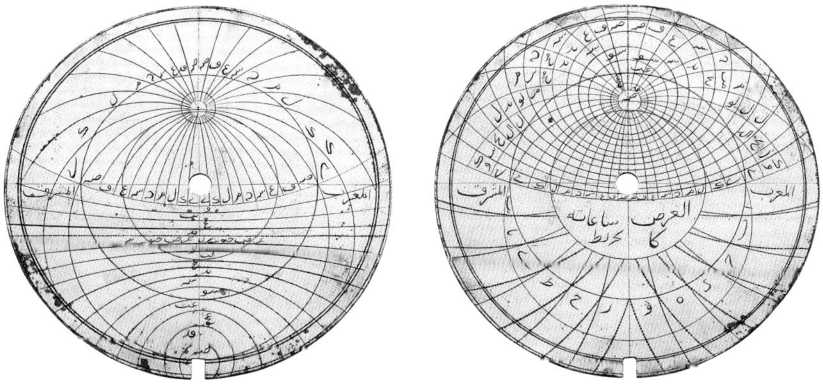

Tympan #5 of No. 34

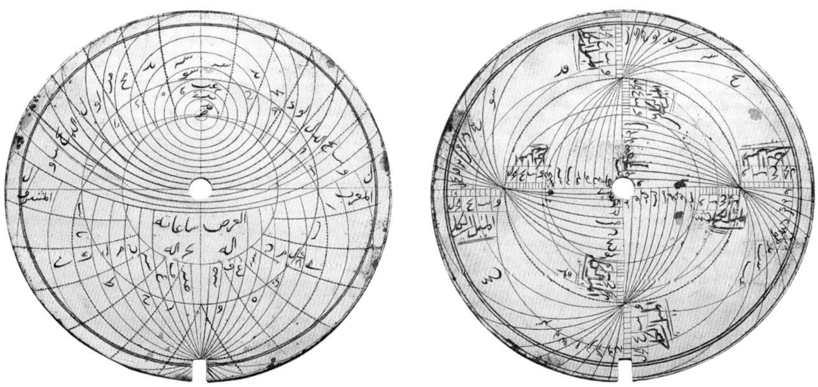

Tympan #6 of No. 34

35 Astrolabe, Imperfect and Incomplete

Muḥammad Ṭabāṭabā'ī
Iran
1792/3
"Muḥammad Ṭabāṭabā'ī, the servant (of God) — may He love him — made it in 1207."
18.7 x 11.3 x 4.9 cm
Mater thickness 1.1 cm
Brass, but tympans silvered
M-40

Face of No. 35

Imperfect and incomplete astrolabe made by Muḥammad Ṭabāṭabā'ī in A. H. 1207 (AD 1792/3) according to the inscription on the edge of the body opposite the throne: *ṣana'ahu muḥammad ṭabāṭabā'ī al- 'abd ḥabbahu fī 1207*. The rest of the edge contains a long inscription enumerating the parts of the astrolabe. This instrument was never finished. Inscriptions in naskh.

Consists of body, rete, uninscribed diopter with sights, pin, horse, handle, ring, and six tympans, which have notches at the bottom fitting over a corresponding tab in the body.

The rim of the front is graduated for every degree; the numbers, running counter-clockwise, are 5, 10, 15, ...360. In the recess in the front is a geographical table in two circles. (❍)(●)

The upper right quadrant is graduated at the top for every degree (the graduation was not applied to the upper left quadrant); the numbers, running clockwise from the left to the top and counter-clockwise from the right to the top, are 5, 10, 15,...90. In the upper left quadrant are the cosines and sines for every 5 of the 60 parts of the Radius and 17 radii. In the upper right quadrant are the noon altitude curves of the Sun for two unspecified latitudes and the azimuths of the qibla for five unnamed cities. At the edge of the lower half are indicated the cotangents. Below the center is an uninscribed shadow-square within which is the skeleton of a table of the elements, zodiacal signs, and planetary lords of the four triplicities. Below this are five semicircles indicating with mistakes the planetary lords of the terms, the names of the zodiacal signs, the planetary lords (denoted by their final letters) of the decans according to the Greek system, and the names of the lunar mansions.

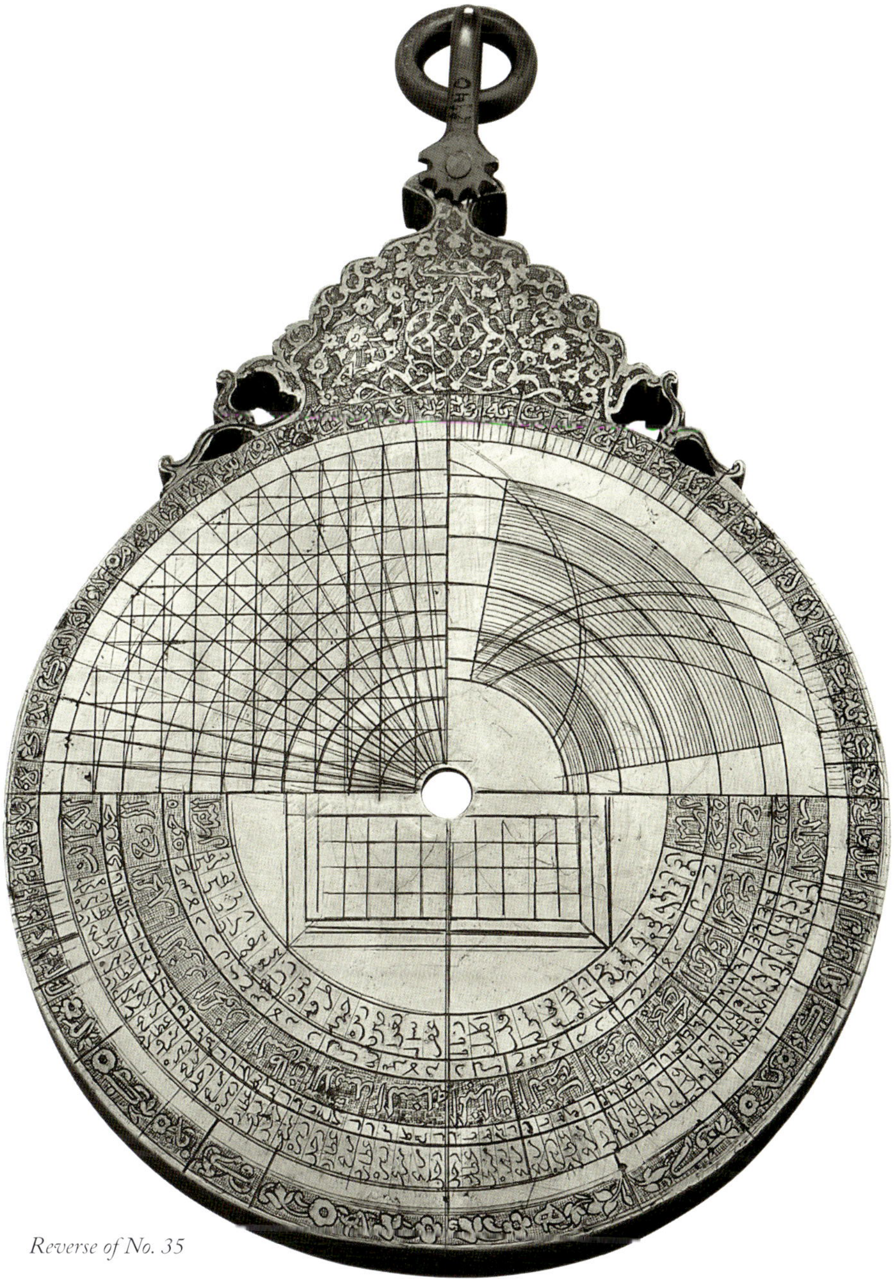

Reverse of No. 35

The rete, though it is well made, is uninscribed except for the names of the zodiacal signs on the ecliptic.

The six tympans have altitude circles for every 10°, azimuths for every 10° (except IIa, b and IIIb have no azimuths), and equal and unequal hour curves; they are not always carefully engraved. They indicate their latitudes and longest daylights with mistakes. (■)

VIb has horizons for unspecified latitudes.

PROVENANCE — A.W.M. Mensing, Amsterdam, 1924; Max Adler, Chicago, 1930; AP gift, 1930.

ICA 2501

REFERENCES — Engelmann Catalogue (1924), p. 16; Stephenson, Bolt, and Friedman (2000), p. 108.

Mater of No. 35

❍ OUTER CIRCLE

City	*Longitude*	*Latitude*	*Inclination*	*Direction*	*Identification*
makkah mu'aẓẓamah	76;10°	21;40°	0;0°	0;0°	Mecca
madīnah ṭayyibah	75;20°	25;10°	36;8°	SE	Medina
baghdād	80;0°	33;25°	12;46°	SW	Baghdād
baṣrah	85;0°	30;0°	36;59°	SW	Basra
shīrāz	88;0°	29;36°	53;23°	SW	Shīrāz
jurfādaqān	85;30°	34;15°	32;21°	SW	Golpayegan
iṣfahān	86;40°	32;25°	40;29°	SW	Eṣfahān
kāshān	84;0°	34;0°	34;31°	SW	Kāshān
qumm	85;40°	34;46°	31;55°	SW	Qom
ṭihrān	86;20°	35;0°	36;26°	SW	Tehrān
qazwīn	85;0°	35;0°	27;34°	SW	Qazvīn
sāwah	85;0°	35;0°	29;16°	SW	Sāveh
hamadān	83;0°	36;10°	22;17°	SW	Hamadan
simnān	88;0°	36;0°	36;17°	SW	Semnān
basṭām	89;30°	36;10°	39;53°	SW	Bastam
nīshābūr	92;30°	36;21°	46;25°	SW	Neyshābūr
mashhad muqaddas	104;30°	36;0°	45;6°	SW	Mashhad
harāt	95;20°	34;30°	54;5°	SW	Herat
marw	106;0°	36;30°	54;42°	SW	Mary
qandahār	107;40°	33;0°	45;5°	SW	Kandahar
miṣr	63;20°	30;20°	48;38°	SE	Cairo
ṭarābalas	45;0°	32;0°	48;17°	SE	Tripoli
iskandariyyah	61;55°	30;58°	59;21°	SE	Alexandria
ṣana'ā	76;0°	14;30°	1;55°	SE	Ṣan'ā'
liḥsā	83;30°	24;0°	69;23°	SW	Aḥsā
bayt al-maqdis	66;30°	31;55°	46;43°	SE	Jerusalem
antākiyyah	71;26°	35;30°	21;36°	SE	Antakya
qusṭanṭah	55;55°	45;0°	38;57°	SE	Istanbul
kanjah	83;0°	41;20°	14;9°	SW	Kirovabad
bāb al-abwāb darband	85;0°	43;0°	19;29°	SW	Darband
ahwāz	85;0°	31;16°	36;14°	SW	Ahvāz
dimashq	70;0°	33;15°	23;49°	SE	Demascus

● INNER CIRCLE

City	*Longitude*	*Latitude*	*Inclination*	*Direction*	*Identification*
tabzārā	82;0°	38;0°	15;1°	SW	Tabrīz
ardabīl	82;30°	38;0°	17;13°	SW	Ardabīl
ayrawān	89;15°	38°	36;22°	SW	Yerevan
yazd	89;0°	32;0°	48;0°	SW	Yazd
marāghah	80;0°	36;20°	16;17°	SW	Marāgheh
kirmān	92;30°	29;35°	47;40°	SW	Kermān
astarābād	59;35°	36;50°	38;48°	SW	Gorgān
āmul	83;0°	38;0°	51;29°	SW	Amol
turshīz	92;0°	36;0°	48;11°	SW	Kāshmar
bulghār	90;0°	99;30°	25;9°	SW	(Bulgars)
kūfah	79;0°	31;30°	12;30°	SW	Kūfah
surramanra'ā	79;0°	34;0°	7;57°	SW	Sāmarrā'
kāshghar	106;30°	94;0°	57;12°	SW	Kashgar
khutan	107;41°	92;0°	41;17°	SW	Khotan
qarāqūram	45;0°	97;0°	65;32°	SW	Karakoram
khutan	107;41°	92;0°	61;4°	SW	Khotan

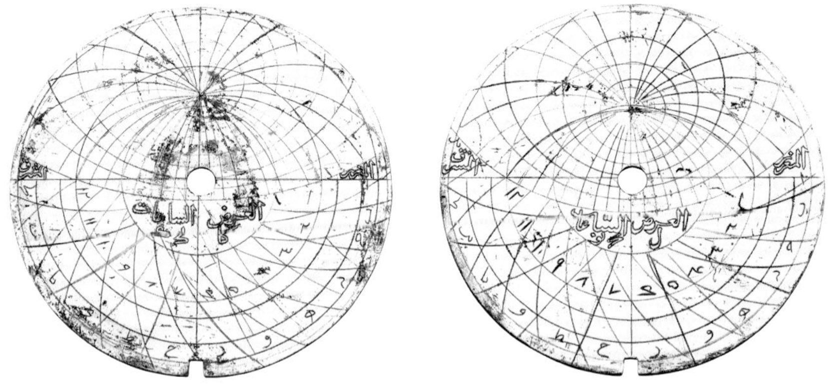

Tympan #1 of No. 35

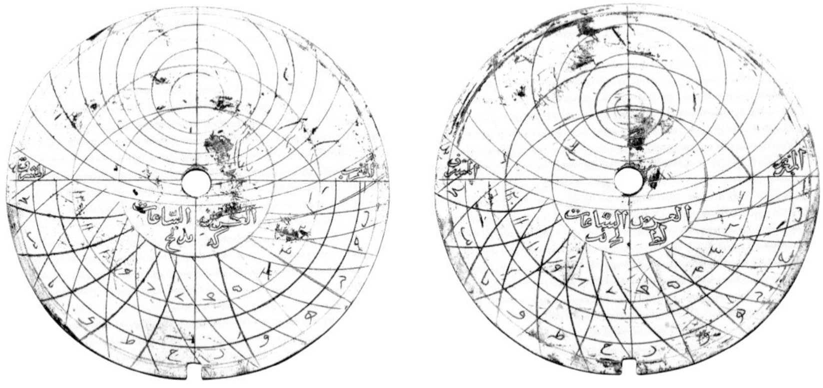

Tympan #2 of No. 35

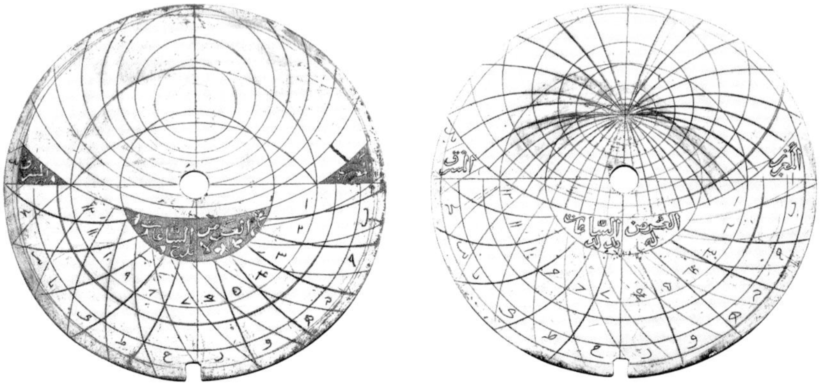

Tympan #3 of No. 35

TYMPANS

Tympan	*Latitude*	*Longest Daylight*
Ia.	21°	13;20 hours
Ib.	30°	13;56 hours
IIa.	25°	14;38 hours
IIb.	39°	13;52 hours
IIIa.	31°	14;8 hours
IIIb.	35°	14;34 hours
IVa.	32°	14;10 hours
IVb.	40°	14;51 hours
Va.	34°	14;36 hours
Vb.	38°	14;39 hours
VIa.	36°	14;28 hours.

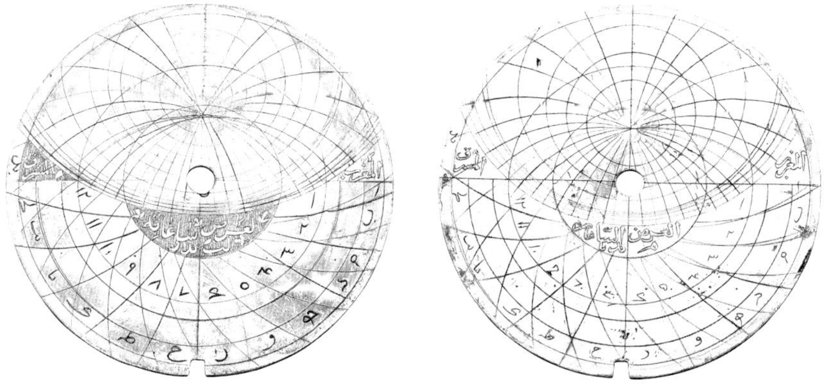

Tympan #4 of No. 35

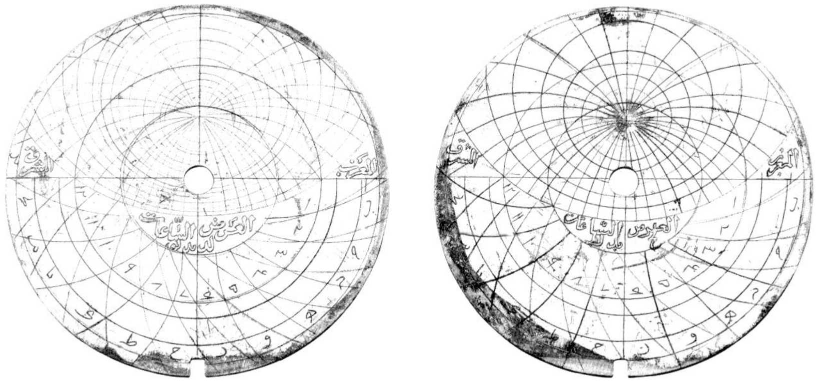

Tympan #5 of No. 35

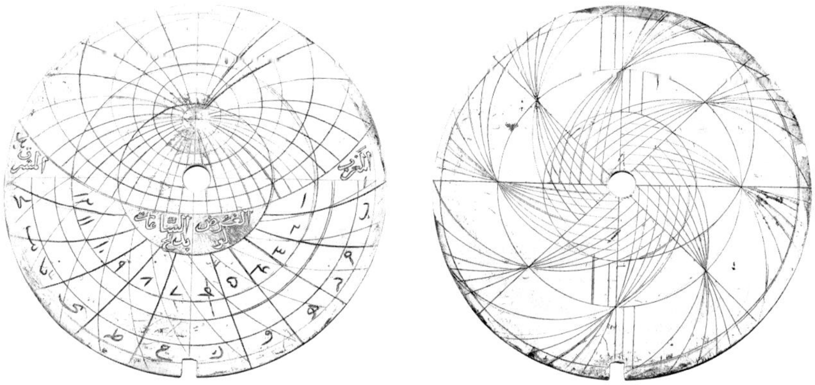

Tympan #6 of No. 35

36 Fake Astrolabe

Maker unknown, but imitates some features of No. 10
Probably Iran
Modern
"Belonging to the treasury of the exalted Sulṭān, the helper of the world and of religion, the unblemished Ismāʿīl — may God make eternal his days."
27.4 x 20.7 x 4.8 cm
Mater thickness 0.9 cm
Brass
A-83

Face of No. 36

Fake astrolabe partially inspired by No. 10. It is falsely stated to have belonged to a Sulṭān Ismāʿīl, who bears the titles of the Maḥmūd of No. 10, in the inscription on the lower part of the back: *li-khizānat al-sulṭān al-muʿaẓẓam mughīth al-dunyā wa al-dīn sālim ismāʿīl. khallad allāh ayyāmahu.* Inscriptions in Kūfic.

Consists of body, rete, inscribed diopter with sights, handle, ring, and five tympans which have tabs at the bottom though the slot in the body is at the top. The pin and horse are replaced by a screw and nut.

The rim of the front is graduated for every degree; the numbers, running clockwise, are 5, 10, 15, ...360. In the recess in the front is an inaccurate projection for latitude 90°.

The back is incorrectly graduated for every degree; the numbers running clockwise from a point 12° below the left, are 5, 10, 15,...360. In the upper left quadrant are 100 sloppy cosines. In the upper right quadrant is a design of arcs, radii, and diagonals. At the bottom are five semicircles inaccurately copying those in No. 10.

The rete is purely decorative; the inscriptions without sense.

The five tympans are identical imitations of tympans I and II of No. 10, except that latitude 22° has been interpreted to be 24°.

PROVENANCE —Ali-Kuli Khan of the House of Persian Art, New York, 1941; AP purchase, 1941.
ICA 2556
EXHIBITIONS — Instrument Exhibit, Purdue University, July 1, 1961 – April 15, 1962

Reverse of No. 36

37 Fake Astrolabe

Maker unknown
Iran
Modern
"The Sulṭān of Sulṭāns, al-Muqtadir Mīrzā Taymūr — may God make eternal his days — in the year 778 of the Hijra."
19 x 14.8 x 3.1 cm
Mater thickness 0.7 cm
Brass
A-73

Face of No. 37

Fake astrolabe. It is claimed to have belonged to the non-existent Sulṭān al-Muqtadir Mīrzā Taymūr in A.H. 778 (A.D. 1376/7) in the inscription on the front of the throne: *sulṭān, al-salāṭīn al-muqtadir mīrzā taymūr. khallad allāh ayyāmahu. fī sanah 778 al-hijrī.* Inscriptions in naskh.

Consists of body, rete, inscribed diopter with imperfect sights, ring fitted to hole at top of throne, and five tympans which have neither notches nor tabs. The pin and horse are replaced by a screw and nut.

The rim of the front is irregularly graduated; the numbers are senseless. In the recess in the front are six circles of meaningless numbers.

The back is graduated and numbered in a manner as senseless as is the front. Within the graduation are pious inscriptions. In the upper left quadrant are 68 crude cosines. The upper right quadrant is like that of No. 36. The lower half of the back is blank.

The rete is purely decorative with senseless inscriptions.

The five tympans are each inscribed on both sides with crude projections for latitude 35° with longest daylight 14;22 hours.

PROVENANCE — E. S. David, New York, 1939; AP purchase 1940.
ICA 2546
EXHIBITIONS — *Special Topics Exhibit: Impostors!*, Adler Planetarium, March 8, 2007 – Sept. 10, 2007

Reverse of No. 37

38 Fake Astrolabe

Maker unknown
Iran.
Modern
"'Abd al-A'immah
made it. 976."
23 x 15.1 x 2.3 cm
Mater thickness 0.9 cm
Brass
N-71

Face of No. 38

Fake astrolabe. It is claimed to have been made by 'Abd al-A'immah in A.H. 976 (A.D. 1568/9) in the inscription below the shadow-square on the back: *'abd al-a'immah ṣana'ahu. 976*. It is presumably indicative of the forger's inspiration that 'Abd al-A'immah the Elder made the famous Shāh 'Abbās astrolabe in A.H. 986 (Arabic 8 is Arabic 7 upside down). Inscriptions in naskh.

Consists of body, rete, graduated and inscribed diopter without sights, pin, horse, handle, ring, tassel, and five tympans which have neither notches nor tabs.

The rim of the front is irregularly graduated; the numbers are senseless. In the recess in the front is a geographical table in two circles. (❍)(●)

The upper half of the back is graduated irregularly, the numbers, running clockwise from the left to the top and counter-clockwise from the right to the top, are 5, 10, 15...90. At the edge of the lower half are indicated the cotangents. In the upper left quadrant are 34 sines and 40 cosines. In the upper right quadrant are 36 arcs concentric with the center of the back; the noon altitude curves of the Sun for latitudes miscalled 34°, 35°, 30°, 28°, and 30°; above is the inscription: *dawā'ir inṣāf al-nuhur fī al-'urūḍ al-marqūmah 'alā aṭrāfahi;* and the azimuths of the qibla for Baghdād, Shīrāz, Iṣfahān, and Ṭūs; above the center is the inscription: *khuṭūṭ sumūr al-qiblah fī al-bilād al-marqūmah 'alā aṭrāfahā bi-irtifā' al-gharbī*. Below the center is a shadow-square mistakenly marked for 5 feet and 5 digits within which is a table indicating erroneously the elements, the zodiacal signs, and the planetary lords of the four triplicities. Below this are five concentric semicircles indicating with errors the planetary lords of the terms, the ends of the terms, the names of the zodiacal signs, a meaningless sequence of

Reverse of No. 38

numbers labeled "decans", and the names of the lunar mansions.

The rete is a meaningless fake.

The five tympans are also worthless.

PROVENANCE — N. Israel; Kenneth Nebenzahl, Glencoe, IL; AP gift, 1986.

REFERENCES — Gingerich, King, and Saliba (1972), pp. 191, 195, 197; Newbury (2004a).

Mater of No. 38

◐ OUTER CIRCLE

City	*Longitude*	*Latitude*	*Inclination*	*Direction*	*Identification*
qandahār	107;40°	18;0°	75;5°	SW	Kandahar
harāh	95;30°	34;30°	54;8°	SW	Herat
marw	97;0°	37;40°	52;30°	SW	Mary
qāyin	93;20°	33;40°	54;1°	SW	Qāyen
tūn	92;0°	34;30°	50;20°	SW	Ferdows
tūrshīz	92;0°	35;0°	48;51	SW	Kāshmar
ṭūs	92;30°	36;0°	45;6°	SW	Ṭūs
nīshābūr	92;30°	36;21°	3525°	SW	Neyshābūr
basṭām	89;30°	36;10°	39;18°	SW	Bastam
dāmghān	83;55°	37;20°	18;55°	SW	Dāmghān
samnān	83;0°	36;0°	36;17°	SW	Semnān
astarābād	89;35°	36;50°	18;48°	SW	Gorgān
ṭāliqān	85;45°	36;10°	29;18°	SW	Taliq-an
rayy	86;10°	35;0	36;26°	SW	Shahr Rey
kāshān	86;0°	34;0°	34;31°	SW	Kāshān
iṣfahān	86;40°	32;25°	29;40°	SW	Eṣfahān
jurfādaqān	85;30°	34;54°	18;0°	SW	Golpayegan
qazwīn	85;0°	36;0°	27;34°	SW	Qazvīn
hamadān	83;0°	35;10°	22;16°	SW	Hamadan
yazd	89;0°	32;0°	48;18°	SW	Yazd
ahwāz	85;0°	31;0°	40;30°	SW	Ahvāz
kūfah	79;30°	31;30°	12;31°	SW	Kūfah
surramanra'ā	79;0°	32;0°	7;56°	SW	Sāmarrā'
baṣrah	84;0°	30;0°	32;19°	SW	Basra
baghdād	82;0°	38;25°	12;45°	SW	Baghdād
miṣr	68;20°	30;20°	48;18°	SE	Cairo
ḥalab	72;10°	35;50°	18;29°	SE	Aleppo
dimashq	70;0°	38;55°	30;31°	SE	Damascus
bayt al-maqdis	66;30°	31;50°	45;56°	SE	Jerusalem
madīnah	75;20°	25;0°	27;10°	SE	Medina
makkah	77;10°	21;40°	0;0°	0;0°	Mecca

● INNER CIRCLE

City	*Longitude*	*Latitude*	*Inclination*	*Direction*	*Identification*
kirmān	92;30°	29;50°	62;11°	SW	Kermān
multān	107;35°	29;40°	80;48°	SW	Multan
lahāwar	109;20°	31;50°	73;26°	SW	Lahore
kābul	104;40°	34;26°	69;35°	SW	Kabul
shīrāz	88;0°	29;36°	58;18°	SW	Shīrāz
kāzirūn	87;0°	29;0°	51;48°	SW	Kāzerūn
shūshtar	84;30°	31;30°	35;24°	SW	Shūshtar
nihāwand	88;55°	34;20°	27;29°	SW	Nahāvand
kirmānshāh	88;0°	34;30°	28;18°	SW	Kermānshān
ayrawān	89;55°	38;30°	36;20°	SW	Yerevan
sabzawār	84;30°	40;50°	9;20°	SW	Sabzevār
nakhjawān	81;55°	38;40°	12;49°	SW	Nakhichevan
marāghah	82;0°	37;20°	16;17°	SW	Marāgheh
ardabīl	82;30°	38;18°	17;18°	SW	Ardabīl
tabrīz	82;0°	38;0°	15;40°	SW	Tabrīz

39 Fake Astrolabe

Maker unknown
Place unknown
Alleged to have been made in 1629/30
"In the year 1039."
14.7 x 9 x 2.3 cm
Mater thickness 0.7 cm
Brass
M-38

Face of No. 39

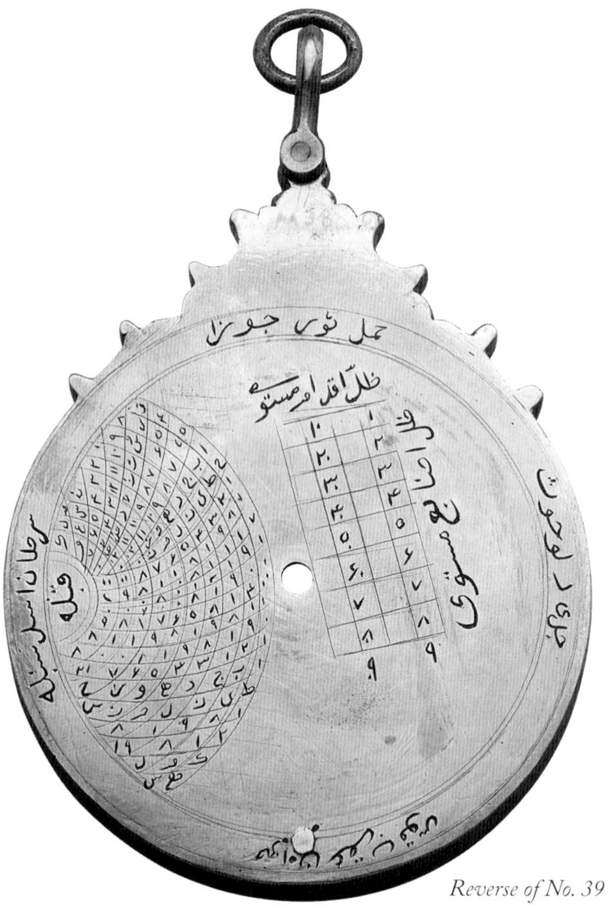

Reverse of No. 39

Fake astrolabe falsely dated A.H. 1039 (A.D. 1629/30) in the inscription on the front of the throne: *sanah 1039*. Inscriptions in kursī.

Consists of body, rete, uninscribed diopter with sights, pin, pointer, horse, handle, ring, and two tympans, with notches at the bottom fitting about a corresponding tab in the body.

The rim of the front is crudely graduated; the numbers are false. In the recess in the front are the names of fifteen localities; *makkah al-mubārrakah, madīnat rusūl, fārs, marṣ* (read *miṣr*), *bayt al-maqdis, ardabīl, baghdād, baṣrah, shābūr, ṭārisān* (?) (read *ṭabāristān* ?), *yazd, hamadan, qazwīn, isfahān, and kāshān*. The longitudes associated with them are generally correct, but the latitudes and distances are meaningless.

The back contains tables that falsely claim to have something to do with the qibla and with shadows. Around the edge are inscribed the names of the zodiacal signs.

The rete is inscribed with a random mix of names of stars and of lunar mansions, numbers, and combinations of letters. Even the names of the zodiacal signs on the ecliptic are in the wrong places and the wrong order.

The two tympans are arbitrarily engraved with circles, curves, numbers, and letters without meaning.

PROVENANCE — A.W.M. Mensing, Amsterdam, 1924; Max Adler, Chicago, 1930; AP gift, 1930.
ICA 113
REFERENCES — Engelmann Catalogue (1924), p. 16; Gunther (1932) 113, p. 246.

40 Fake Astrolabe

Alleged to have been made by 'Abd al-'Alī
Probably Iran
Modern
'Abd al-'Alī made it."
"According to the salvation (?) of Shaykh Luṭfallāh in the year 1048..."
"Mullā Ḥasanī."
18.5 x 14.6 x 3.2 cm
Mater thickness 0.6 cm
Brass
A-85

Face of No. 40

Reverse of No. 40

Fake astrolabe alleged to have been made by 'Abd al-'Alī in an inscription on the back of the throne: *ṣana'ahu 'abd al-'alī.* 'Abd al-'Alī made No. 29 in the early eighteenth century. A corrupt inscription on the front of the throne claims that it was made for Shaykh Luṭfallāh in A.H. 1048 (A.D. 1638/9): *ḥabahba* (read *ḥaṣb) al-islām* (?) *shaykh luṭfallāh fī sanah 1048...* A final inscription, in the cartouche below the triplicity-table on the back, mentions Mullā Hasanī; Muḥammad Ibrāhīm al Ḥasanī made No. 25 in 1699/1700. Inscriptions in naskh.

Consists of body, rete, inscribed diopter with unperforated sights, pin, horse, handle, and three tympans, which have neither tabs nor notches.

The rim of the front is not graduated; the numbers are senseless. In the recess in the front are circles of numbers, names of the zodiacal signs and planets, and senseless words.

The back is inaccurately graduated; the numbers are senseless. In the upper half, below the graduation, are two semicircles with names of the planets and meaningless words. In the upper left quadrant is a pattern of vertical and horizontal lines. In the upper right quadrant is a graduation arc, an arc of numbers, and eight arcs concentric with the center of the back; below is a most corrupt version of the usual inscription for azimuths of the qibla. Below the center is a table of the planetary lords of the four triplicities; it is labeled as if it were a shadow-square. Below this are five semicircles giving false or corrupt information regarding terms and decans.

The rete is purely decorative. Three tympans are meaningless, labeled each with the names of two zodiacal signs.

PROVENANCE — Ali-Kuli Khan of the House of Perian Art, 1941; AP purchase, 1941.

41 Fake Astrolabe

Alleged to have been decorated by Muḥammad Mahdī al-khādim (the servant), engraved by ʿAbdallāh (?)
Probably Iran
Modern (after 1856)
"Muḥammad Mahdī the servant — the least, the repentant — decorated it in 1070."
"Engraving of ʿAbdallāh (?) the helpers."
"In the court of Shāh Sulṭān Ḥ<us>ayn Ṣafawī."
7 x 11.1 x 3.7 cm
Mater thickness 0.5 cm
Brass
A-72

Face of No. 41

Fake astrolabe said to have been decorated by Muḥammad Mahdī al-khādim in A.H. 1070 (A.D. 1659/60) in an inscription below the shadow-square on the back: *namaqahu ʿ. li-aqall* (read *al-aqall*) *al-nādim muḥammad mahdī al-khādim.*" Muḥammad Mahdī made the last instrument described by Morley (1856) pp. 48-49. Instrument No. 41 was engraved by ʿAbdallāh (?) according to the inscription at the bottom of the back: *naqshī ʿabd allāh* (?) *anṣār.* The cartouches with these names are in the same position as those of the maker (ʿAbd al-ʿAlī ibn Muḥammad) and decorator (Muḥammad Bāqir) of the Shāh Sulṭān Ḥusayn Safawī astrolabe, made in 1712, which is the first instrument described by Morley (1856) pp. 21-31; moreover, the general layout and decoration of the backs of Shāh Ḥusayn's (see Morley (1856) plate XVII) and Adler astrolabes are identical. To clinch the argument that the maker of this fake instrument was inspired by Morley's descriptions is the inscription on either side of the cartouche of ʿAbdallāh (?) which states that it was in the court of Shāh Sulṭān Ḥ<us>ayn Ṣafawī: *dar darbār shāh sulṭān ḥayn ṣafawī.* Since Shāh Ḥusayn's reign began only in 1694 (he resigned in 1722), the maker could not have been Muḥammad Mahdī who was active four or five decades earlier. Pious inscriptions on the front and back of the throne. Inscriptions in naskh.

Consists of body, rete, graduated and inscribed diopter with sights, pin, horse, handle, rin, and four tympans, which have notches at the bottom fitting loosely about a corresponding tab in the body.

The rim of the front of the body is graduated for every degree; the numbers are arbitrary and meaningless. In the recess in the front is a geographical table in two circles. (❍)(●)

Reverse of No. 41

The upper half of the back is graduated for every degree; the numbers, running clockwise from the left to the top and counter-clockwise from the right to the top, are 5, 10, 15, ...90. At the edge of the lower right quadrant are indicated the cotangents. In the upper left quadrant are the cosines for every one of the 60 parts of the Radius and six hour curves labeled : *khuṭūṭ sāʿāt muʿawwajah.* In the upper right quadrant are the noon altitude curves of the Sun for latitudes 24°, 27°, 30°, 33°, 36°, 39°, and 42°; at the top is the inscription: *dawāʾir inṣāf al-nuhur fī al-ʿurūḍ al-marqūmah ʿalā aṭrāfahā.* In the same quadrant are the azimuths of the qibla for Yazd, Ṭūs, Iṣfahān, Baṣrah, Baghdād, and Shīrāz; below is the inscription: *khuṭūṭ sumūt al-qiblah fī al-bilād al-madhkūrah ʿalā aṭrāfahā bi-irtifāʿ al-gharbī.* Below the center is the shadow-square for 7 feet and 12 digits, within which is a table indicating the elements, zodiacal signs, and the planetary lords of the four triplicities. Below this are five semicircles indicating, with many mistakes, the planetary lords of the terms, the ends of the terms, the names of the zodiacal signs, the planetary lords (denoted by their final letters) of the decans according to the Greek system, and the names of the lunar mansions.

The rete is decorative rather than useful. The ecliptic is not graduated. Some 19 pointers have names of stars, but usually of the wrong ones; many pointers have only flowers.

The tympans are also crude forgeries with no astronomical meaning.

PROVENANCE — E.S. David, New York, 1939; AP purchase, 1940.
ICA 2545
EXHIBITIONS — St. John's College, Feb. 13, 1960 – July 7, 1961. "Imposters!" Special Topic Exhibit, Adler Planetarium & Astronomy Museum, Chicago, IL, March 8, 2007 – Sept. 10, 2007.

Mater of No. 41

❍ OUTER CIRCLE

City	Longitude	Latitude	Inclination	Direction	Identification
makkah	77;10°	21;40°	0;0°	0;0°	Mecca
madīnah	75;20°	25;0°	37;10°	SE	Medina
miṣr	63;20°	30;20°	58;13°	SE	Cairo
ṣana'ā	77;0°	14;1°	1;55°	SW	Ṣan'ā'
liḥsā	83;30°	24;0°	69;30°	SW	Aḥsā
bayt al-maqdis	66;30°	31;50°	45;17°	SE	Jerusalem
dimashq	70;0°	33;15°	30;31°	SE	Damascus
ḥalab	72;10°	35;50°	18;29°	SE	Aleppo
mawṣil	77;0°	34;30°	17;12°	SW	Mosul
najawān	81;15°	38;40°	12;15°	SW	Nakhichevan
marāghah	82;0°	37;20°	16;17°	SW	Marāgheh
tabrīz	82;0°	38;0°	15;40°	SW	Tabrız
ardabīl	82;30°	38;0°	17;13°	SW	Ardabīl
surramanra'ā	79;0°	34;0°	7;56°	SW	Sāmarrā'
(illegible)	79;30°	31;30°	12;31°	SW	Kūfah
(illegible)	70;15°	33;25°	12;45°	SW	Baghdād
baṣrah	84;0°	30;0°	37;19°	SW	Basra
shūsh	84;30°	31;30°	35;24°	SW	Shūsh
kāzirūn	87;0°	29;15°	51;47°	SW	Kāzerūn
shīrāz	88;0°	29;0°	53;18°	SW	Shīrāz
yazd	89;0°	32;0°	48;28°	SW	Yazd
hamadān	83;0°	35;10°	22;16°	SW	Hamadan
qazwīn	85;0°	36;0°	27;34°	SW	Qazvīn
jurbādaqān	85;30°	34;14°	38;41°	SW	Golpayegan
iṣfahān	86;40°	32;25°	40;28°	SW	Eṣfahān
kāshān	86;0°	34;0°	34;31°	SW	Kāshān
qumm	85;40°	34;45°	31;54°	SW	Qom
ṭāliqān	85;45°	36;10°	29;33°	SW	Taliq-an
astarābād	89;35°	36;50°	38;48°	SW	Gorgān
simnān	88;0°	36;0°	36;17°	SW	Semnān
dāmghān	83;55°	36;20°	38;0°	SW	Dāmghān
akrī'	114;0°	26;53°	90;58°	SW	Agra
āmul	88;20°	36;5°	34;37°	SW	Amol

● INNER CIRCLE

City	Longitude	Latitude	Inclination	Direction	Identification
mashhad	92;30°	37;0°	45;6°	SW	Mashhad
nīshābūr	92;30°	36;21°	46;25°	SW	Neyshābūr
tūn	92;30°	34;30°	55;20°	SW	Ferdows
sabzawār	91;30°	36;0°	44;12°	SW	Sabzevār
harāh	94;30°	34;30°	34;8°	SW	Herat
qāyin	93;20°	33;40°	54;36°	SW	Qāyen
balkh	101;0°	36;41°	65;36°	SW	Wazirabad
hurmūz	92;0°	25;0°	74;30°	SW	Hormoz
kirmān	92;30°	29;50°	62;11°	SW	Kermān
qandahār	107;40°	23;0°	75;5°	SW	Kandahar
lahāwar	109;20°	31;55°	73;26°	SW	Lahore
kashmīr	108;0°	31;0°	75;20°	SW	Srinagar
kanjah	83;0°	41;20°	15;49°	SW	Kirovabad
tiflīs	83;0°	43;0°	14;41°	SW	Tbilisi
shīrwān	91;30°	36;0°	84;13°	SW	Shīrvān
multān	106;30°	29;40°	80;48°	SW	Multan

42 Fake Astrolabe

Ascribed to Muḥammad Amīn ibn Mīrzā, but perhaps made by the maker of No. 43
Probably India (Bombay mentioned)
Modern
"The servant (of God), Muḥammad Amīn ibn Mīrzā, victorious in the faith, made it in the year one thousand and ninety-nine, 1099."
"According to the order of the exalted support of the dynasty, for the victory of the majority (?), Shaykh 'Alīkhān Ḍifjāhī (?)."
"Muḥammad Amīn."
37.2 x 22.2 x 4.1 cm
Mater thickness 1.1 cm
Brass
G-36

Face of No. 42

Fake astrolabe said to have been made by Muḥammad Amīn ibn Mīrzā in A.H. 1099 (A.D. 1687) according to the inscription to the right of the shadow-square on the back: *ṣana'ahu al-'abd muḥammad amīn ibn mīrzā muẓaffar al-dīn fī sanah tasa' wa tasa'ayn wa alf. 1099*. It was made at the command of Shaykh 'Alīkhān Ḍifjāhī (?) according to the inscription to the right of the shadow-square: *ḥasb al-amr 'ālījāh mu'tamad al-dawla li-ghalabat al-ghālibīyyah* (?) *shaykh 'alīkhān ḍifjāhī* (?). Muḥammad Amīn has also signed his name at the top of the back of the throne. Below the shadow-square is a Persian inscription, beginning with a line mentioning the engraving found in several other instruments. About the shadow-square is a fourth inscription in Persian listing the parts of the astrolabe. Inscriptions in naskh. On the front of the throne are two angels holding a compass set into the body; between them is a pot of flowers. On the back of the throne a bull standing on a fish; above them a terrestrial globe.

Consists of body, rete, imperfectly graduated and inscribed diopter with sights, pin, horse, handle, ring, and five tympans, which have notches fitting a tab in the body.

The rim of the front of the body is graduated into 576 parts divided into 144 sections, each of which is assigned a planet, zodiacal sign, or letter, and a number. In the recess in the front are two circles listing cities associated with meaningless numbers (for longitudes, latitudes, and inclinations). In the center is a circle of figures of the zodiacal signs surrounding circles with two wedges for each zodiacal sign with the name of a planet and five numbers.

The upper part of the back is graduated for every degree; the numbers are nonsensical to the left, but, running clockwise from the top are 5, 10, 15, ...90. At the edge of the lower half are indicated the

Reverse of No. 42

cotangents. In the upper left quadrant are inaccurate sines and cosines. In the upper right quadrant are the noon altitude curves of the Sun for latitudes 28°, 30°, 32°, 34°, 36°, 38°, and 40°; above is the inscription: *dawā'ir inṣāf al-nahār fī al-'urūḍ 'alā aṭrāfahā*; and a quatrain. In the same quadrant are the azimuths of the qibla for Iṣfahān, Bahgdād, Shīrāz, Nawshahr, and Būmbāī; below is the inscription: *khuṭūṭ al-sumūt al-qiblah fī al-bilād 'alā aṭrāfahā al-marqūmah bi-irtifā' min al-jarūb ilā al-gharbī*. Below the center is the shadow-square for 7 feet and 12 digits, within which is a table indicating the elements, the zodiacal signs, and the planetary lords of the four triplicities. Below the inscriptions mentioned above are five semicircles indicating the planetary lords of the terms, the ends of the terms, the names of the zodiacal signs, the planetary lords (denoted by their final letters) of the decans according to the Greek system, and the names of the lunar mansions.

The rete has a symmetrical design with two ecliptics with the names of the zodiacal signs in the wrong places and various obscure words. There are no pointers, but rather scattered among the flower pattern boxes inscribed with the names of the planets and of the lunar mansions. The rim of the rete is inscribed randomly.

The five tympans are adorned with curves, numbers, and the names of planets and zodiacal signs, sometimes representing the nested planetary spheres.

PROVENANCE — Alexander Karim, 1947; Edgar Berliner, Beverly Hills, 1947; AP gift, 1947.

ICA 2565

EXHIBITIONS — First Federal Savings and Loan, Chicago, IL, Jan. 4 – Feb. 1, 1966; DePaul University, Chicago, IL, March 14 – April 1, 1967; *History of the Occult Sciences*, Hallmark Gallery, New York, N.Y., Aug. 18 – Nov. 21 1969

43 Fake Astrolabe

Ascribed to Muḥammad Amīn but perhaps made by the maker of No. 42
Probably India (Bombay mentioned)
Modern
"Muḥammad Amīn in the year 1136."
"Order of Ḥājjī Muḥammad Muḥsin ...; the work of Muḥammad Amīn."
28.3 x 17.4 x 2.9 cm
Mater thickness 1.1 cm
Brass
A-153

Face of No. 43

Fake astrolabe falsely said to have been made by Muḥammad Amīn in A.H. 1136 (A.D. 1723/4) according to the inscription in the center of the recess in the front: *muḥammad amīn fī sanah 1136.* A Persian inscription at the right edge of the back claims that Muḥammad Amīn made it for Ḥājjī Muḥammad Muḥsin: *farmāyish ḥājjī muḥammad muḥsin yaran* (?) (*bar ān*?) *'amal muḥammad amīn.* On the front of the throne is an inscription: *quṭb mu'addal niṣf al-nahār* ("pole of the equalizer of noon"); on the back of the throne is a picture of a crowned, human-headed lion shooting an arrow at its own serpent-headed tail beneath another meaningless inscription: *janūb al-rajul.* In the upper left quadrant is another inscription: *ikhtiyārāt 'alā al-ittiṣālāt* ("elections according to conjunctions"). Inscriptions in naskh.

Consists of body, rete, inscribed diopter without sights, pin, horse, handle, ring, and five tympans, without notches as the body is without a tab.

The rim of the front is graduated for every degree; the numbers, running counter-clockwise in each quadrant, are 5, 10, 15, ...90. In the recess in the front is a geographical table in three circles; between the second and the third are pictures of the zodiacal signs similar to those in No. 42. Meaningless numbers meant to resemble longitudes and latitudes accompany the names of the cities; the list of cities, which is close to that on No. 42, includes London.

The back is graduated for every degree; the numbers, running clockwise from the bottom to the left and counter-clockwise from the bottom through the right and the top to the left, are in each quadrant 5, 10, 15, ...90. Below the graduation in the upper left quadrant are the names of 18 cities, including London and Berlin, with false inclinations. In the upper left quadrant is a pattern of vertical, horizontal, and diagonal

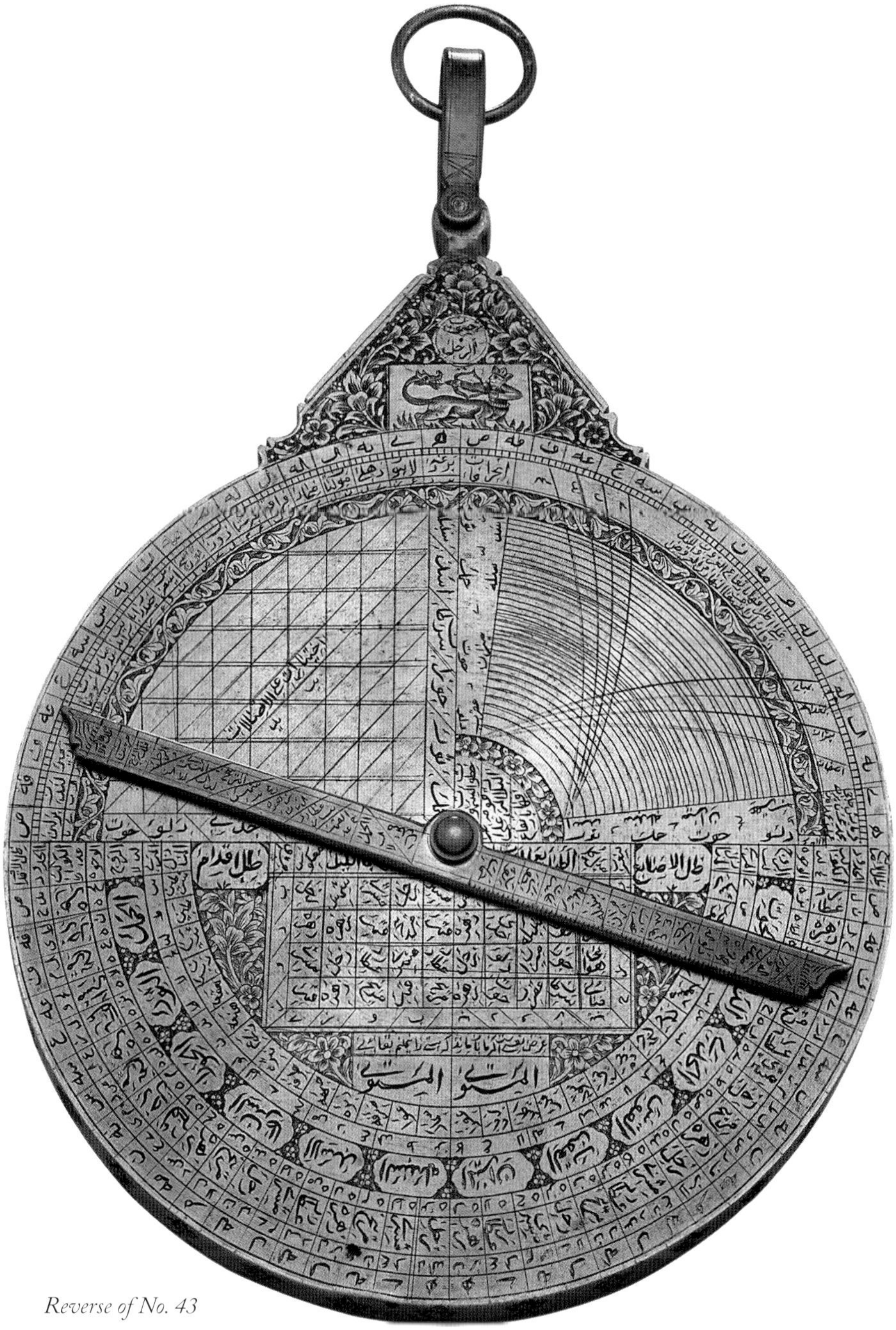

Reverse of No. 43

lines. In the upper right quadrant are four noon altitude curves of the Sun for falsely identified latitudes; above is the corrupt inscription: *dawā'irāt inṣifāf al-nahār min al-'urūḍ 'alā aṭrafahā bi-irtifā' al-ghurūb wa al-layl.* In the same quadrant are the azimuths of the qibla for Iṣfahān, Shīrāz, Baghdād, and Bambā'ī. Below the center is an inaccurate shadow-square for 7 feet and 12 digits, within which is a table of the elements, the zodiacal signs, and the planetary lords of the four triplicities. Below this, with many false entries, are six semicircles meant to indicate the terms and their planetary lords, the zodiacal signs, the planetary lords of the decans, and the lunar mansions.

The rete is purely decorative, as are the inscriptions that embellish it.

The five tympans have designs similar to those on No. 42, but less elaborate.

PROVENANCE — Karl Wede, Inc., Saugerties, NY, 1960; AP purchase, 1961.

EXHIBITIONS — *History of the Occult Sciences*, Hallmark Gallery, New York, N.Y., Aug. 18 - Nov. 21 1969; Abrams Planetarium, Michigan State University, East Lansing, April 29 - Nov. 30, 1977.

REFERENCES — Adler Planetarium (1970) p.24; Wede (n.d.), cover; Wede (1960), lot 56.

44 Fake Astrolabe

Signed Abū Isḥāq
Probably Iran
Modern
"Abū Isḥāq."
18 x 11.2 x 3.1 cm
Mater thickness 1 cm
Brass
A-80

Face of No. 44

Reverse of No. 44

Fake astrolabe. On the front of the throne is the inscription: *abū isḥāq*. Inscriptions in naskh.

Consists of body, rete, diopter with scales, but lumps instead of sights, pin, horse, handle, ring, and five tympans which have notches at the bottom fitting about a corresponding tab in the body.

The rim of the front is crudely graduated; the numbers are senseless. In the recess is a geographical table in two circles; the longitudes, latitudes and inclinations are almost all incorrect.

The back is crudely graduated; the numbers, where legible, are senseless. In the upper left quadrant are 67 inaccurate cosines. In the upper right quadrant is a table of the planetary lords of the four triplicities. Below the center is a shadow-square for 7 feet and 12 digits.

The rete bears names that are senseless and repetitive.

The five tympans are crude and useless.

PROVENANCE — E.S. David, New York, 1939; AP purchase, 1939.
ICA 2553

45 Fake Astrolabe

Maker unknown
Probably Iran
Modern
13.2 x 9.1 x 2.5 cm
Mater thickness 0.9 cm
Brass
A-89

Face of No. 45

Reverse of No. 45

Fake astrolabe. Inscriptions in naskh.

Consists of body, rete, inscribed diopter with scales and sights, pin, bent bar for a horse, handle, ring, and eight tympans with notches at the bottom fitting about a corresponding tab in the body.

The rim of the front is graduated for every degree; the numbers, running clockwise, are 10, 20, 30,...100, 10,...200, 10,...300, 10,...60. The bounds of the zodiacal signs and some of their names are engraved on the side of the body. In the recess in the front is a partial projection with the pole (marked *maghrib*, "west") to the right, and a geographical table; the latitudes, longitudes, inclinations, and directions given are generally incorrect.

The upper half of the back is graduated for every degree; the numbers, running clockwise from the left to the top and counterclockwise from the right to the top, are 10, 20, 30,...90. Below the graduation are the names of the Syrian months: *nīsān, ayyār, khazīrān, tammūz, ab, īlūl, tashrīn awwal, tashrīn thānī, kānūn awwal, kānūn thānī, shubāṭ*, and *adhār*. In the upper left quadrant are eight arcs concentric with the center of the back and seventeen radii. In the upper right quadrant is a pattern of vertical and horizontal lines. Below the center is a shadow-square for 7 feet and 12 digits.

On the rete the ecliptic is graduated for every 6°; there are no numbers. Names of stars and lunar mansions are sloppily engraved.

The eight tympans have inaccurately engraved altitude circles for every 6°, azimuths for every 10°, and unequal hour curves. Their cities and latitudes are indicated. (■)

PROVENANCE — E.S. David, New York, 1943; AP purchase, 1943.
ICA 2561

TYMPANS

Tympan	City	Latitude
Ia.	Makkah	21;20°
Ib.	Madīnah	24°
IIa.	Iskandariyyah	30°
IIb.	Bayt maqdis	32°
IIIa.	Baṣrah	31;8°
IIIb.	Baghdād	33;25°
IVa.	Mawṣil	36°
IVb.	Diyārbakr	38°
Va.	Ḥalab	37°
Vb.	Karkūk	37°
VIa.	Iṣfahān	39°
VIb.	Shirwān	40°
VIIa.	Qusṭanṭiyyah	41°
VIIb.	Adurnah	42°
VIIIa.	Bulghār	49°
VIIIb.	gives horizons for many different latitudes.	

Tympan #1 of No. 45

Tympan #2 of No. 45

Tympan #3 of No. 45

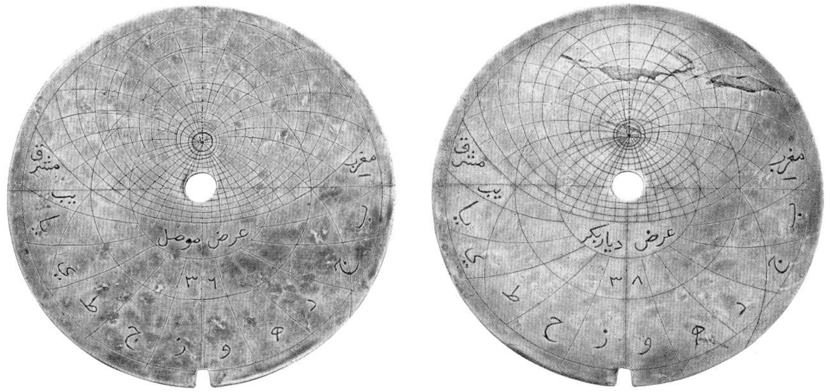

Tympan #4 of No. 45

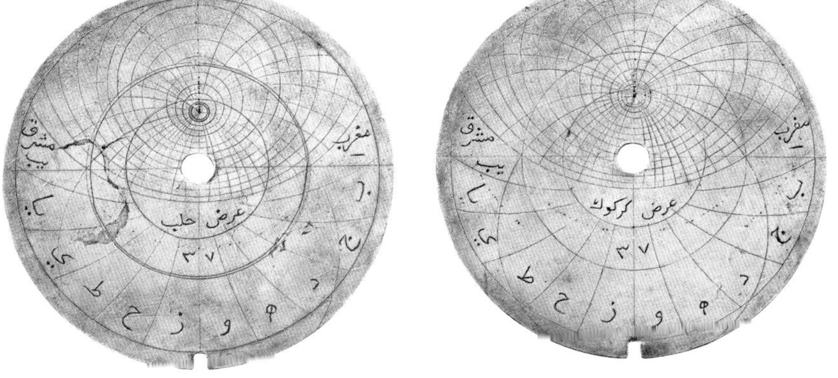

Tympan #5 of No. 45

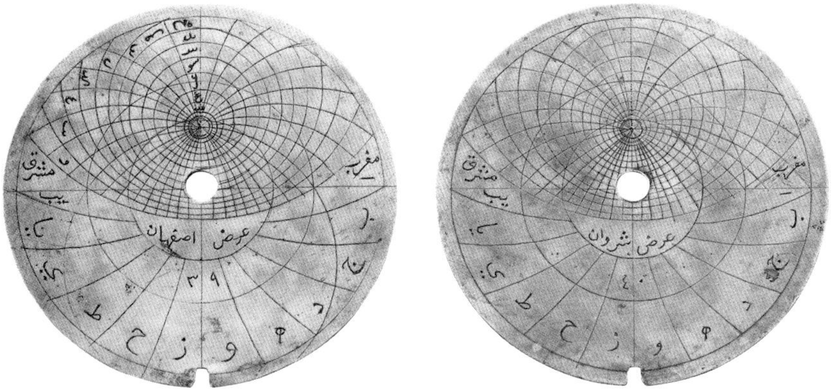

Tympan #6 of No. 45

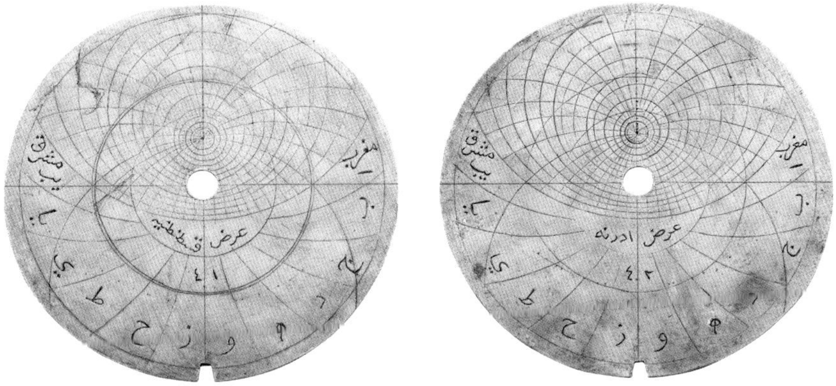

Tympan #7 of No. 45

Tympan #8 of No. 45

46 Fake Astrolabe

Maker unknown
Iran
Modern
14.1 x 8.7 x 1.8 cm
Mater thickness 0.8 cm
Brass
A-212

Face of No. 46

Reverse of No. 46

A modern Iranian fake astrolabe. Inscriptions in naskh.

Consists of body, rete, inscribed diopter without sights, pin, horse, handle, crude ring, and four tympans without notches as the body has no tab.

The rim of the front is not graduated. In the recess in the front are two circles of the names of predominantly Iranian cities, accompanied by meaningless numbers.

The back is adorned with circles of the names of the planets and the names and pictures of the zodiacal signs.

The rete is without pointers or star-names.

The four tympans are meaningless.

PROVENANCE — Charlotte Olson, Glencoe, IL, before 1965; AP gift, 1975.

47 Fake Astrolabe

Maker unknown
Perhaps Israel
Modern
12 x 8.2 x 2.2 cm
Mater thickness 0.5 cm
Brass
A-204

Face of No. 47

Reverse of No. 47

A modern fake astrolabe perhaps made in Israel. See CCA Nos. 4002 and 4003 in Gibbs and Saliba, (1984) pp. 187-188. Inscriptions in Hebrew and Arabic.

Consists of body, rete, inscribed diopter without sights, pin, horse, crude handle, and three tympans without notches as the body has no tab.

The rim of the front is unevenly graduated; there are no numbers. In the recess in the front are three circles of Hebrew letters; the central circle has the names of the planets.

The back is unevenly graduated; there are no numbers. There are three circles containing the names of the zodiacal signs in Hebrew, pictures of the zodiacal signs (Scorpius and Sagittarius are transposed), and Hebrew letters. In the center is a 4 x 4 square of Hebrew letters.

The rete is purely decorative. The names of the zodiacal signs in Arabic are written in the wrong places on the ecliptic.

The three tympans are meaningless. One is decorated with Hebrew letters, the other two with the Arabic names of the planets, the zodiacal signs, the directions, etc.

PROVENANCE — Gordon F. Hull, c/o Marika's Antiques, Boston; Roderick S. Webster, Winnetka, IL, 1969; AP gift.

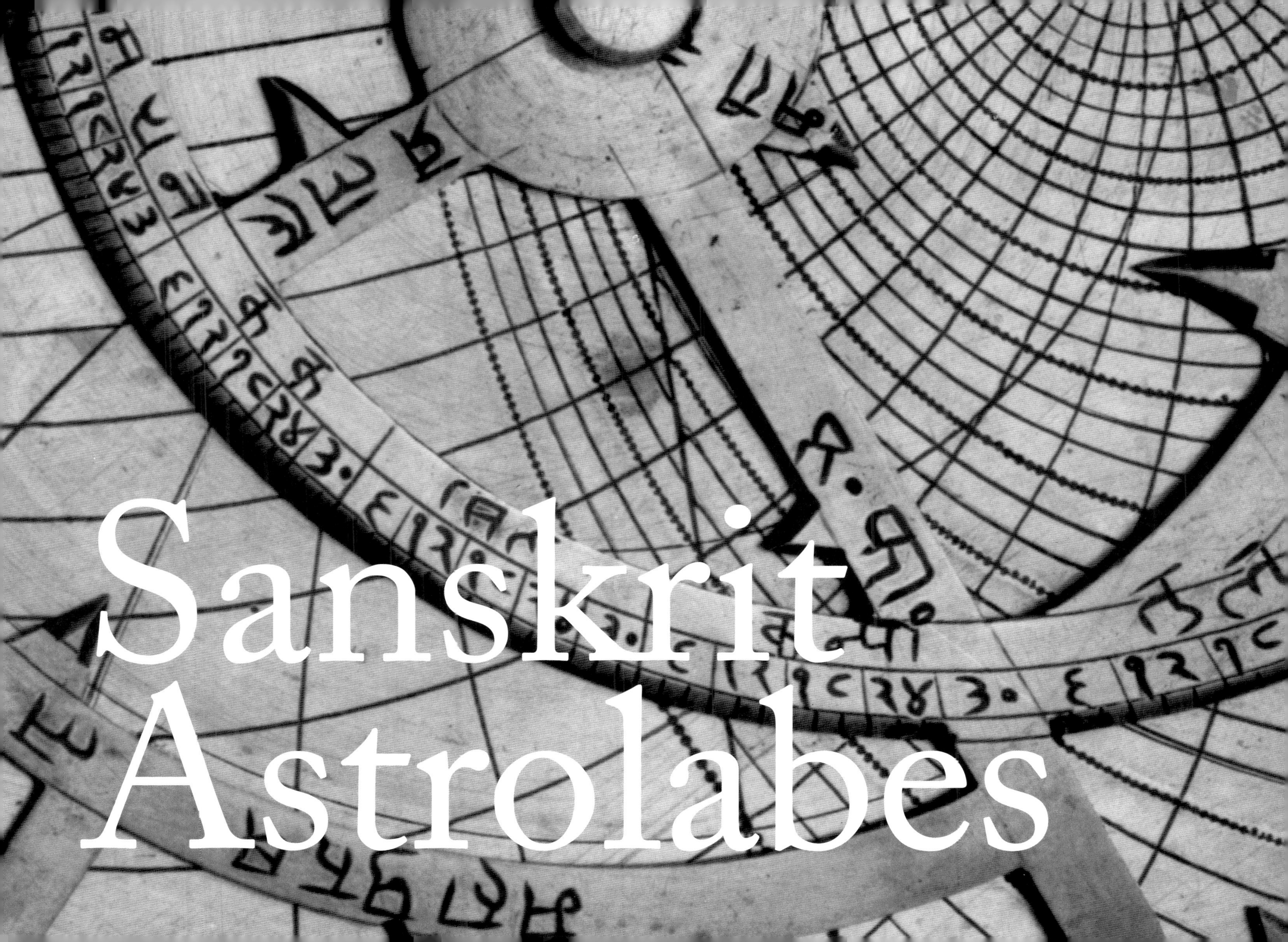
Sanskrit
Astrolabes

48 Yantrarāja

Maker unknown
India
18th century (?)
23.7 x 17.7 x 1.7 cm
Mater thickness 0.1 cm
Brass
A-68

Face of No. 48

Astrolabe (called *antrarājaḥ* on the front of the throne) made probably in the eighteenth century in southern Rajasthan (the latitude is that of central Mewar). Inscriptions in Nāgarī.

Consists of body, rete, uninscribed diopter whose sights have been lost, pin, handle, and ring. In place of the horse is a modern bent bar. A piece has been riveted to the handle that prevents the ring from rotating.

The edge of the front is graduated for every degree; the numbers, running clockwise from the left to the top and counter-clockwise from the right to the top, are 6, 12, 18, ...90. On the front is an inaccurately drawn projection for latitude 25° (noon equinoctial shadow: 5;35 digits). The altitude circles are for every 2°, the azimuths for every 3°. There are no hour curves.

The back is graduated and numbered exactly as is the front. At the edge of the lower right quadrant are indicated the cotangents. In the upper left quadrant are cosines and sines for every 2 of the 60 parts of the Radius. In the upper right quadrant are 27 arcs concentric with the center of the back. Below the center is the shadow-square for 7 feet and 12 digits.

On the rete, the ecliptic is graduated for every degree; the numbers for each zodiacal sign are 6, 12, 18, 24, 30. The names of the stars are listed. (★)

PROVENANCE — Purchased by Leo Metzenberg in China, 1933; AP purchase, 1933.
ICA 2575

★ STAR LIST

Name	*Identification*
abhijit	α Lyr.
aśvamukha	ε Peg.
ādrāḥ	α Ori.
uttarā bhādrapadā	γ Peg.
kukudapuccha	α Cygn.
citrāh	α Virg.
naharakoṭāroḥ	θ Erid.
pati	β Pers.
pūrvā bhādrapadā	β Peg.
maghāḥ	α Leo
māhāpuruṣa	α Hydr.
mithunapādaḥ	β Ori.
rohinī	α Tau.
lubdhaka	α Can. Mai.
lubdhakavadhaḥ	α Can. Min.
viśākhā pītarāvaca	α Cor. Bor.
vrahmāhṛda	α Aur.
śatabhiṣakaḥ	λ Aquar.
śravaṇa	α Aquil.
samudrapakṣiḥ	ι Ceti
sarpadhāriśira	α Oph.
svātiḥ	α Boot.
hastaḥ	δ Corv.

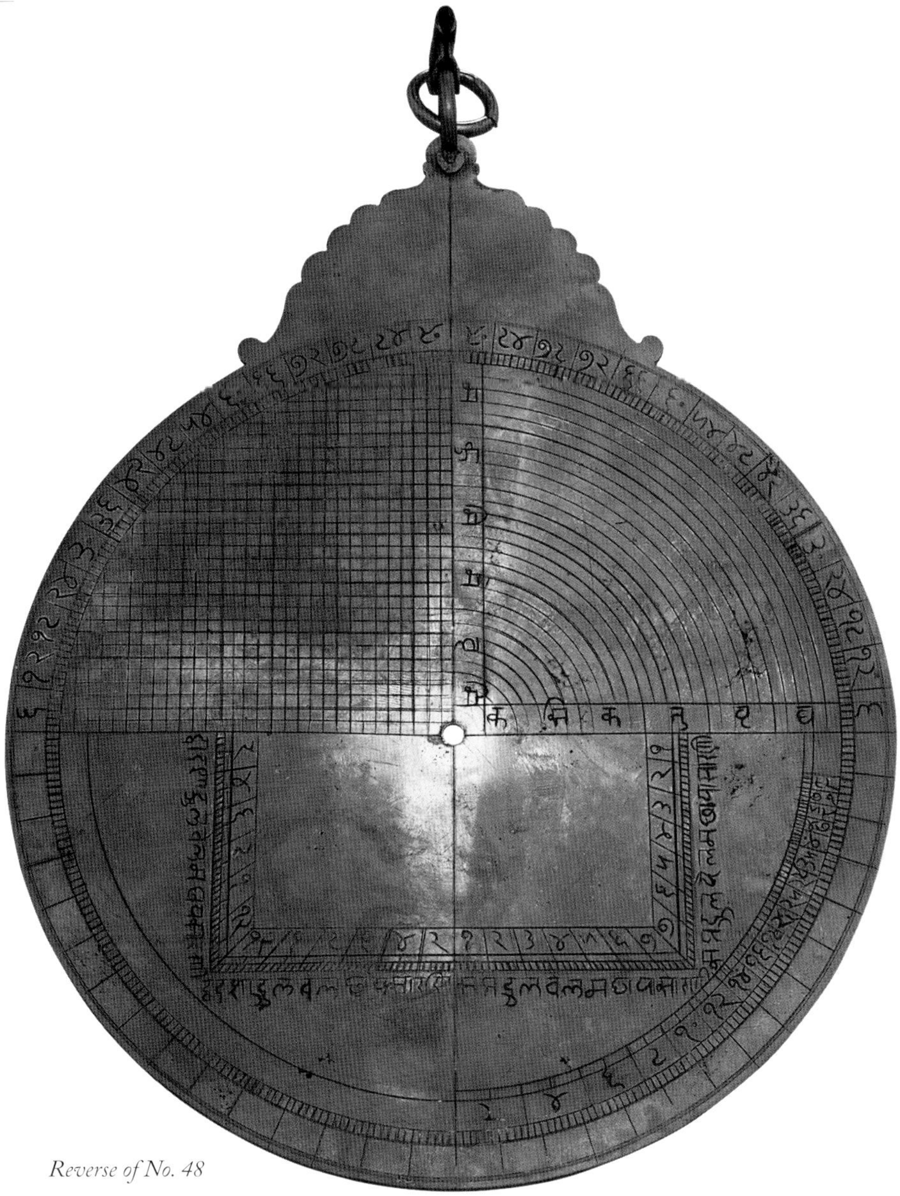

Reverse of No. 48

Rete of No. 48

49 Yantrarāja, incomplete

Maker unknown
India
19th century (?)
23.3 x 17.4 x 2.6 cm
Mater thickness 0.1 cm
Brass
W-102

Face of No. 49

Imperfect astrolabe made for the area near Multān or Thanesar, apparently never completed. Inscriptions in Nāgarī.

Consists of plate, rete, pin, washer, horse, sighting tube, handle, and ring.

The front of the plate is graduated for every degree; the numbers, indicating arcs of 6° and running clockwise, are 1, 2, 3, ...60. The front is an imperfectly executed projection for latitude *ca.* 30°; the altitude circles are for every 3°, the azimuths for every 10°, and there are unequal hour curves. The back of the plate is graduated every degree; the numbers, running clockwise from the left to the top and counter-clockwise from the right to the top, are (with some mistakes) 6, 12, 18, ...90. Otherwise the back is blank.

On the rete the ecliptic is graduated for every 2°; the numbers in each zodiacal sign are 6, 12, 18, 24, 30. The names of the stars are listed. (★)

PROVENANCE — Saul Moskowitz, 1984; Roderick S. Webster, Winnetka, IL, 1985; AP gift, 1985.
EXHIBITIONS — Museum of Science and Industry, Chicago, IL, June 1, 1986 – March 1, 1989.
REFERENCES — Moskowitz (1984), p. 21, lot 88.

★ STAR LIST

Name	*Identification*
abhiji	α Lyr.
aśvamuṣa	ε Peg.
ārdrā	α Ori.
citrā	α Virg.
pū phā	δ (?) Leo
pū phā	β Peg.
pre	β Pers.
matsyodara	β Andr.
ma. pā	α Pers.
marīca	η (?) Urs. Mai.
mahāpuruṣa	α Hydr.
mi. vā. pā	β Ori.
rohaṇī	α Tau.
la badhu	α Can. Min.
lubdhaka	α Can. Mai.
vralahṛ	α Aur.
śatabhiṣā	λ Aquar.
śravaṇa	α Aquil.
samudrapakṣī	ι Ceti
sarpadhārā	α Oph.
svāti	α Boot.
hasta	δ Corv.

Reverse of No. 49

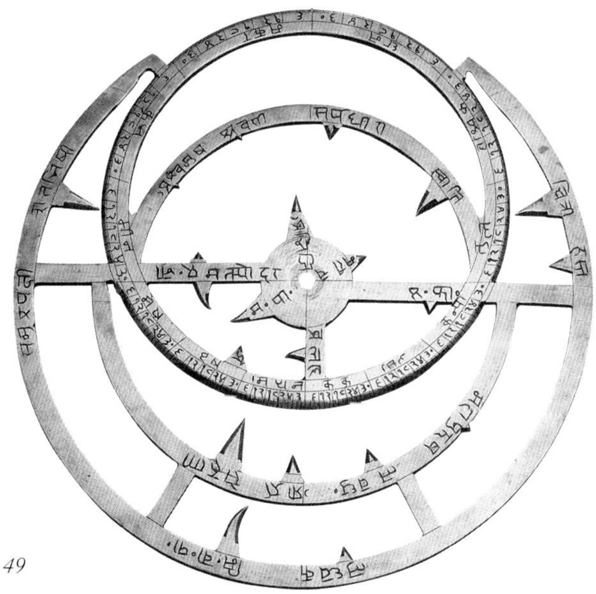

Rete of No. 49

Other Instruments

A Quadrant

Maker unknown.
North Africa.
Radius 15.9 cm
18.5 x 22.6 x 1.3 cm
Brass
W-94

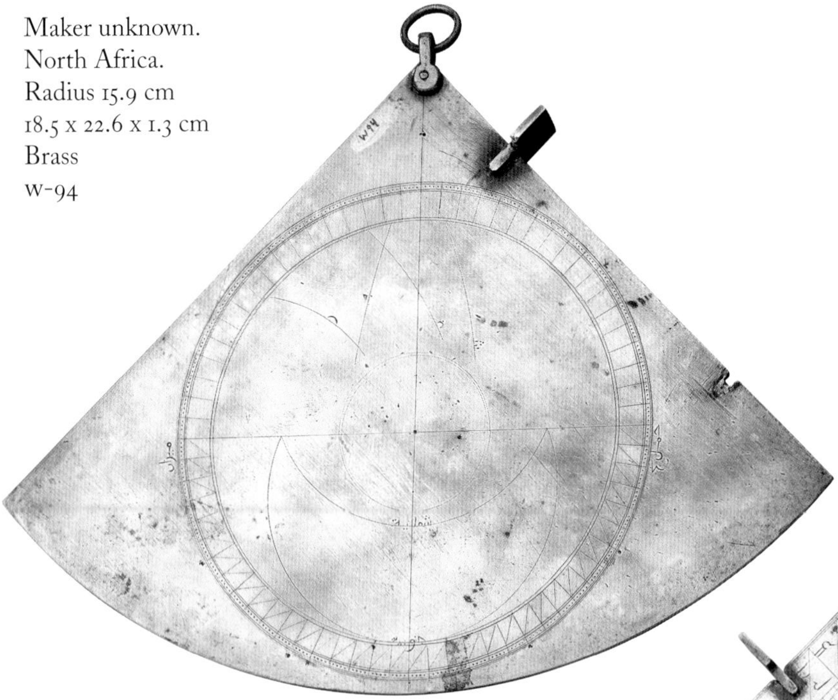

LEFT: *Face of A*
RIGHT: *Reverse of A*

A quadrant originally with two sights on the upper edge (one is now lost). Inscriptions in kūfic (both 60 and 90 represented by ṣād).

On the front is a circle graduated for every degree, within which are various curves and circles. The center is labeled north; Aries is to the right and Libra to the left. Intended for use at Marrakesh.

On the back the quadrant is divided by 60 sines and 60 cosines; along both straight edges in both directions are the numbers 5, 10, 15, ...60. The edge of the arc is graduated for every degree; the numbers in both directions are 5, 10, 15, ...90. The ruler is missing.

PROVENANCE — Roderick S. Webster, Winnetka, IL, 1983; AP gift, 1984.
EXHIBITIONS — *The Heritage of Islam,* Islam Centennial Fourteen, Washington, D.C., Feb. 1, 1982 – Sept. 19, 1983; *Special Topics Exhibit: Islamic Astronomy,* Adler Planetarium, Aug. 23 – Nov. 27, 2006.
REFERENCES— Edwards and Signell (1982), p. 52; King (1991), p. 182; Turner, H. R. (1995), p. 100.

B Horary Quadrant

Maker unknown
Eastern
Radius 8.5 cm
9 x 1.1 x 12 cm
Wood
N-53

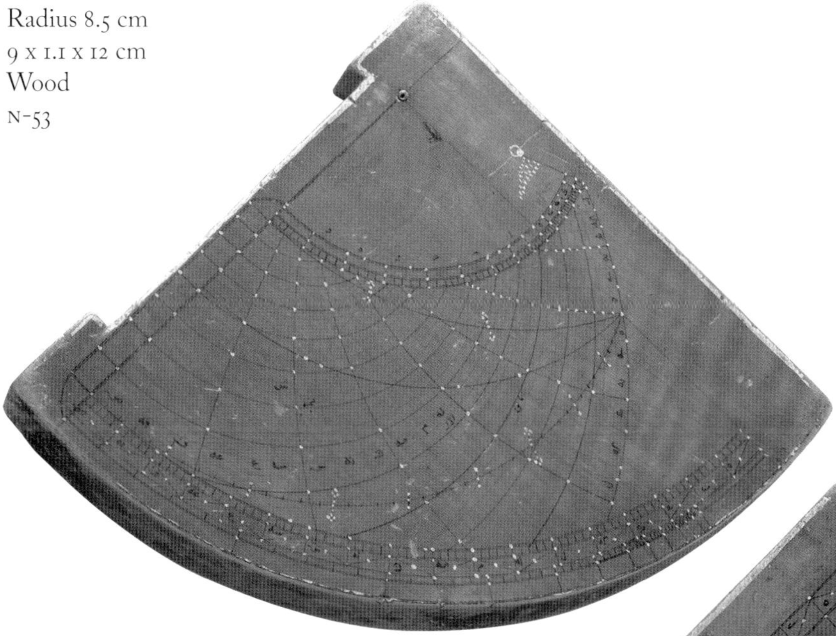

LEFT: *Face of B*
RIGHT: *Reverse of B*

A quadrant which appears not to have had a diopter. Inscriptions in naskh; numerals follow the Eastern style.

On one side the arc is divided into 90°; each fifth is numbered 5, 10, 15, ...90. There are various other curves whose significance is unclear to me, but which are similar to those on a quadrant in the Museo Arqueológico in Madrid. Photos of the Madrid instrument are published by Franco (1945), Nos. 54 and 55 between pp. 366 and 367; the instrument is described on pp. 371-372.

On the other side is a quadrant for finding sines and cosines. Again the arc is divided into 90° and every fifth degree is numbered 5,10, 15, ...90. The straight sides are each divided by twelve perpendicular straight lines numbered 5, 10, 15, ...60.

PROVENANCE — Acquired by Kenneth Nebenzahl, Istanbul bazaar; AP gift, 1986.

C Yantracintāmaṇi

Hṛṣikeśa.
India; perhaps in Multān.
18th century.
"A yantracintāmaṇi measured by the astronomer Hṛṣikeśa for the son of Caṇḍīdatta."
Radius 13.7 cm
14.1 x 20.5 x 1.5 cm
Brass
W-103

Face of C

A yantracintāmaṇi (quadrant) made by Hṛṣikesa for the son of Caṇḍīdatta according to the inscription below the center on the front:

śrījyotirviddhṛṣīkeśanirmito
yantracintāmaṇiḥ
śrīcaṃḍīdattātmajasyārtham

The noon equinoctial shadow is 7;1 digits from a 12 digit gnomon; the latitude is therefore *ca.* 30;20°. Inscriptions in Nāgarī.

On the front are three concentric circles. The first is divided into 180 equal parts (time-degrees); the numbers, running counter-clockwise, are 1, 2, 3, ...30. The middle circle gives the thirty degrees (the numbers are 6, 12, 18, 24, 30) in the zodiacal signs; above the circle are listed counter-clockwise Aries to Virgo. The third circle gives the same thirty degrees for each zodiacal sign, but above it are listed clockwise Libra to Pisces. The oblique ascensions are listed. (■)

Above the center is written: *palabhā* (noon equinoctial shadow) 7;1, and *carakhaṇḍāni* (half-equations of daylight) 70, 56, 23. These last numbers are increments to the half-equation of daylight when the Sun is at Taurus 0°, Gemini 0°, and Cancer 0° respectively; therefore, the half-equation of daylight for the longest day is 149 = 2;29 ghaṭikās, and the longest daylight is 34;58 ghaṭikās, the shortest 25;2 ghaṭikās. There is a pointer on the front that facilitates reading the oblique ascension for each degree of the ecliptic.

The back is divided by a sine and a cosine for each of the 30 parts of Radius; the numbers along both edges are 1, 2, 3, ...30. The arc is graduated at the edge for every degree; the numbers are 6, 12, 18, ...90. The boxes for 7 and 12 as sine and cosine, are numbered. There is a ruler whose edge is divided like the Radius into 30 equal parts (numbered 1, 2, 3, ...30) with a cursor which swings over the quadrant. The 90° point of the arc is labeled *kuja* (horizon); the 0° point of the arc *khamadhya* (midheaven),

OBLIQUE ASCENSIONS

Aries	21°	Pisces
Taurus	24°	Aquarius
Gemini	29°	Capricorn
Cancer	35°	Sagittarius
Leo	35°	Scorpius
Virgo	36°	Libra

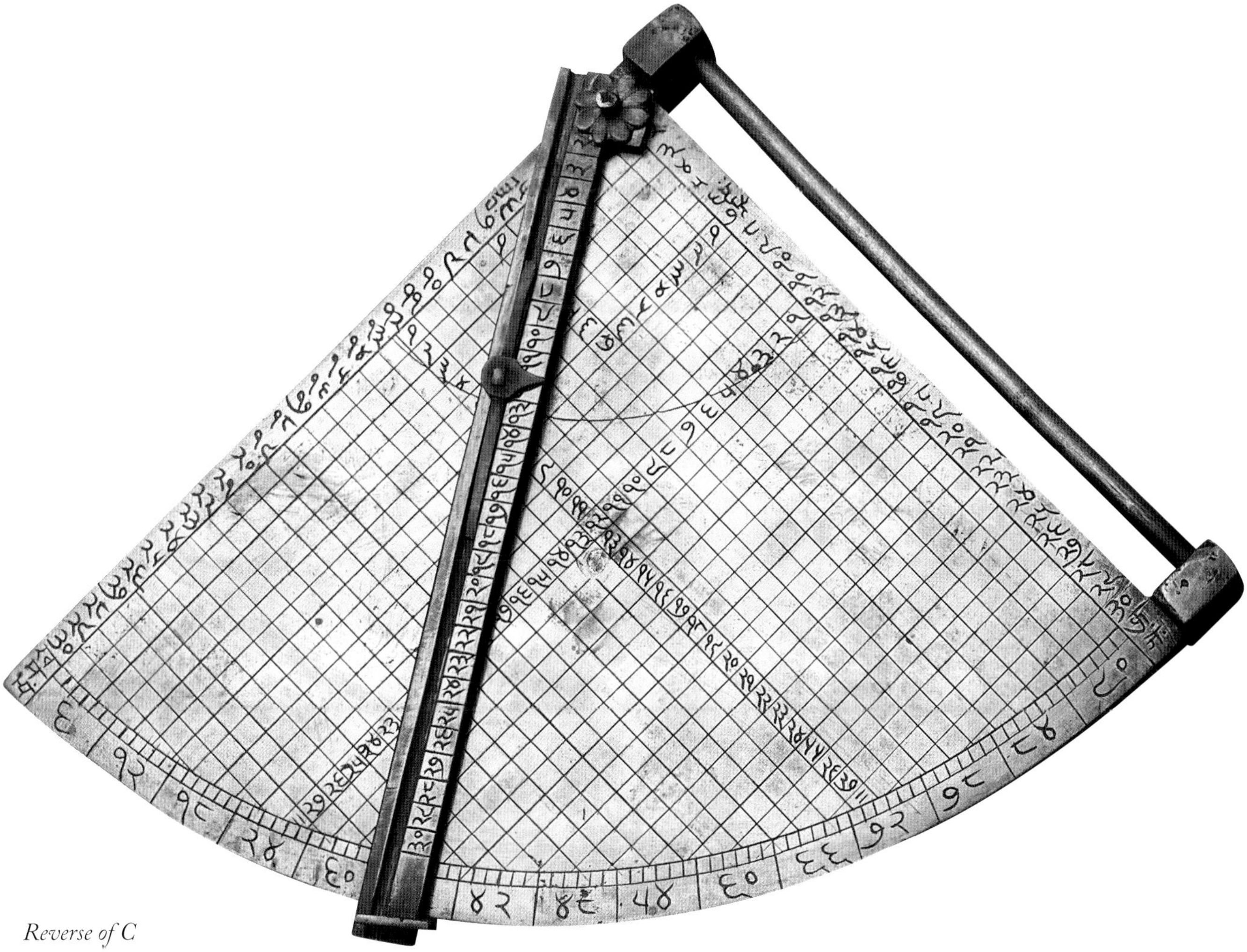

Reverse of C

the 7 on the right edge *bhūmi* (earth), and the 7 on the left edge, in Arabic, *nasab* (proportion).

PROVENANCE — Roderick S. Webster, Winnetka, IL, 1985; AP gift, 1985.
EXHIBITIONS — *Mystery Objects,* Adler Planetarium, June 1, 1992 – May 31, 1994.

D Quadrant

Maker unknown.
Turkey (?)
19th century (?)
13.5 x 15 x 2 cm
Wood
W-93

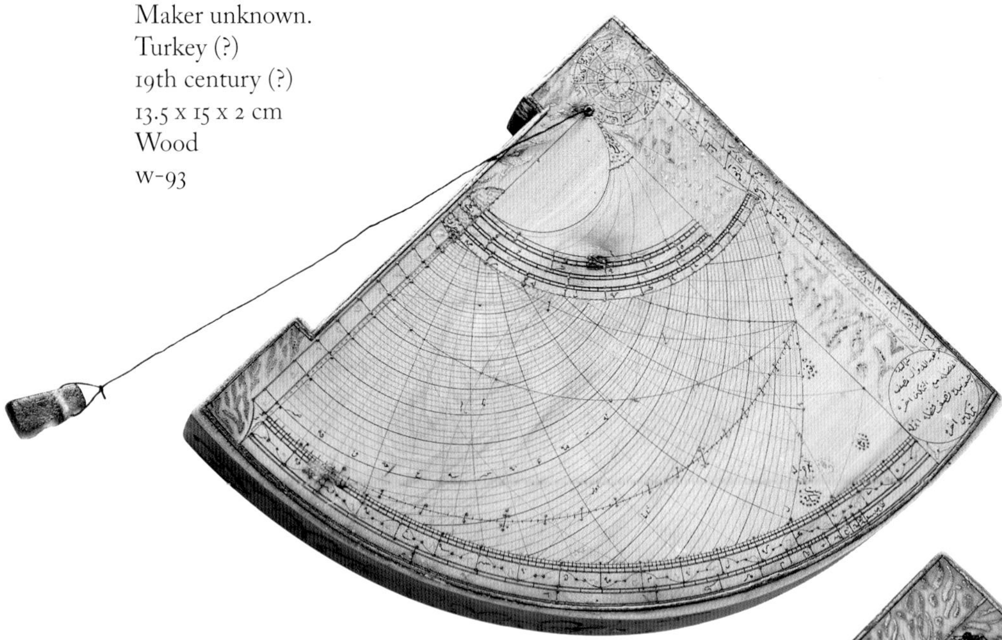

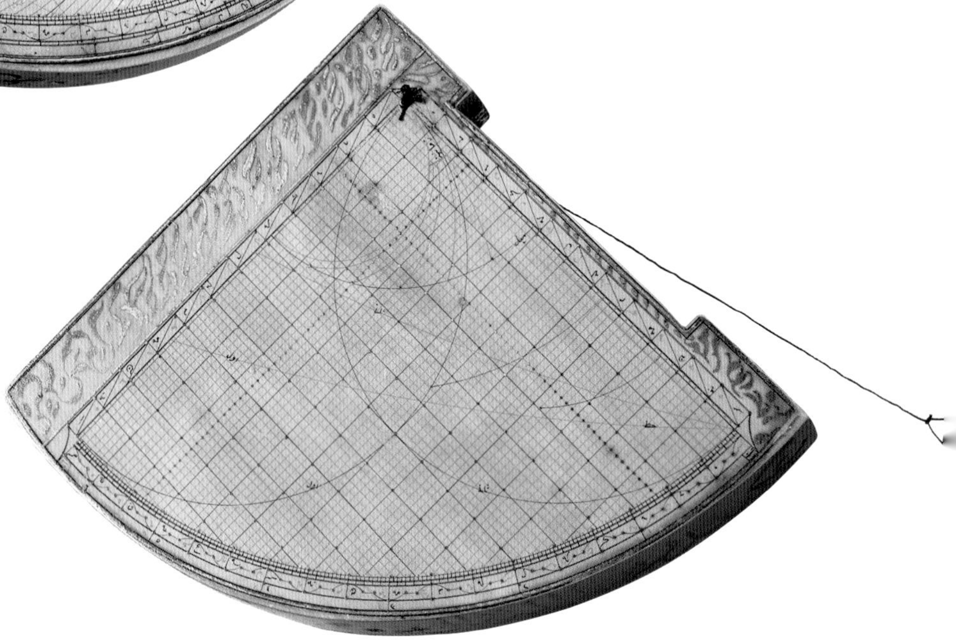

LEFT: *Face of D*
RIGHT: *Reverse of D*

Quadrant with a plumb. Inscriptions in naskh.

On the front are curves for determining the hours, etc. At the top in the center is a circle indicating the zodiacal signs, the months (mixed French and Syrian), and a number. The months (the first corresponds to Aries) are: *nīsān, mārs, khazirān, tammūz, aghtūs* (read *aghūst*), *īlūl, tishrīn awwal, tishrīn thānī, kānūn awwal, kānūn thānī, shubāṭ*, and *that* (?). Along the top edge are listed the seven week-days and twelve months; the latter now are: *mārs, nīsān, māyyah, khazirān, tammūz, aghtūs, īlūl, tishrīn awwal, tishrīn thānī, kānūn awwal, kānūn thānī,* and *shubāṭ*. The arc at the edge is graduated for every degree; the numbers are 5, 10, 15, ...90. Below these numbers are indicated the cotangents. At the right end of the arc is an inscription, partially illegible, in a circle; it gives directions for using this face of the instrument similar to those given along the side of the instrument and the quadrant illustrated in Turner (1985) 28, pp.212–214. In the center of the front is a scale comparing three functions; their limits are 24, 26.5 and 45. The entries in the third, if interpreted as decimals, are close to but not precisely the tangents of the degrees in the first.

The back is divided by 60 sines and 60 cosines; the two straight edges are numbered 5, 10, 15, ...60. The arc at the edge is graduated for every degree; the numbers are 5, 10, 15, ...90.

PROVENANCE — Roderick S. Webster, Winnetka, IL, 1983; AP gift, 1984.
EXHIBITIONS — *The Heritage of Islam,* Islam Centennial Fourteen, Washington, D.C., Feb. 1, 1982 – Sept. 19, 1983.
REFERENCES — Edwards and Signell (1982), p. 52.

E Diopter

Majd al-Dīn Ḥasan
Iran, probably Yazd
1675/6
"Al-'Abd al-faqīr Majd al Dīn Ḥasan of Yazd — peace be upon him! — made it in the year 1075."
19.6 x 15.8 x 0.4 cm
Brass
M-466

Face of E

An instrument made by Al-'abd al-faqīr Majd al-Dīn Ḥasan of Yazd in A.H. 1075 (A.D. 1675/6), according to the inscription on the plate within the circle of the zodiacal signs: *ṣana'ahu al-'abd al-faqīr majd al-dīn ḥasan al-yazdī 'afā 'anhu fī sanah 1075*. Inscriptions in nashk.

Consists of a circular plate perpendicular to whose surface are two rings: one a fixed solsticial ring, and the other a revolving meridian ring. Pinned to the center of the plate is a diopter; its sighting holes are plugged up.

The rim of the plate, which represents the plane of the ecliptic, is graduated for every degree. In an inner circle, divided into twelve equal parts, are written the Arabic names of the zodiacal signs. Between the graduation circle and the zodiacal circle lies a circle in which is numbered each 5° arc of each zodiacal sign: 5, 10, 15, ...30.

The fixed solsticial ring passes through the ecliptic solsticial points. It is notched at the points where an equatorial ring might have passed (no provision is made for it, however, in the zodiacal plate); but the notches are located only about 17° from the ecliptic. The solsticial ring is graduated for every degree; those numbered represent declinations from the equator and depressions from the poles: 5, 10, 15, ...30.

The meridian ring is attached to the solsticial ring with its axis at the North and South poles of the equator. It also is graduated for every degree. Every 5° in either direction from each of the two poles is numbered: 5, 10, 15, ...90.

PROVENANCE — A.W.M. Mensing, Amsterdam; Max Adler, Chicago, 1930; AP gift, 1930.
EXHIBITIONS — *Special Topics Exhibit: Islamic Astronomy,* Adler Planetarium, Aug. 23 – Nov. 27, 2006.

F **Qibla Indicator**

Iran
1800/1
8.6 x 5.8 x 2.7 cm
Brass
W-152

TOP: *Face of F*
BOTTOM: *Reverse of F*

A compass embedded in an oblong tablet having four feet. Made in A.H. 1215 = A.D. 1800/1 if the number 1215 inscribed between the two lozenges in the lower right corner of the top is a date. Inscriptions in nashk.

The top of the tablet is labeled *janūb* (south) above the compass, mashriq (east) to the left, *shamāl* (north) at the botton, and *maghrib* (west) to the left. There is a quadrant of a circle with intervals numbered 5, 10, 15, ...in the upper right, with a rotating pointer. Straight lines emanating from the region of the compass terminate in numbers written in Persian. Proceeding clockwise from the lower left they are: *yak* (one), *dū* (two), *sih* (three), *chahār* (four), *panj* (five), *shash* (six), *haft* (seven), *hasht* (eight), *nuh* (nine), *dah* (ten), *yāzdah* (eleven), *duwāzdah* (twelve), *sīzdah* (thirteen), and *chahārdah* (fourteen). In the lozenges in the lower right is written a Persian quatrain:

īn quṭb namā chū qarṣ mihr anwar
ṭāli' shuda āz mashriq dast khāwar
itmām chū yāft ṭab' ḥashamat guftā
āz īn quṭb namā qibla wa sā'at

This pivot, indicating when the disc of the Sun, splendid,
Becomes the ascendant from the East, is the hand of the West.
When the nature of magnificence obtained completion, he said:
From this pivot point out the qibla and the hour.

On the side of the tablet is a quotation in Arabic, which I have not found in the *Qur'ān*:

qul ṣabbara yūsuf li-nūr 'aynayhi
aṣ-ṣafḥa waqfan bi-man iksāb ilāhi
a'nī al-walād al-arshad mā min sa<l>ilihi
man baddalahu fala'nat allāh 'alāhi

"Say: Yūsuf preserves for the light of his two eyes
The sheet, an endowment to him on behalf of whom he acquired it.

■ COLUMN I

City	*Inclination*	*Identification*
(illegible)	27;1°	
ṭā'if	1;9°	Ṭa'if
liḥsā	69;30°	Aḥsā
miṣr	58;38°	Cairo
ṣana'ā	1;55°	Ṣan'ā'
mawṣil	17;0°	Mosul
surramanra'	2;56°	Sāmarrā'
baghdād	52;45°	Baghdād
baṣrah	37;19°	Basra
wāsiṭ	20;55°	Wasit
shushtar	34;24°	Shustar
kāzirūn	51;57°	Kāzerūn
shīrāz	58;58°	Shīrāz
yazd	78;29°	Yazd
iṣfahān	40;29°	Eṣfahān
kāshān	34;31°	Kāshān
qumm	31;54°	Qom
qazwīn	27;34°	Qazvīn
hamadān	22;17°	Hamadan
sāwah	29;17°	Sāveh
mazdaqān (?)	32;41°	
samīrum	38;0°	Semirum
kirmān	62;51°	Kermān
mashhad	85;6°	Mashhad

■ COLUMN II

City	*Inclination*	*Identification*
nīshābūr	46;24°	Neyshābū
shirwān	44;2°	Shīrvān
simnān	36;17°	Semnān
dāmghān	38;0°	Dāmghān
basṭām	39;28°	Bastam
mazīnān	40;14°	Kalāteh-ye Mazinan
tūn	80;20°	Ferdows
qāyin	54;1°	Qāyen
zūzan	51;58°	Zūzan
harāt	54;0°	Herat
tabrīz	55;2°	Tabrīz
shimākhī	20;9°	Shemakha
ardabīl	17;18°	Ardabīl
marāghah	16;17°	Marāgheh
nakhjawān	18;0°	Nikhichevan
barda'a	17;0°	Barda
amul	32;0°	Amol
sārī	34;0°	Sārī
astarābād	38;48°	Gorgān
rayy	37;26°	Shahr Rey
ḥūrāh	35;0°	
turshīz	48;51°	Kāshmar
marw	52;30°	Mary
sarakhs	51;14°	Sarakhs

■ COLUMN III

City	*Inclination*	*Identification*
bardachān (?)	55;0°	
ṭāliqān	29;38°	Taliq-an
shahraūr	22;0°	Shahraz'r
nihāwand	24;0°	Nahāvand
abhar	25;0°	Abhar
rīmar (?)	34;0°	
jurjān	40;0°	Jurjān
marand	12;0°	Marand
lāhījān	23;0°	Lāhījān
hurmuz	74;0°	Hormoz
ghaznīn	70;37°	Ghazna
kunjūr	30;0°	Kalardasht
ṭabas	40;15°	Ṭabas
burūjird	34;30°	Burūjird
firūzābād	70;0°	Firūzābād
iṣṭakhr	70;0°	Sivand
rāmuz	42;0°	Rāmhurmuz
kābul	69;35°	Kabul
qandahar	75;5°	Kandahar
multān	80;48°	Multan
kirmānshāhān	20;1°	Kermānshāh
karbalā	12;0°	Karbalā'
najaf	15;0°	Najaf
ḥillah	17;1°	Ḥilla

I mean, the mature children from
among his descendants.
Whoever alters it, the curse of God upon
him!"

On the bottom of the tablet is a list of cities in three columns indicating their names, their inclinations, and the direction: the last is always j.gh. (southwest): (■)

In the lozenges to the left of the geographical table is another Persian quatrain:

khwāhī ki shūd ba-har balad qibla
bayān
āz janūb khaṭṭrā tābi' murghak dān
pas madd aliph bar far ān shahr buna
tā ān ki shūd qibla ān mulk 'iyān

"Should you wish that for each city the
qibla should be manifest,
From the south know the line following
the little bird.
Afterwards extend an aliph; above that
city [is its] place
So that that which is the qibla of that
kingdom may be clear."

PROVENANCE — Ahuan Islamic Art, 1985; Roderick S. Webster, Winnetka, IL, 1985; AP gift, 1985.
EXHIBITIONS —Arabic Influences, Robert F. Longway Planetarium, Flint, MI, Sept. 10 - Dec. 18, 2002; *Special Topics Exhibit: Islamic Astronomy,* Adler Planetarium, Aug. 23 – Nov. 27, 2006.

G Qibla Indicator

Maker unknown
Iran
6 x 6 x 1.8 cm
Brass
A-27

■ UPPER BAND

City	Inclination	Identification
kanja	15°	Kirovabad
bādkūba	21°	
zanjān	28°	Zanjan
urūmiyyah	6°	Urmiya
marand	12°	Marand
nakhjawān	13°	Nakhichevan
marāghah	36°	Marāgheh
ardabīl	18°	Ardabīl
tabrīz	16°	Tabrīz
darband	18°	Derbent
bardha'a	36°	Barda
khuway	18°	Khvoy
salmās	1°	Shāpūr
shīrwān	20°	Shīrvān
ayrawān	36°	Yerevan
shimākhī	20°	Shemakha

Compass of G

A circular box, with a lid, inside of which is a compass. Inscriptions in naskh.

The rim of the compass is graduated for every degree. Every five degrees from north and south to east and west is numbered 5, 10, 15, ...90. The four cardinal directions are named on tabs. A piece of paper on which are two mutually perpendicular diameters bearing the names of the four directions has been inserted under the pointer of the compass; it is not aligned with the tabs.

Around the side of the circular box are two bands with lists of cities, their inclinations, and the direction to Mecca; the direction is always gh.j. (southwest). (■)

The bottom, lid, and underside of the lid each have two circles of geographical lists giving the name of the city, its inclination, and the direction to Mecca. On the bottom of the box (plate 1), all the directions are gh.j. (southwest). (○)(●)

PROVENANCE — Ali-Kuli Khan, 1934; AP purchase, 1934.
REFERENCES — Savage-Smith (1992), pp. 47, 67; Yonge (1968), p. 83.

LOWER BAND

City	*Inclination*	*Identification*
harāt	62°	Herat
qandahār	75°	Kandahar
balkh	63°	Wazirabad
bukhārā	71°	Bukhara
samarqand	55°	Samarkand
turshīz	48°	Kāshmar
shūshtar	53°	Shustar
bujnaward (?)	47°	
sarakhs	52°	Sarakhs
kābul	71°	Kabul
qāyin	52°	Qāyen
marw	53°	Mary
tūn	57°	Ferdows
ṭabas	53°	Ṭabas
qūchān	50°	
bulghār	15°	(Bulgars)
badakhshān	64°	Faizabad

PLATE I: OUTER CIRCLE

City	*Inclination*	*Identification*
qaṣr shīrīn	19°	
hamadān	24°	Hamadan
najaf āshraf	12°	Najaf
kūfah	12°	Kūfah
karbalā	12°	Karbalā'
kākhalīn (?)	12°	
baghdād	13°	Baghdād
surramanra'	7°	Sāmarrā'
ḥillah	12°	Ḥilla
nahrawān	15°	
baḥrayn	57°	Manāmah
madā'in	23	Ctesiphon
baṣrah	37°	Basra
liḥsā	59°	Aḥsā
masayyāb (?)	18°	
ya'qūbiyyah	2°	
shahrbān	21°	
zuyhāb	23°	
kirmānshāh	23°	Kermānshāh
kangawar	3°	Kangavar

PLATE I: INNER CIRCLE

City	*Inclination*	*Identification*
burūjird	24°	Burūjird
nihāwand	24°	Nahāvand
jurfādaqān	38°	Golpayegan
yazd	48°	Yazd
kirmān	62°	Kermān
ṭihrān	37°	Tehrān
qumm	32°	Qom
sāwah	29°	Sāveh
kāshān	34°	Kāshān
iṣfahān	40°	Eṣfahān

Bottom of box of G, Plate 1

PLATE 2: OUTER CIRCLE

City	Inclination	Direction	Identification
burhānbūr	32°	NW	Burhanpur
dawlatābād	80°	NW	Daulatabad
sinjār	36°	SE	Sinjār
ra's al-'ayn	37°	SE	Ra's al-'ayn
mawṣil	34°	SE	Mosul
qusṭanṭiyyah	39°	SE	Costanta
miṣr	19°	SE	Cairo
qayrawān	82°	SE	Kairouan
ṭarābulus	78°	SE	Tripoli
arbīl	1°	SE	Arbīl
ṭanjah	79°	NE	Tangier
ṣana'ā	1°	NE	Ṣan'ā'
ḥabashah	9°	NE	(Ethiopia)
'adan	5°	NE	Aden
nūbah	68°	NE	Dongola
sarandīb	70°	NW	(Śrīlankā)

Underside of the lid of G, Plate 2

PLATE 2: INNER CIRCLE

City	Inclination	Direction	Identification
arzan al-rūm	40°	SE	Erzurum
raqqah	36°	SE	Raqqa
madīnah	25°	SE	Medina
bayt al-	32°	SE	Jerusalem
maqdis	33°	SE	Damascus
dimashq	34°	SE	Aleppo
ḥalab	45°	SE	Istanbul
istāmbūl	41°	SE	Konya
qūnyah			

Lid of the box of G, Plate 3

PLATE 3: OUTER CIRCLE

City	Inclination	Direction	Identification
lāhījān	29°	SW	Lāhījān
qazwīn	26°	SW	Qazvīn
zasht	30°	SW	
sulṭāniyyah	26°	SW	Solṭānīyeh
mahd maqdis	40°	SW	
nīshābūr	45°	SW	Neyshābūr
sabzawār	42°	SW	Sabzevār
basṭām	39°	SW	Bastam
simnān	36°	SW	Semnān
dāmghān	38°	SW	Dāmghān
mazīnān	32°	SW	Kalāteh-ye Mazinan
ahwān	40°	SW	Ahvāz
mabāqī (?)	32°	SW	
astarābād	39°	SW	Gorgān
bārfurūsh	32°	SW	Babol
amul	30°	SW	Amol
(Henceforth the directions are not indicated)			
sārī	33°		Sārī
kajūr	26°		Kalardasht
khuwār	30°		Garmsar
satākūh	29°		

PLATE 3: INNER CIRCLE

City	Inclination	Identification
kalbūr (?)	23°	
abarqū	45°	Abarqū
naṭanz	39°	Naṭanz
damāwand	33°	Damāvand
ṭāliqān	29°	Taliq-an
qamshaḥ (?)	41°	
lārsarār (?)	66°	
jahram (?)	29°	
kāzirān	52°	Kāzerūn
shīrāz	13°	Shīrāz

H Sundial

Maker unknown
Probably Europe
18th–19th century
2.5 x 10 x 9.7 cm
Brass
A-148

Face of H

A universal inclining portable sundial for the 'Turkish Market.' The sundial is kept in a padded and decorated wooden case, which was not intended to contain it. The inscriptions, except for the hours on the chapter ring and the latitude scale on the circular arc, in naskh.

The compass is housed in a circular box with hinged chapter ring. On the chapter ring there is a folding gnomon with a fixed 60° inclination to the ring. The dial is set for the correct latitude by means of a brass circular arc that is used to incline the chapter ring to the correct angle so that the gnomon is parallel to the earth's axis. The hours are engraved with the typical adaptation of numerals found on European clocks, watches, and sundials that were made for the 'Turkish market' [see Kurtz (1975) p.30 and plates XII – XIV.] The hours on the chapter ring run, beginning from the upper right, 4, 5,…12, 1, 2,…8. The circular arc is graduated for every 10°, starting at the top, 10, 20,…60. Set at 60, the chapter ring is flush with the compass box. An almost identical sundial is illustrated in Higton (2001) p.113.

On the compass dial the four cardinal directions are noted. The rim is graduated for every 2°; every 10° from north and south to both east and west are numbered: 10, 20, 30,…90.

There are many geographical lists giving the names of cities, their latitudes, and their inclinations on the underside of the chapter ring, on the rim on the top of the compass box, and in two circles on the bottom of the compass box. (■)(■)(❍)(●)

Around the side of the compass box is a difficult Persian inscription concerning the qibla.

by David Pingree and Bruce Chandler

PROVENANCE — AP purchase, before 1931.
EXHIBITIONS — Purdue University, Lafayette, IN, July 1, 1961 – April 15, 1962.

UNDERSIDE OF TEH CHAPTER RING

City	Latitude	Inclination	Identification
iṣfahān	32;35°	40;29°	Eṣfahān
kāshān	34;0°	34;0°	Kāshān
qumm	34;45°	31;55°	Qom
ṭihrān	35;0°	36;26°	Tehrān
khuwar	35;40°	29;50°	Garmsar
sāwah	35;0°	29;36°	Sāveh
qazwīn	35;0°	29;36°	Qazvīn
abhar	36;45°	24;18°	Abhar
ṭāliqān	37;10°	29;10°	Taliq-an
zanjān	36;30°	22;4°	Zanjan
ardabīl	38;0°	17;3°	Ardabīl
marāghah	37;20°	16;17°	Marāgheh
marand	37;50°	51;50°	Marand
urūmiyyah	37;0°	9;1°	Urmiya
salmās	37;40°	7;0°	Shāpūr
khuway	37;40°	7;0°	Khvoy
tabrīz	38;0°	15;40°	Tabrīz
nakhjawān	38;40°	12;49°	Nakhichevan
barda'a	40;30°	16;17°	Barda
kanjah	41;20°	15;49°	Kirovabad
tiflīs	43;0°	43;0°	Tbilisi

RIM OF THE TOP OF THE COMPASS BOX

City	Latitude	Inclination	Identification
harāt	34;30°	54;8°	Herat
qandahār	33;0°	75;5°	Kandahar
kābul	34;30°	77;27°	Kabul
kashmīr	35;0°	68;44°	Srinagar
ghaznīn	33;35°	70;37°	Ghazna
bukhārā	39;50°	49;18°	Bukhara
marw	37;40°	52;30°	Mary
mashhad	37;0°	45;7°	Mashhad
nīshābūr	36;21°	46;21°	Neyshābūr
sabzāwar	36;5°	44;12°	Sabzevār
mazīnān	36;0°	40;11°	Kalāteh-ye Mazinan
basṭām	36;10°	39;53°	Bastam
dāmghān	36;5°	38;5°	Dāmghān
simnān	36;0°	36;17°	Semnān
jurjān	36;50°	39;18°	Jurjān
astarābād	36;50°	28;48°	Gorgān
sārī	37;0°	34;28°	Sārī
amul	36;15°	35;0°	Amol
lāhījān	37;15°	22;40°	Lāhījān
darband	43;0°	19;29°	Derbent
qarābā' (?)	40;0°	18;42°	
bākawiyyah	39;30°	21;29°	Baku
shimākhī	40;50°	20;9°	Shemakha

BOTTOM OF THE COMPASS BOX: OUTER CIRCLE

City	Latitude	Inclination	Direction	Identification
madīnah	29;0°	37;28°	SE	Medina
ḥalab	35;50°	18;29°	SE	Aleppo
dimashq	33;35°	30;31°	SE	Damascus
miṣr	30;20°	18;38°	SE	Cairo
qās	22;0°	78;24°	SE	Fès
jiddah	14;10°	1;5°	NE	Jiddah
ṣana'ā	14;30°	1;15°	NE	Ṣana'ā
ṣuḥār	14;10°	30;10°	NW	Sohār
baḥrayn	25;15°	57;36°	SW	Manāmah
liḥsā	20;0°	69;30°	SW	Aḥsā
qaṭīf	25;0°	63;24°	SW	Qaṭīf
baṣrah	30;0°	37;59°	SW	Basra
shūshtar	31;30°	35;24°	SW	Shustar
rāmuz	31;0°	41;30°	SW	Rāmhurmuz
kāzirūn	29;15°	51;43°	SW	Kāzerūn
bihbihān	30;40°	37;40°	SW	Behbehān
shirāz	29;0°	53;58°	SW	Shīrāz
hurmūz	25;0°	74;0°	SW	Hormoz
kirmān	29;50°	65;51°	SW	Kermān

BOTTOM OF THE COMPASS BOX: INNER CIRCLE

City	Latitude	Inclination	Direction	Identification
najwāsharf (?)	31;30°	31;12°	SW	
surramanra'	34;0°	7;16°	SW	Sāmarrā'
baghdād	33;25°	7;12°	SW	Baghdād
madā'in	33;10°	13;30°	SW	Ctesiphon
nahrawān	36;10°	15;36°	SW	
kirmānshāh	35;30°	23;9°	SW	Kermānshāh
kangāwar	34;30°	30°	SW	Kangavar
burūjird	35;10°	24;30°	SW	Burūjird
hamadān	35;10°	22;17°	SW	Hamadan
nahāwand	34;20°	24;22°	SW	Nahāvand
jurfādaqān	34;15°	32;21°	SW	Golpayegan

Box of H

I Qibla Indicator

Maker unknown
Iran
10.2 x 6.1 x 3 cm
Brass
W-101

Face of I

An oblong box with a compass set into it. Inscriptions in naskh.

There is an irregular graduation around the rim of the compass. On either side of the compass is a column of names of cities and their inclinations: (■)

Below the compass is a dial with a graduated rim. The circle is irregularly divided into segments by radii. In some of these segments are written the names of places: *shīrāz, birlin, bīq (?), nīshābūr, afghān, turkiyya, rūsiyya, al-ṭaliqān, ḥalab, chīn,* and *'adan.* The upright piece is labeled shākhiṣ, "rigid".

On the back is a list of cities and their inclinations divided into six columns. (●)

PROVENANCE — Roderick S. Webster, Winnetka, IL, 1985; AP gift, 1985.

EXHIBITIONS — *The Heritage of Islam,* Islam Centennial Fourteen, Washington, D.C., Feb. 1, 1982 – Sept. 19, 1983; *Special Topics Exhibit: Islamic Astronomy,* Adler Planetarium, Aug. 23 – Nov. 27, 2006.

REFERENCES — Edwards and Signell (1982), p. 52.

■ COLUMNS

Left		*Right*	
iṣfahān	46°	samsān	43°
ṭihrān	37°	dāmghān	45°
baghdād	19°	basṭām	42°
baṣrah		sabzawār	

COLUMN 1

City	Inclination	Identification
miṣr	51°	Cairo
shīrāz	57°	Shīrāz
ābād (?)	52°	Anar
barda'ah	16°	Barda
bulghār	31°	(Bulgars)
baghdād	19°	Baghdād
baṣrah	28°	Basra
madīnah	4°	Medina
makkah		Mecca

COLUMN 2

City	Inclination	Identification
ṭūs	56°	Ṭūs
basṭām	62°	Bastam
rāyy	36°	Shahr Rey
khīkath	16°	
maknah (?)	85°	
karbalā	18°	Karbalā'
'abaq (?)	12°	
jiddah	4°	Jiddah
kamrah (?)		

COLUMN 3

City	Inclination	Identification
qāyin	56°	Qāyen
kirmān	64°	Kermān
tabrīz	22°	Tabrīz
abhar	22°	Abhar
khuway	8°	Khvoy
shām	17°	(Syria)
kawil	17°	Aligarh
ḥalab	11°	Aleppo
rāmlā (?)		Ramla

Reverse of I

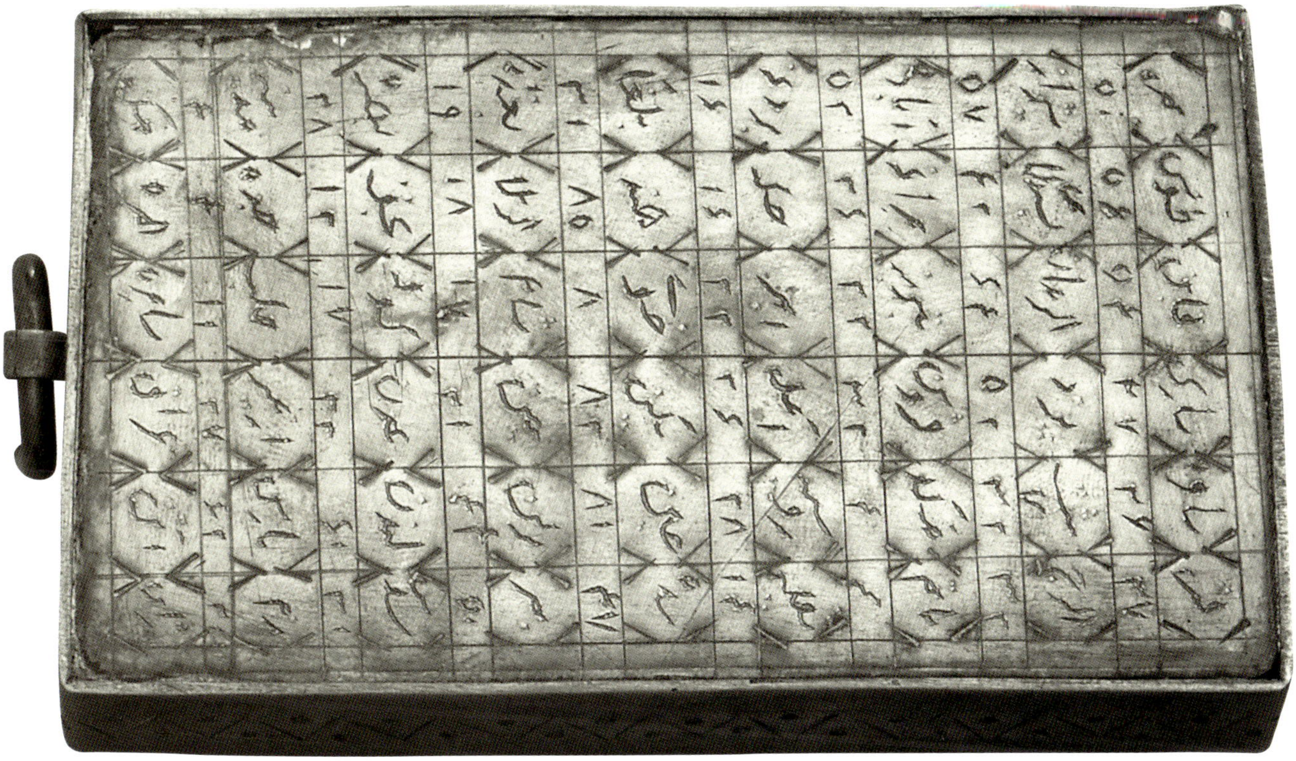

COLUMN 4

City	Inclination	Identification
sāri	27°	Sārī
yazd	52°	Yazd
qazwīn	31°	Qazvīn
amul	24°	Amol
yakan (?)	83°	
yaman	21°	(Yemen)
baḥran	23°	Manāmah
ajmīr	27°	Ajmer
'irāq		(Iraq)

COLUMN 5

City	Inclination	Identification
sāwah	26°	Sāveh
zasht	36°	
mahhak	0°	Mecca
rūsiyya	28°	(Russia)
chīn	81°	(China)
birlin	40°	Berlin
landan	60°	London
bāris	62°	Paris
athan		Athinai

COLUMN 6

City	Inclination	Identification
qumm	37°	Qom
al-raj	32°	
bartim (?)	?	
tiflīs	16°	Tbilisi
sah (?)	67°	
baḥr	50°	
ra's (?)	10°	
zam	21°	Kerki
rayy		Shahr Rey

J

Dialing Instrument

'Alā' al-Dīn ibn Ḥasan Sanandaj
Iran
1864/5
"According to the command of the chief, the light of the king Amān Allāh Khān, the eternal, the most learned, al-dā'ī 'Alā' al-Dīn ibn Ḥasan Sanandaj finished it in 1281."
12.2 x 12.2 x 1 cm
Brass
A-43

Face of J

Disk with hole in center and four orthogonal holes drilled through sides parallel to the horizontal plane; made by 'Alā' al-Dīn ibn Ḥasan Sanandaj for the king Amān Allāh Khān in A.H. 1281 (A.D. 1864/5) according to the inscription of the front: *ḥaṣb al-farmān sar kār ḍiyā' al-malik amān allāh khān bāqī 'allāmī rasīd sanandaj al-dā'ī 'alā' al-dīn ibn ḥasan 1281.* Inscriptions in naskh.

The top is graduated in three segments: 0 to 90, 0 to 180, and 0 to 90. In the first segment is the inscription given above, a quadrant with the cosines for 6, 12, 18, ...60 parts of the Radius, and another quadrant with the sines (but inaccurate, since Radius 60 occurs at arc 63) for 3°, 9°, 15°, ...87°, 90°. In the second segment are the ascensional differences of the zodiacal signs: 28, 30, and 32, as well as both the first and the second declination (the first is measured along a great circle passing through the poles of the equator, the second on one passing though the poles of the ecliptic). The third segment contains tangents for Radius 60 and for Radius 7 and sines for Radius 60.

The side of the instrument, starting at the hole below the beginning of the first segment, is graduated clockwise and counter-

Reverse of J

clockwise in each semicircle from 0° to 180°.

The bottom of the instrument is divided into 4 quadrants, each of which is graduated from 0° to 90°. In each there are also three sequences of numbers keyed to the graduation. In each the first sequence is at intervals of 4: 1, 5, 9,...61; 2, 6, 10,...62; 3, 7, 11,...63; 4, 8, 12,...64. The three sequences in the first quadrant reach 61 at 35°, 45°, and 55°; 62 at 35°, 50°, and 60°; 63 at 35°, 50°, and 60°; and 64 at 40°, 45°, and 60° (but 60 at 35°, 40°, and 55°). I do not know what the meaning of these patterns might be.

There is another disk like this made by the same person at the same date in the History of Science Museum, Oxford.

PROVENANCE — E.S. David, New York, 1940; AP purchase, 1940.
EXHIBITIONS — "The Science and Technology of Islam", Science Museum, London, Feb. 13, 1976 – Sept. 9, 1976.
REFERENCES — Maddison and Turner (1976) p. 166; Brieux and Maddison (forthcoming).

K Pillar Sundial

Maker unknown
Turkey (?)
17.3 x 4.9 x 4.9 cm
Gnomon 4.6 cm
Wood
W-209

Two views of K

A cylinder. Inscriptions in naskh.

A wooden cylinder, on which is written an inscription in a language (Turkish) unknown to me, but employing Arabic technical terms. Thus it begins: *irtifāʿ waqtihi a la* ("The elevation of its time 1;31"). And the last line begins: *ʿarḍ ma h* ("latitude 41;5°"). The actual latitude of Istanbul is 41;2°.

PROVENANCE — Ahuan Islamic Art, London; Roderick S. Webster, Winnetka, IL; AP gift, 1986.

L Artillery Level

Painted by Ḥusayn
Probably North Africa
1783/4
"Ḥusayn painted it in 1198."
33.7 x 11.9 x 1.7 cm
Wood painted red with gold designs
Case (not illustrated) 36 x 14.2 x 2.5 cm
Leather
A-63

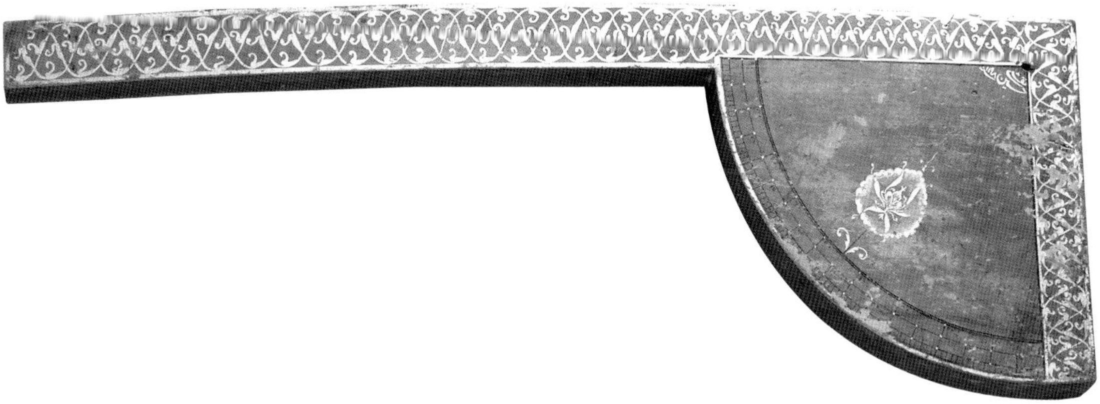

Artillery level painted by Ḥusayn in A.H. 1198 (A.D. 1783/4.) according to the inscription at the top of the front: *naqashahu ḥusayn 1198*. Incription in naskh.

On the front the arc is graduated for every degree; the numbers, running clockwise, are 5, 10, 15, ...90.

TOP: *Face of L*
BOTTOM: *Reverse of L*

PROVENANCE — Jacob Zeitlin, Inc., Los Angeles, 1937; AP purchase, 1937.
EXHIBITIONS — "The Science and Technology of Islam", Science Museum, London, Feb. 13, 1976 – Sept. 9, 1976.

M Artillery Level

Muḥammad ibn al-'Arab (maker or owner)
Probably North Africa
18th or 19th century
25.7 x 8.9 x 0.6 cm
Brass
A-67

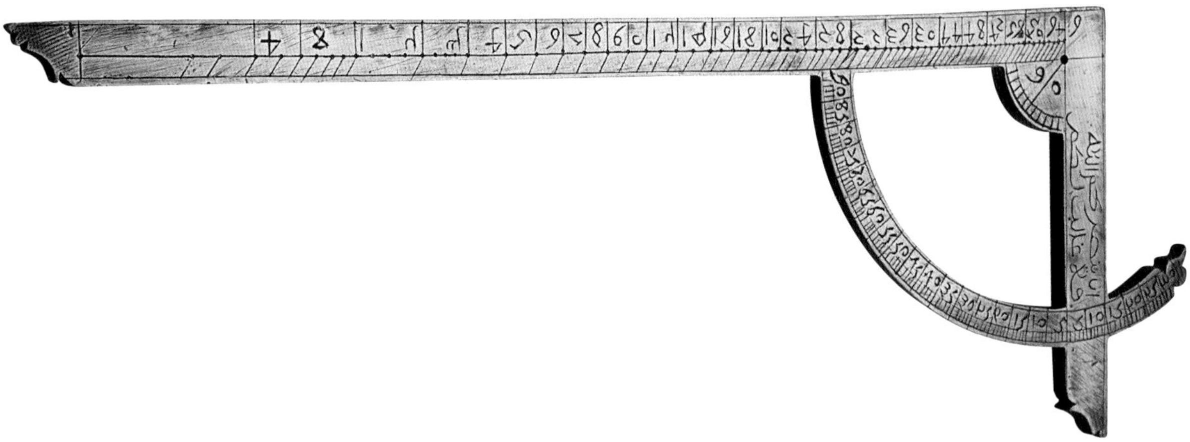

Artillery level. On the back of the top horizontal is the name of the maker or owner: *muḥammad ibn al-'arab*. On the front of the same horizontal is the inscription: *in yanṣura najm ilāyhi fa lā ghālib lakama* ("if a star helps him, the victor does not strike a blow"). Words in inscription in naskh, numbers European.

The arc is graduated for every degree. The numbers from the horizontal clockwise are 5, 10, 15, ...90; from the horizontal counterclockwise are 5, 10, 15, ...30.

PROVENANCE — Jacob Zeitlin, Inc., Los Angeles, 1937; AP purchase, 1937.

EXHIBITIONS — "The Science and Technology of Islam", Science Museum, London, Feb. 13, 1976 – Sept. 9, 1976; *Special Topics Exhibit: Islamic Astronomy,* Adler Planetarium, Aug. 23 – Nov. 27, 2006.

TOP: *Face of M*
BOTTOM: *Reverse of M*

N Level

Muḥammad ibn 'Alī
North Africa
22 February 1788
"Its maker was Muḥammad ibn 'Alī.in Sayl (?) 15 of Jumādī first...1202."
"Belonging to the servant of his master, 'Abd al-'azīz al-Ḥusayn Fanīsh."
12.6 x 14.1 x 0.9 cm
Brass
A-66

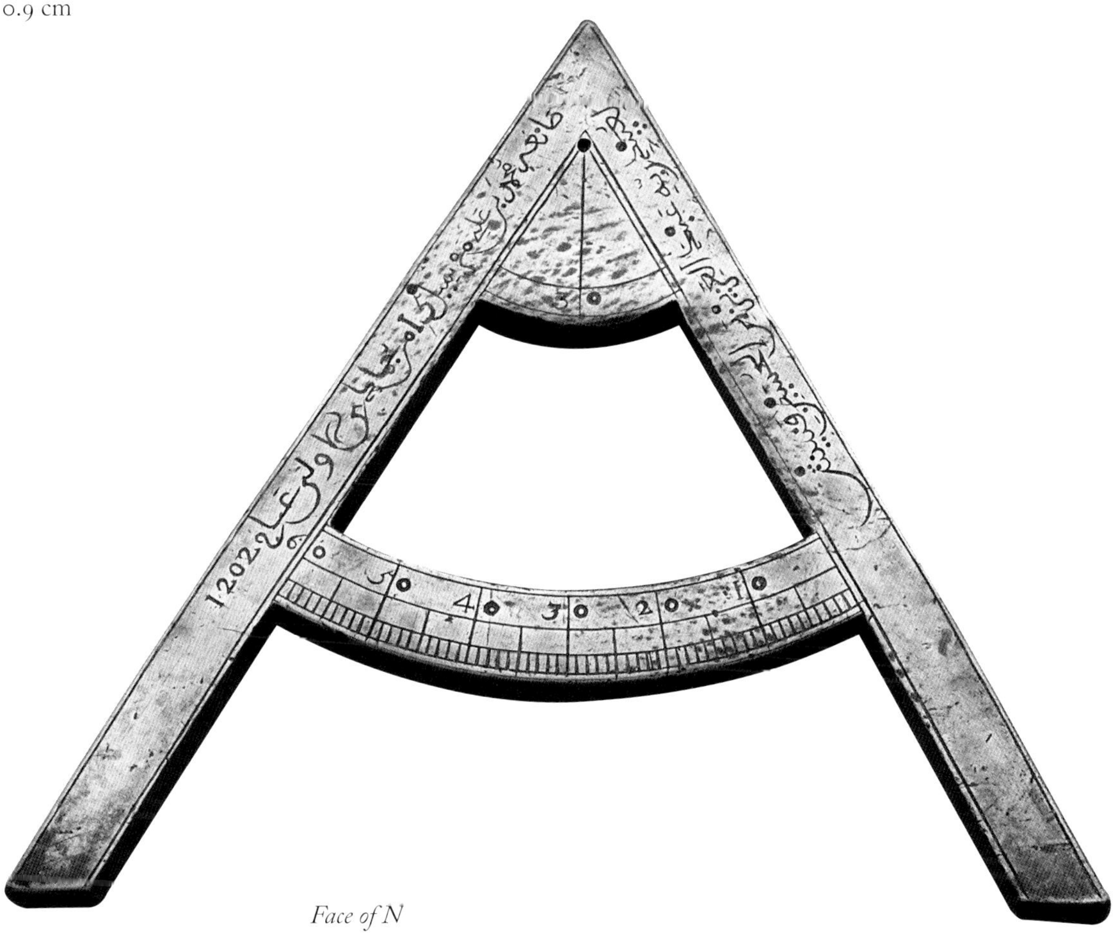

Face of N

Level made by Muḥammad ibn 'Alī.in Sayl (?) on 15 Jumādī I A.H. 1202 (22 February 1788) according to the inscription on the front of the left leg: *ṣāni'hu muḥammad ibn 'alī fī sayl (?) 15* min Jumādī ūl*ā 'āḥ (?) 1202*. It belonged to 'Abd al-'azīz al-Ḥusayn Fanīsh according to the inscription on the front of the right leg: *li-'abd rabbihi 'abd al-'azīz al-ḥusayn fanīsh*. Words in inscriptions in naskh, numbers European.

The crossbar on the front is graduated into 60 parts; the numbers are 10, 20, 30, ...60.

PROVENANCE — Jacob Zeitlin, Inc., Los Angeles, 1937; AP purchase, 1937.
EXHIBITIONS — "The Science and Technology of Islam", Science Museum, London, Feb. 13, 1976 – Sept. 9, 1976. "Arabic Influences", The Robert F. Longway Planetarium, Flint, MI, Sept. 10, 2002 – Dec. 18, 2002.; *Special Topics Exhibit: Islamic Astronomy*, Adler Planetarium, Aug. 23 – Nov. 27, 2006.

O Level

Maker unknown
North Africa
18th or 19th century
11.4 x 12.7 x 0.4 cm
Brass
A-64

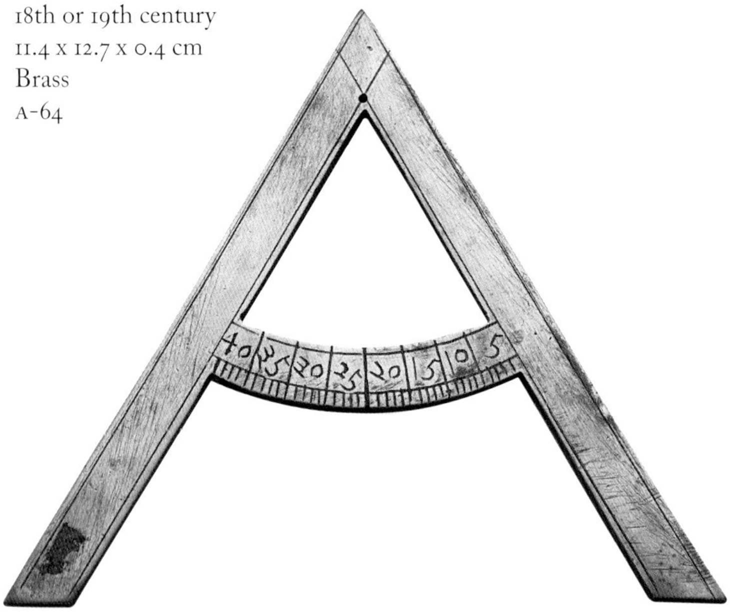

LEFT: *Face of O*
RIGHT: *Reverse of O*

Level. On the back of the crossbar is the inscription: *banayka al-jahd in shā' allāh* ("effort will establish you if God wishes"). Inscription in naskh, numbers European.

The front of the crossbar is graduated into 40 parts; the numbers are 5, 10, 15, ...40.

PROVENANCE — Jacob Zeitlin, Inc., Los Angeles, 1937; AP purchase, 1937.
EXHIBITIONS — "The Science and Technology of Islam", Science Museum, London, Feb. 13, 1976 – Sept. 9, 1976; *Special Topics Exhibit: Islamic Astronomy,* Adler Planetarium, Aug. 23 – Nov. 27, 2006.

P Fake Instrument

Maker unknown
Europe (?)
Modern
19.5 x 15.6 x 0.3 cm
Brass
M-47

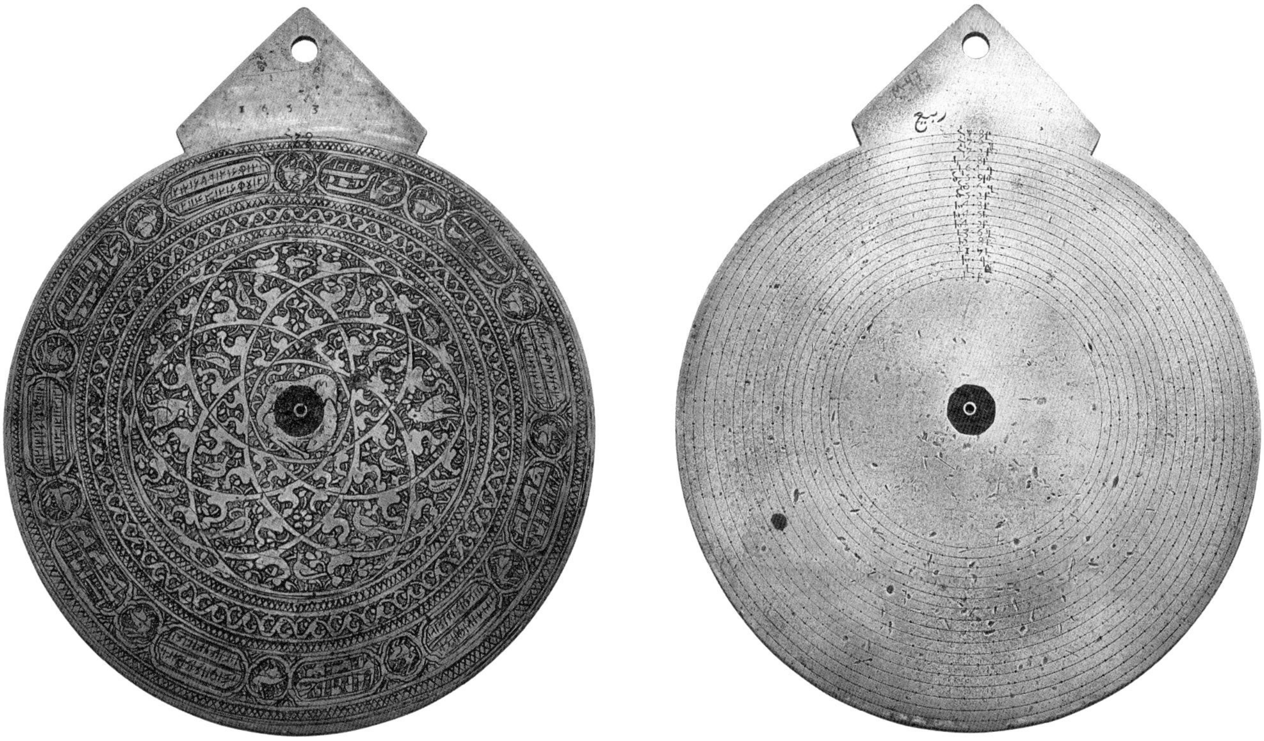

LEFT: *Face of P*
RIGHT: *Reverse of P*

A plate of no apparent significance. There is a hole in the center for attaching something else to it, and a hole in the triangular throne for suspension. Inscribed in naskh with Arabic and European numerals as well as unidentified numerals. On the front of the throne is the date (?) 1653 in European numerals; on the back of the throne is written: *rabīʿ* ("Spring"). On the front is an elaborate design of animals, birds, and flowers; around the rim are cartouches containing alternately Arabic words and numbers. On the back are 17 concentric circles, which are numbered according to two systems (see the photo for the second system, which I transcribe with dotted numbers). These numbers are:

.7 7· 8 .8
.7 7· 6 .6
.7 7· 2 .2
.6 6· 8 .8
.6 6· 4 .4
.5 5· 9 .9
.5 5· 6 .6
.5 5· 2 .2
.4 4· 8 .8
.4 4· 5 .5
.4 4· 2 .2
.3 3· 6 .6
.2 2· 8 .8
.1 1· 7 .7
.4· .4
.3· .8
1· .2

Note that .2, .4, .5, and .7 each has two distinct forms.

PROVENANCE — A.W.M. Mensing, Amsterdam, 1924; Max Adler, Chicago, 1930; AP gift, 1930.
EXHIBITIONS — *Special Topics Exhibit: Islamic Astronomy,* Adler Planetarium, Aug. 23 – Nov. 27, 2006.
REFERENCES — Engelmann Catalogue (1924), p. 17.

Q Celestial Globe

Maker unknown
[Lahore, Pakistan]
17th century
Stand with globe 29 x 20.5 x 20.5 cm
Diameter 16.9 cm
Brass inlaid with silver
M-14

Celestial globe with constellations. Star positions relative to the equinoxes conform to those of the seventeenth-century Lahore workshop. Inscriptions in naskh.

No seam observable; several plugs visible including one large oblong one; rattles; clearly polished on lathe. Weight 1835.7 grams. Full set of constellation figures with about 1018 stars indicated by inlaid silver points. Ecliptic latitude-measuring circles. Ecliptic and equator graduated by single degrees with every fifth labeled; ecliptic repeats every 30°; equator is numbered continuously from the vernal equinox. Zodiacal names engraved along ecliptic; equatorial and ecliptic poles, constellations, and many stars labeled; naskh. Has several small dark engraved circular patches ringed by dark dots (one in Serpentarius and three in Leo) whose purpose is not known. Clearly two different engravers worked on globe, for Boötes, Hercules, and the Northern Crown are rather crudely drawn and lightly incised with some parts redrawn, while other figures are well executed and deeply engraved. Mughal iconography.

Quadruped metal stand on eight-lobed ring base resting on four spherical legs, with two cross bars holding a central support for a semi-circular arc beneath the globe. This semicircular arc bears the names of 42 localities with their corresponding longitudes and latitudes; there are 23 on one side and 19 on the other, each side beginning with the words *asmāʾ bilād, ṭūl*, and *ʿurḍ* (names of cities, longitude, and latitude). The cities are: Mawṣil (Mosul), al-Zubayr, Marand, Nakhjawān (Nakhichevan in Azerbaijan), Marāgha, Tabrīz, Kūfah, Baghdād, Baṣrah, ʿAbādān, Firūzābād (the earlier Gor in Persia, not the one in India), Shīrāz, Yezd, Shahrazūr, Nahāwand (Nehawend), Abhar, Hamadān, Narakh, Qazwīn (Kazvin), Iṣfahān, Qum, Rayy, Astarābād (Astrabad), Medina

(Medinah Rasūl), Mecca (Meccah Mu'aẓẓamah), Hajar in Baḥrayn, Jerusalem (Bayt al-muqaddas), Mahdia, Qayruwān (Kairowan), Ṭarāblus Maghrib (Tripoli in North Africa), Andalus (Southern Spain), Alexandria, Miṣr (Cairo-Fusṭāt), Aden, Yemen, Zabīd, Asqalān (Ascalon), Tiberias, Ṭarāblus Shām (Tripoli in the Lebanon), Damascus, Baalbek, Ḥalab (Aleppo). The horizon ring is graduated by single degrees with every fifth labeled and bears the names of the four cardinal points and the four intermediate directions near the outer edge of the ring; numbering is in non-consecutive segments of 90° beginning at the north-south points, where there are also two notches for the meridian ring, and converging at the east-west points. The outside edge of the horizon ring is divided into about 16 groupings of three segments each, one large and two smaller, separated by engraved double lines; in one grouping there is engraved *asmā' al-buldān* (the names of cities) in the large section, *ṭūl* (longitude), and *'urḍ* (latitude) in the two smaller sections; the other spaces are left blank, indicating that the ring and stand were never completed. On the underside of the horizon ring are also the words *asmā' bilād, ṭūl,* and *'urḍ* (names of cities, longitude, and latitude).

The meridian ring is graduated on one quadrant by single degrees with every fifth labeled beginning at a point marked North and proceeding clockwise. From the point labeled North, there are a series of holes anti-clockwise along the ring labeled 10°, 18°, 24°, 27°, 32°, 40°, 48°, 54°, 66°30', 72°, and 90°, the latter hole having a special dark inset and being labeled "up" (*fawq*). Continuing anti-clockwise from "up" the next quadrant is ungraduated and contains no holes. Continuing with the remaining quadrant, the zero point is labeled South, and there are holes labeled 10°, 18°, 24°, 27°, 32°, 40°, 48°, 54°, 66°30', and 72°. Thus if the ring is positioned on the horizon ring with the North point on the north point of the horizon ring (and the South point on the south point) and the point marked "up" at the top to mark the zenith, the axis of the globe can then be set at 10 specified northern geographical altitudes; there is no accommodation for southern latitudes. A pin passes through the meridian ring and the equatorial (celestial) poles.

by Emilie Savage-Smith

PROVENANCE — A. W. M. Mensing, Amsterdam, 1924; Max Adler, 1930; AP gift, 1930.

EXHIBITIONS — St. John's College, Annapolis, MD, Feb. 13, 1960 – July 7, 1961; *Man, His Planet and Space,* World Exhibition Feb. 1 – Dec. 1, 1967.

REFERENCES — Adler Planetarium (1970), p. 25; Adler Planetarium (1973), p. 29; Adler Planetarium (1980), p. 28; Engelmann Catalogue (1924), p. 12; Fox (1932), pp. 328-30; Fox (1933), pp. 31-32; McAllister (1970); Savage-Smith (1985), pp. 238-39, 292; Savage-Smith (1990), p. 25; Stephenson, Bolt, and Friedman (2000), p. 104; Yonge (1968), p. 83.

Two views of Q

R Celestial Globe

Maker unknown
Lahore, Pakistan
17th century
Stand with globe 23.8 x 21.2 x 19.7 cm
Diameter 16.6 cm
Brass inlaid with silver
A-140

Three views of R

Celestial globe with constellations. Star positions relative to the equinoxes conform to those of the seventeenth-century Lahore workshop. Probably Indo-Persian product of seventeenth-century Lahore. Inscriptions in naskh.

No seam observable; five plugs visible and one figure-eight hole where there was once a plug. Weight 1942 grams; wall thickness 2 mm. Full set of constellation figures with about 1018 stars indicated by inlaid silver points. Ecliptic latitude-measuring circles. Ecliptic and equator graduated by single degrees with every fifth labeled; ecliptic repeats every 30°; equator numbered continuously from vernal equinox. Zodiacal names engraved along ecliptic; constellations and major stars labeled; naskh, engraving filled in white. While star placement, engraving of circles and graduations are precise, the constellation figures are rather crudely drawn. Surface of globe badly scratched and beaten.

Metal quadruped stand on ring, eight-lobed base with cross bar, but with no undersupport for globe. Stand and horizon ring clearly made to go with a different globe, for the ring is too small for this globe and does not allow globe to rest with ring at its midpoint. Horizon ring graduated by single degrees with every

fifth labeled in non-consecutive segments of 90° converging at the two notches for the meridian ring which are labeled north and south; bears the names of the four cardinal points near outside edge of horizon ring. A 90° graduated arc (same pattern as ring) is attached and hinged to the top of the horizon ring to serve as a marker for zenith and prime vertical. Ungraduated (recent) meridian ring.

by Emilie Savage-Smith

PROVENANCE — Michel N. Maloof, Wilmington, DE, 1941; AP purchase, 1941.

EXHIBITIONS — *The Sphere of the Cartographer,* Golda Meir Library, University of Wisconsin-Milwaukee, Oct. 14 – Dec. 6, 1996; *Nostalgia and Technology,* Museum of Art, Brigham Young University, Provo, UT, Nov. 7, 2005 - May 15, 2006; *Special Topics Exhibit: Islamic Astronomy,* Adler Planetarium, Aug. 23 – Nov. 27, 2006.

REFERENCES — Savage-Smith (1985), pp. 239-240; Yonge (1968), p. 83.

s Celestial Globe

Maker unknown
[Lahore, Pakistan]
17th century
Diameter 11.5 cm
Brass inlaid with silver
A-115

Celestial globe with constellations. Star positions relative to the equinoxes conform to those of the seventeenth-century Lahore workshop. Inscriptions in naskh.

No seam observable; three plugs visible including two large round plugs in line on opposite sides of the sphere; rattles slightly. Weight 921.4 grams. Full set of constellation figures and about 1018 stars indicated by very small inlaid silver points. Ecliptic latitude-measuring circles; it appears that a circle with the south equatorial pole as center was begun but not finished; the circle would have been between the south tropic circle and the south polar circle had it been completed. For this reason work was probably stopped on the globe and no holes were drilled at either the equatorial or ecliptic poles, and the signature of the maker was not put on it. Ecliptic and equator graduated by single degrees with every fifth labeled; ecliptic repeats every 30°; equator is numbered continuously from the vernal equinox. Zodiacal names engraved along ecliptic; constellations and major stars labeled; naskh. The twelve zodiacal constellation figures are engraved with dotted lines. No hole drilled at either the equatorial or ecliptic poles; however, two (recent?) large holes drilled opposite each

Four views of S

other about 25° east of the equinoxes; purpose unknown. Mughal iconography. Fairly carefully executed engraving.

Stand and rings missing.

by Emilie Savage-Smith

PROVENANCE — Provenance unknown.
EXHIBITIONS — *IBM*, Hayden Planetarium, New York, N.Y., June 11, 1964 – May 3, 1974; *Episodes in the Story of Time*, Adler Planetarium, April 4 - Sept. 30, 2001.
REFERENCES — Savage-Smith (1985), pp. 43, 95, 241, 305.

T Celestial Globe

Maker unknown
Ṣafavid Iran
1603/4
"By the command of the sulṭān of sulṭāns, the honorable and powerful Bandah Shāh Walāyat 'Abbās, completed 1012."
Diameter 14.9 cm
Brass inlaid with silver
A-114

Celestial globe without constellations. Unsigned, but bears patron's name: "*haṣb al-amr/sulṭān al-sulāṭīn/mu'aẓẓam wa-al-muqtadir bandah/shāh walāyat 'abbās/ khātim-yāft/1012*". Inscriptions in naskh.

Hemispheres with seam at ecliptic. Weight 1389.1 grams. Only twelve figures representing the zodiacal houses are engraved; they are not actually constellation outlines but only decorative medallions. There are thirty-eight stars indicated by inlaid silver points surrounded by a circle. Ecliptic latitude-measuring circles; equatorial tropic and polar circles; ecliptic tropic circles; equinoctial colure and one circle 30° to the west passing through the equatorial poles (a meridian). There are also circles about 12° inside each of the tropic circles, parallel to both the ecliptic and equatorial tropic circles. Equator represented by a simple line with no graduations; ecliptic represented by the line of the seam. The figures representing the Houses are not the usual constellation figures (e.g., Libra is a cross-legged human figure holding the scales like a yoke over the shoulders) and there are no stars indicated in them; each figure is enclosed in a circle and is of a decorative style found on some other forms of Islamic metalwork. Each medallion figure is labeled with the name of the zodiacal house outside the circle; equinoxes, solstices, ecliptic latitude circles, meridian, both colures, both sets of poles, ecliptic, equator, *mayl kullī* ("full obliquity," referring to the greatest distance of a point on the ecliptic from the celestial equator), tropic and polar circles are labeled; naskh in floriated double-outlined writing on recessed stippled ground. A unique feature of the globe is a large crescent moon engraved on it between the north equatorial pole and the House of Aries. The inscription, with the patrons name, is within a cartouche located between

the houses of Aries and Pisces and the south ecliptic pole. The patron was Shāh ʿAbbās I, who ruled A.D. 1587-1629 (for the formula used for his title see Borgomale (1945), pp. 32–33.

Stand and rings are missing.

by Emilie Savage-Smith

PROVENANCE — Ali-Kuli Khan of the House of Persian Art, 1941; AP purchase, 1941.
EXHIBITIONS — Abrams Planetarium, Michigan State University, East Lansing, Jan. 28, 1964 - June 20, 1965; *The Sphere of the Cartographer,* Golda Meir Library, University of Wisconsin-Milwaukee, Oct. 14 - Dec. 6, 1996; *Arabic Influences,* Robert F. Longway Planetarium, Flint, MI, Sept. 10 - Dec. 18, 2002; Special Topics Exhibit: Islamic Astronomy, Adler Planetarium, Aug. 23 - Nov. 27, 2006.
REFERENCES — Savage-Smith (1985), pp. 45-47, 58, 61, 68, 80, 215, 249-250, 293, 295, 305, 306; Savage-Smith (1992), pp. 47, 67; Yonge (1968), p. 83.

Four views of T

U Celestial Globe

Muḥammad Samʿī Tarī
Iran or India
1822/3
"The work of Muḥammad Samʿī Tarī in 1238"
Globe 10.1 x 10.3 x 10.3 cm
Diameter 10.3 cm
Brass
A-116

Two views of U

Celestial globe without stars and constellations. *"bandah-i mubarrā(?)/muḥammad samʿī (?) tarī (?)/ʿamal bar (?)/1238"*. Inscriptions in naskh.

No seam observable; no plugs visible. Weight 411.1 grams. Ecliptic latitude-measuring circles; equatorial tropic and polar circles; equinoctial colure (with the ecliptic latitude-measuring circle intersecting the equinoctial colure omitted). Ecliptic and equator indicated by simple lines with no graduations. Zodiacal names engraved along ecliptic; solstices, solstitial colure, *mayl kullī* ("full obliquity," referring to the greatest distance of a point on the ecliptic from the celestial equator), ecliptic, equatorial poles, equinoxes, and solstices labeled; naskh, written with floriated double-outlined writing on dark recessed ground with a band around each inscription except for the cartouche near the vernal equinox in which the date and the name of the maker are given. The latter is lightly engraved with a stippled background, possibly added later. The inscription in the cartouche could be interpreted in several ways, perhaps reading: "the work of (ʿamal) Muḥammad Samʿī Tarī in 1238." The top part of the cartouche seems to read *bandah-i mubarrā* (a freed slave), though this is by no means clear. Globe is probably of Persian origin, though India is also possible. Stand and rings missing.

by Emilie Savage-Smith

PROVENANCE — AP acquisition, before 1945.
EXHIBITIONS — *The Sphere of the Cartographer,* Golda Meir Library, University of Wisconsin-Milwaukee, Oct. 14 – Dec. 6, 1996; *Special Topics Exhibit: Islamic Astronomy,* Adler Planetarium, Aug. 23 – Nov. 27, 2006.
REFERENCES — Savage-Smith (1985), pp. 56, 263, 296, Yonge (1968), p. 84.

V Celestial Globe

Maker unknown
Iran or India
18th century or later
Stand with globe 17.4 x 9.2 x 9.4 cm
Diameter 6.7 cm
Brass
A-39

Two views of V

Celestial globe without stars and constellations. Inscriptions in naskh.

No seam observable; no plugs visible; rattles. Weight 127.6 grams including brass meridian ring. Ecliptic latitude-measuring circles; equatorial tropic and polar circles; ecliptic tropic and polar circles; equinoctial colure. Ecliptic and equator graduated by 2° intervals with every tenth degree labeled; ecliptic repeats every 30°; equator numbered continuously from vernal equinox. Zodiacal names engraved along ecliptic; the two *mayl kullī* ("full obliquity," referring to the greatest distance of a point on the ecliptic from the celestial equator) labeled; naskh. Some of the engraving is in broad double-outlined writing and some is filled in in white. Graduations are uneven and circles somewhat wobbly.

Metal quadruped stand on ring base with top, center and bottom hexagonal knobs on the straight legs. The semicircular support under the globe attached to horizon ring has a wing-headed screw at bottom that can tighten to hold meridian ring in place. Horizon ring graduated by 2° intervals with every sixth degree labeled in non-consecutive segments of 90° beginning at notches for the meridian ring. Meridian ring graduated by 2° intervals with every sixth degree labeled in non-consecutive 90° segments beginning at the poles where it is attached to the globe by a pin. Rings and stand appear contemporary with globe.

by Emilie Savage-Smith

PROVENANCE — Ali Kuli Khan, 1934; AP purchase, 1934.
EXHIBITIONS — Exhibit, Michigan State University, Jan. 28, 1964 – June 30, 1965; Special Topics Exhibit: Islamic Astronomy, Adler Planetarium, Aug. 23 – Nov. 27, 2006.
REFERENCES — Savage-Smith (1985), pp. 56, 267; Yonge (1968), p. 80.

W Celestial Globe

Maker unknown
Iran or India
18th century or later
Stand with globe 17.3 x 9.8 x 9.9 cm
Diameter 7 cm
Brass
A-38

Two views of W

Celestial globe without stars and constellations. Inscriptions in naskh.

No seam observable; no plugs visible; rattles. Weight 134.7 grams including metal meridian ring. Ecliptic latitude-measuring circles; equatorial tropic and polar circles; ecliptic tropic and polar circles; equinoctial colure. Ecliptic and equator graduated by 2° intervals with every tenth degree labeled; ecliptic repeats every 30°; equator numbered continuously from the vernal equinox. Zodiacal names engraved along ecliptic; the two *mayl kullī* ("full obliquity," referring to the greatest distance of a point on the ecliptic from the celestial equator) are labeled; naskh with some large double-outlined writing. Graduations and circles uneven.

Metal quadruped stand on ring base with square knobs at the top, center, and bottom of each tall straight leg. The semicircular support under globe attached to horizon ring has a very large wing-headed screw at the bottom (mid-point), which can tighten to hold meridian ring in place; the semicircular support terminates in a four-sided finial at mid-point. Horizon ring graduated by 2° intervals with every sixth degree labeled in non-consecutive segments of 90° beginning at notches for meridian ring. Meridian ring graduated by 2° intervals; every sixth degree labeled in non-consecutive segments of 90° beginning at the poles where it is mounted to the globe by a pin. Stand has been repaired. Probably contemporary with globe.

by Emilie Savage-Smith

PROVENANCE — E. S. David, 1940; AP purchase, 1940.
EXHIBITIONS — *Special Topics Exhibit: Islamic Astronomy,* Adler Planetarium, Aug. 23 – Nov. 27, 2006.
REFERENCES — Savage-Smith (1985), pp. 56, 267-68; Yonge (1968), p. 80.

x Magic Bowl

Maker unknown
Middle East
Modern
2.8 x 20.3 x 20.6 cm
A-112

Interior of X

Shallow bowl of no apparent significance. Inscriptions in Arabic.

Shallow copper bowl with signs of silvering. Covered inside with pious phrases and tables of numbers in Arabic. Signs of the zodiac with Gemini as a two-headed Siamese twin holding an addition head in his right hand. Sun and two human figures in center of bowl. Outside of bowl has tables of numbers, two cartouches, and a device resembling a wind rose.

PROVENANCE — AP acquisition, before 1945.
EXHIBITIONS — *Special Topics Exhibit: Islamic Astronomy,* Adler Planetarium, Aug. 23 – Nov. 27, 2006.

Y Magic Bowl

Maker unknown
Middle East
Modern
4 x 18.8 x 18.8 cm
A-113

Shallow bowl of no apparent significance. Inscriptions in Arabic.

Shallow copper bowl covered with pious phrases in Arabic. Center of interior very worn. Exterior has signs of the zodiac with Gemini represented by a two-headed Siamese twin holding a fish in each hand. Two human figures on bottom surrounded by numbers.

PROVENANCE — AP acquisition, before 1940.
EXHIBITIONS — *Special Topics Exhibit: Islamic Astronomy*, Adler Planetarium, Aug. 23 – Nov. 27, 2006.

LEFT: *Interior of Y*
RIGHT: *Exterior of Y*

z Magic Bowl

Maker unknown
Middle East
Modern
4.2 x 12.6 x 12.6 cm
A-142

LEFT: *Interior of Z*
RIGHT: *Exterior of Z*

Bowl of no apparent significance. Inscriptions in Arabic.

Bronze bowl with raised conical center. Cracked. Covered inside and out with pious phrases in Arabic. Signs of zodiac on outside includes Gemini represented by a two-headed Siamese twin.

PROVENANCE — James Keggie, London, 1940; AP purchase, 1940.
EXHIBITIONS — *Special Topics Exhibit: Islamic Astronomy*, Adler Planetarium, Aug. 23 – Nov. 27, 2006.

AA. Inscribed Stone Shaped like an Astrolabe

Maker unknown
Amber, Jaipur, India
1742/3
"1799. Ābarā, observation room."
25.3 x 21.8 x 1.2 cm
Black stone
w-273

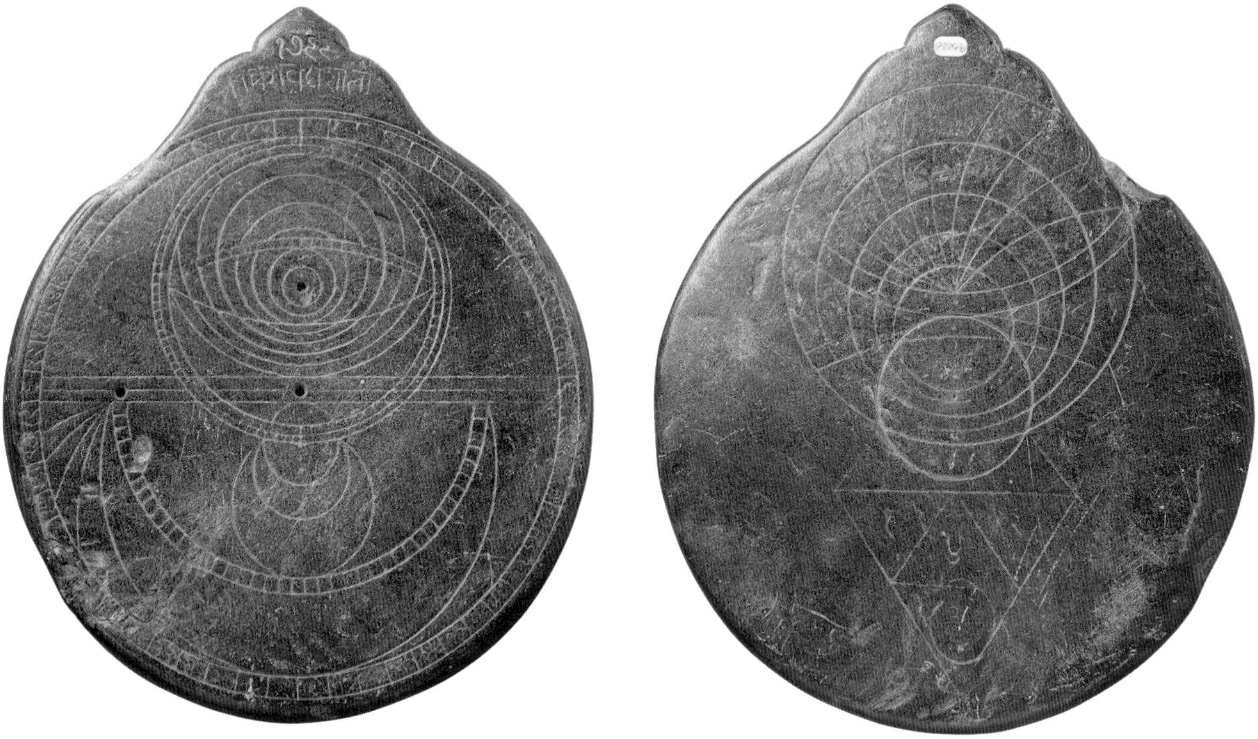

Object of uncertain purpose made, if the inscription has been properly read as *1799 ābarā <e>dhasālā*, at Amber, the orginal capital of the present Jaipur region of Rājasthān, in <Saṃvat> 1799 or A.D. 1742/3 (the Mahārāja Jayasiṃha died in 1743). Inscriptions in Nāgarī.

The upper two sections of the rim of the front are graduated for every 6°; the numbers, running clockwise on the left and counter-clockwise on the right, are 6, 12, 18, ... 90. The numbers in the lower part of the rim are too obscure to be read.

The front and upper part of the rear contain drawings of circles, curves, and straight lines similar to but not identical with the drawings on a tympan.

On the front, along the North-South and East-West lines, are drilled shallow holes, whose purpose can only be guessed at.

On the rear, below the tympan-like design, is a triangle divided into four more or less equal triangles. The three top triangles, from left to right, are numbered in Nāgarī 1, 4, and 2; the bottom triangle, which has a circle crudely inscribed in it, perhaps was intended to contain a 3.

PROVENANCE — David Berstein Fine Art, New York, 1993; Roderick S. Webster, Winnetka, IL, 1993; AP gift, 1993.
EXHIBITIONS — *Special Topics Exhibit: Islamic Astronomy*, Adler Planetarium, Aug. 23 – Nov. 27, 2006.
REFERENCES — Dershowitz and Reingold (1997), p. 118.

Appendices

Catalogue of Arabic Star-Names

ANDROMEDA		Cat. No.
β	baṭn al-ḥūt	3,13,16,33
	baṭn ḥūt	20
	baṭn	19
γ	rijl al-musalsalah	16,23
	rijl musalsalah	33

AQUARIUS		
δ	sāq ayman sākib al-mā'	23
	sāq sākib al-mā'ī janūbī	16
	sāq sākib al-mā'ī	17
	sāq sākib al-mā'	13,14,18,19,20,33
	sāq ayman	15,24,28,30,31
	sāq	21,25

AQUILA		
α	al-nasr al-ṭā'ir	1
	nasr ṭā'ir	13,15,16,17,18,19,22,23,24,25,27,28,30,31,32,33,34
	al-ṭā'ir	6,7,8,10
	ṭā'ir	2,3,9,11,14,20,21
ζ	dhanab al-'uqāb	23,24,30,31

ARIES		
β	muqaddam sharaṭān	19
γ	mu'akhkhar sharaṭān	19

AURIGA		
α	al-'ayyūq	1,6,10
	'ayyūq	2,3,7,8,9,11,13,14,15,16,19,21,22,23,24,25,27,28,30,31,33,34

BOOTES		
α	al-simāk al-rāmiḥ	1
	simāk rāmiḥ	13,15,18,19,22,23,24,31,32,33
θ	al-rāmiḥ	6,7,8,10
	rāmiḥ	2,3,9,11,14,16,17,21,25,27,28,30,34

CANCER		
α	shafat al-saraṭān	1

CANIS MAIOR		
α	al-shi'rā al-abūr	1
	al-'abūr	2,6,7,8
	shi'rā yamāniyyah	13,15,17,19,20,33
	shi'rā yamāni	22,24,25,27,28,30,31,32,34
	al-yamāniyyah	10
	yamāniyyah	14,23
	yamāni	21
ζ (?)	al-furūdayn	13

CANIS MINOR

		Cat. No.
α	al-ghumayṣā	6,7,8
	ghumayṣā	2,3
	al-shiʿrā al-ghumayṣā	1
	shiʿrā shaʾāmiyyah	13,16,17,32,33
	shiʿrā shaʾāmī	18,19,23,34
	al-shaʾāmiyyah	10
	shaʾāmiyyah	11,14,20
	shaʾāmī	15,21,24,25,27,28,30,31

CAPRICORNUS

β	qaran al-jady al-thānī	19
γ	saʿd nāshirah	1
δ	dhanab al-jady	2,6,7,10,13,16,17,18,19,20, 22,33,34
	dhanab jady	8,23

CASSIOPEIA

β	kaff al-khaḍīb	10,13,14,15,16,17,18,22,23, 24,27,28,30,31,32,33,34
	kaff	11,21
	al-khaḍīb	1,25
	khaḍīb	2,3,20,34

CETUS

α	kaff al-jadhmāʾ	19,33
β	dhanab al-qayṭīs al-janūbī	16
	dhanab al-qayṭus janūbī	32
	dhanab al-qayṭus ja	34
	dhanab al-qayṭīs	33
	dhanab al-qayṭus	18,19
	dhanab qayṭūs	1
	dhanab qayṭus	13,23,25
	dhanab qayṭus ja	19
	janūbī qayṭus	22
γ	fam al-qayṭīs	20
	fam al-qayṭus	13,16,17,18
	fam qayṭus	14,19,23,24,27,28,30,33
ζ	baṭn qayṭūs	2,6,7,8,34
ι	dhanab al-qayṭus al-shamālī	16
	dhanab al-qayṭus shamā	34
	dhanab al-qayṭīs	20
	dhanab al-qayṭus	17,18,33
	dhanab al-qayṭus	2,6,7,8
	dhanab qayṭus	10,11,14,15,24,27,28,30,32
	dhanab qayṭus sha	19
	qayṭus	21

CETUS *(cont)*

		Cat. No.
π	ṣadr al-qayṭus	16,17,18,20,33,34
	ṣadr qayṭus	19,23,24,27,28,30,31
	ṣadr	15,25
τ	min jasad qayṭus	10

CORONA BOREALIS

α	nayyir al-fakkah	13,14,22
	nayyir fakkah	15,17,18,19,23,24,25,27, 28,30,31
	nayyir	11
	al-fakkah	1,6
	fakkah	2,3,7,8,9,20,21,32,33
	fakk	34
	al-munīr al-fakkah	10

CORVUS

γ	janāḥ al-ghurāb	2,6,8,10,13,15,16,17,18,20, 24,27,28,30,31,32,33
	janāḥ ghurāb	19,34
	janāḥ	11,21,25
	al-ghurāb	7

CRATER

α	qāʿidat al-bāṭiyyah	13,16,17,18,19,20,23,33
	qāʿidat bāṭiyyah	19,34
	qāʿidah	15,24,25,27,28,30,31
	qāʿid	11

CYGNUS

α	dhanab al-dajājah	10,17,18
	dhanab dajājah	15,19,20
	al-dajājah	32
	al-ridf	1,6
	ridf	2,3,7,8,9,14,21,23,24,25, 27,28,30,31,34
β	minqār al-dajājah	13
	minqār dajājah	19,20,23,33,34

DELPHINUS

ε	dhanab al-dulfīn	10,13,20
	dhanab al-dhālfīn	16
	dhanab dulfīn	18,19,33
	al-dulfīn	6,7,8
	dulfīn	2,9,17

DRACO

α	ʿayn al-tinnīn	13,16,33
	ʿayn tinnīn	19

ERIDANUS

		Cat. No.
θ	tālī masaf	19
	tālī masāfat al-nahr	18,33
	tālī masāf al-nahr	13,16,19,20,34
	masāfat al-nahr	15,23,24,31
	masāfah	27,28,30
	misfah	25
	ākhir al-nahr	21
	nahr	3

HERCULES

		Cat. No.
α	ra's al-jāthī	16,17,19
	al-jāthī	33
	jāthī	20,34

HYDRA

		Cat. No.
α	fard al-shujā'	13,14,15,16,17,18,19,20,22, 23,24,25,27,28,30,31, 32,33
	fard shujā'	19
	fard	11,21
	'unq al-shujā'	10
	'unq	3
	munīr al-shujā'	1,2
	al-shujā'	6,8,18
	shujā'	7
σ	mu'akhkhar al-shujā'	18
	mu'akhkhar shujā'	13,34

LEO

		Cat. No.
α	qalb al-asad	2,6,7,10,13,14,17,23,24, 28,30,31,32,34
β	ṣarfah	19
δ	ẓahr al-asad	13
	ẓahr	18
ε (?)	zubānā al-asad	6,13

LIBRA

		Cat. No.
α	kiffah janūbī	13,16,17,23
β	kiffah shā'	19
	kiffah shamālī	13

LYRA

		Cat. No.
α	al-nasr al-wāqi'	13
	nasr wāqi'	15,17,18,22,23,24,25,27, 28,30,31,32,33
	al-wāqi'	1,6,10
	wāqi'	2,3,7,8,9,11,14,16,19,20, 21,34

OPHIUCHUS

		Cat. No.
α	ra's al-ḥawwā'	1,7,10,11,13,15,17,19,23,24, 27,28,30,31,32,33
	ra's ḥawwā'	18,19
	ra's	21
	al-ḥawwā'	6,8,20,34
	ḥawwā'	2,9,14
δ	yad al-ḥawwā'	13,16,19
	yad ḥawwā'	19
ζ	yad yamīn	20
	rukbat al-ḥawwā'	13
	rukbat ḥawwā'	19
	rukbah yusrā	20

ORION

		Cat. No.
α	mankib al-jawzā'	1,6,7,8
	mankib	2
	yad al-jawzā' yumnā	13,18,33
	yad al-jawzā'	10,14
	yad al-yumnā	23
	yad yumnā	15,19,24,25,27,28,30,31,34
	yad yamīn	20
	yad	11
β	rijl al-jawzā' al-yusrā	16
	rijl al-jawzā' yusrā	20
	rijl jawzā' yusrā	18,33
	rijl al-jawzā'	2,8,10,14
	rijl al-yusrā	13,15,23,24,25,27,28,31,32
	rijl yusrā	30,34
	rijl	3,11,19,21
	qadamā al-jawzā'	6,7
	qadam al-jawzā'	1
γ	yad al-jawzā' al-yusrā	16,33
	yad al-jawzā' yusrā	13
	yad jawzā' yusrā	18
	yad jawz al-yusrā	17
	yad yusrā	19,23,34
κ	rijl al-jawzā' al-yumnā	16
	rijl al-jawzā' yumnā	18,33
	rijl jawzā' yumnā	13
	rijl al-yumnā	23
	rijl yumnā	19,20
	rijl yumānī	34
	jawzā' yumnā	17

PEGASUS

		Cat. No.
α	matn al-faras	23
	matn faras	32
β	mankib al-faras	1,13,15,17,18,23,24,27,28,30,31,32
	mankib faras	6,7,8,19,20,33
	mankib	3,10,14,34
	al-faras	16
	faras	2,9,21,34
γ	janāḥ al-faras	13
δ	surrat al-faras	13
	surrat faras	19,33
	surrat	34
ε	fam al-faras	13,15,16,17,18,23,24,25,27,28,30,31,32
	fam faras	19,20,33
	fam	14,21

PERSEUS

β	al-ra's	19
	ra's al-ghūl	6,7,10,13,33
	ra's ghūl	18
	ghūl	3,8,9,15,19,20,27,28,34
o (?)	ʻātiq thurāyyah	21

PUPPIS

ζ	ra's al-shirāʻ	10
ρ	ṭaraf al-sufīnah	13,19,23,34
	ṭaraf sufīnah	19

SAGITTARIUS

ρ (?)	ʻiṣābat al-rāmī	13,19

SCORPIUS

α	qalb al-ʻaqrab	1,2,6,7,10,13,14,15,17,18,23,24,25,27,28,30,31
	qalb ʻaqrab	8

SERPENS

α	ʻunq al-ḥayyah	7,10,15,16,23,24,27,28,30,31,32,33
	ʻunq ḥayyah	19,20
	ʻunq	18,21,25
	al-ḥayyah	6,8,17
	ḥayyah	2,9,14
ρ (?)	minkhār (?) mutaqaddim	33

TAURUS

		Cat. No.
α	al-dabarān	6,7,8
	dabarān	2,3,18,19
	ʻayn al-thawr	1,10,13,14,15,16,17,18,23,24,25,27,28,31,32,33
	ʻayn	11,19,21

URSA MAIOR

α	ẓahr al-dubb al-aʻẓam	16
	ẓahr al-dubb	13,15,24,27,28,30,31
	ẓahr dubb al-akbar	33
	ẓahr dubb akbar	34
	ẓahr dubb	17,18
	yad al-dubb	1,8
	yad dubb	9
	al-dubb	10
	dubb	2
ε	jawn	21
ζ	al-ʻanāq	16,34
	ʻanāq	19,20,23,33
η	nayyir qā'id	32
	al-qā'id	1
	qā'id	2,3,23
	ṭaraf dhanab al-dubb	10
μ	rijl al-dubb	1,6,7
	rijlihi	8
	rijl	9

VIRGO

α	al-simāk al-aʻzal	1,10
	simāk al-aʻzal	20
	simāk aʻzal	13,14,15,17,18,19,22,23,24,25,27,28,30,31,32,34
	al-aʻzal	6,7,8,16
	aʻzal	2,11,21,33
ε	al-mutaqaddim	13

Catalogue of Sanskrit Star-Names

		Cat. No.
ANDROMEDA		
β	matsyodara	49
AQUARIUS		
λ	(śatabhiṣak)	
	śatabhiṣakaḥ	48
	śatabhiṣā	49
AQUILA		
α	śravaṇa	48,49
AURIGA		
α	(brahmahṛdaya)	
	vrahmāhṛda	48
	vralahṛ	49
BOOTES		
α	svātiḥ	48
	svāti	49
CANIS MAIOR		
α	lubdhaka	48,49
CANIS MINOR		
α	(lubdhakabandhu)	
	lubdhakavadhaḥ	48
	la badhu	49
CETUS		
ι	samudrapakṣīḥ	48
	samudrapakṣī	49
CORONA BOREALIS		
α	viśākhā pītarāvaca	48

CORVUS

		Cat. No.
δ	hastaḥ	48
	hasta	49

CYGNUS

		Cat. No.
α	kukudapuccha	48

ERIDANUS

		Cat. No.
θ	naharakoṭāroḥ	48

HYDRA

		Cat. No.
α	mahāpuruṣa	49
	mahāpuruṣa	48

LEO

		Cat. No.
α	maghāḥ	48
δ (?)	(pūrvaphālgunī)	
	pū. phā	49

LYRA

		Cat. No.
α	abhijit	48
	abhiji	49

OPHIUCHUS

		Cat. No.
α	sarpadhāriśira	48
	sarpadhārā	49

ORION

		Cat. No.
α	ārdrāḥ	48
	ārdrā	49
β	(mithunavāmapāda)	
	mithunapādaḥ	48
	mi. vā pa.	49

PEGASUS

		Cat. No.
β	pūrvā bhādrapadā	48
	pū. bhā	49
γ	uttarā bhādrapadā	48
ε	aśvamukha	48
	aśvamuṣa	49

PERSEUS

		Cat. No.
α	(manuṣyapārśva)	
	ma. pā.	49
β	(pati; pretaḥ)	
	pati	48
	pre.	49

TAURUS

		Cat. No.
α	rohiṇī	48
	rohaṇī	49

URSA MAIOR

		Cat. No.
η (?)	(marīci)	
	marīca	49

VIRGO

		Cat. No.
α	citrāḥ	48
	citrā	49

Astrological Tables

Lords of the Decans

		Greek	*Indian*
Aries	1	Mars	Mars
	2	Sun	Sun
	3	Venus	Jupiter
Taurus	1	Mercury	Venus
	2	Moon	Mercury
	3	Saturn	Saturn
Gemini	1	Jupiter	Mercury
	2	Mars	Venus
	3	Sun	Saturn
Cancer	1	Venus	Moon
	2	Mercury	Mars
	3	Moon	Jupiter
Leo	1	Saturn	Sun
	2	Jupiter	Jupiter
	3	Mars	Mars
Virgo	1	Sun	Mercury
	2	Venus	Saturn
	3	Mercury	Venus
Libra	1	Moon	Venus
	2	Saturn	Saturn
	3	Jupiter	Mercury
Scorpius	1	Mars	Mars
	2	Sun	Jupiter
	3	Venus	Moon
Sagittarius	1	Mercury	Jupiter
	2	Moon	Mars
	3	Saturn	Sun
Capricorn	1	Jupiter	Saturn
	2	Mars	Venus
	3	Sun	Mercury
Aquarius	1	Venus	Saturn
	2	Mercury	Mercury
	3	Moon	Venus
Pisces	1	Saturn	Jupiter
	2	Jupiter	Moon
	3	Mars	Mars

Lords of the Terms according to Dorotheus

(Catalogue Numbers: 10,11,18,19,21,22,23,24,25,26,27,28,29,30,31,32,33)

Sign	*Length*	*Bounds*	*Lord*
Aries	6°	6°	Jupiter
	6°	12°	Venus
	8°	20°	Mercury (Jupiter in *18*)
	5°	25°	Mars
	5°	30°	Saturn
Taurus	8°	8°	Venus
	6°	14°	Mercury
	8°	22°	Jupiter
	5°	27°	Saturn (4 in *11*)
	3°	30°	Mars (4 in *11*)
Gemini	6°	6°	Mercury
	6°	12°	Jupiter
	5°	17°	Venus
	7°	24°	Mars (27 in *25*)
	6°	30°	Saturn
Cancer	7°	7°	Mars (6 in *18*)
	6°	13°	Venus
	6°	19°	Mercury (15 in *21, 25*)
	7°	26°	Jupiter (25 in *18, 32*)
	4°	30°	Saturn
Leo	6°	6°	Jupiter
	5°	11°	Venus
	7°	18°	Saturn (13 in *25*; 28 in *30*)
	6°	24°	Mars (25 in *18, 28, 31*; Mercury in *11, 30, 33*)
	6°	30°	Mercury (Mars in *11, 30, 33*)
Virgo	7°	7°	Mercury
	10°	17°	Venus
	4°	21°	Jupiter
	7°	28°	Mars (23 in *25, 26*)
	2°	30°	Saturn

Sign	*Length*	*Bounds*	*Lord*
Libra	6°	6°	Saturn (7 in *24*)
	8°	14°	Mercury
	7°	21°	Jupiter
	7°	28°	Venus (23 in *26, 32*; 27 in *25*)
	2°	30°	Mars (3 in *11*)
Scorpius	7°	7°	Mars (8 in *11*)
	4°	11°	Venus
	8°	19°	Mercury (7 in *11*)
	5°	24°	Jupiter (25 in *25*; 27 in *31*; 29 in *32*)
	6°	30°	Saturn
Sagittarius	12°	12°	Jupiter
	5°	17°	Venus (14 in *21*)
	4°	21°	Mercury
	5°	26°	Saturn
	4°	30°	Mars
Capricorn	7°	7°	Mercury
	7°	14°	Jupiter
	8°	22°	Venus
	4°	26°	Saturn
	4°	30°	Mars
Aquarius	7°	7°	Mercury
	6°	13°	Venus
	7°	20°	Jupiter (Mars in *32*)
	5°	25°	Mars (Jupiter in *32*; 23 in *32*; 26 in *25, 31*)
	5°	30°	Saturn
Pisces	12°	12°	Venus
	4°	16°	Jupiter (15 in *18, 21*; 6 in *23*)
	3°	19°	Mercury (21 in *18*)
	9°	28°	Mars (23 in *32*)
	2°	30°	Saturn

Lords of the Triplicities

			First		*Second*		*Third*
			Diurnal	*Nocturnal*	*Diurnal*	*Nocturnal*	
I	Aries Leo Sagittarius	(fire)	Sun	Jupiter	Jupiter	Sun	Saturn
II	Taurus Virgo Capricorn	(earth)	Venus	Moon	Moon	Venus	Mars
III	Gemini Libra Aquarius	(air)	Saturn	Mercury	Mercury	Saturn	Jupiter
IV	Cancer Scorpius Pisces	(water)	Venus	Mars	Mars	Venus	Moon

Lunar Mansions

(Catalogue Numbers: 2,6,13,14,15,16,17,18,19,21,23,24,25,26,28,29,30,31,32,33,34)

1. saraṭayn (naṭḥ in *2,6*)
2. buṭayn
3. thurayyā
4. dabarān
5. haq'ah (han'ah in *24*)
6. han'ah (haq'ah in *24*)
7. dhirā'
8. nathrah (barah in *30*)
9. ṭarfah (ṭarf in *2, 19*; jabhah in *30*)
10. jabhah (zubrah in *30*)
11. zubrah (kharthān in *5*; ṣarfah in *30*)
12. ṣarfah (ṭarfah in *30*)
13. 'awā'
14. simāk
15. ghafr (ghafrah in *14, 15, 16, 17*)
16. zubānā' (zubānyān in *6*)
17. iklīl (om. in *2*)
18. qalb
19. shawlah
20. na'ā'im
21. baldah
22. dhābiḥ
23. bala'
24. su'ūd (su'ūr in *14*)
25. akhbiyyah
26. muqaddam (fargh muqaddam in *6*)
27. mu'akhkhar
28. rishā' (baṭn al-ḥūt in *6*; ḥūt in *2*)

Metallurgical Analysis

by Bruce Stephenson

While this volume was in preparation the Adler Planetarium undertook a collaborative research project (with Argonne National Laboratory and Lehigh University) to study the metallurgy of historical astrolabes, using high-energy X-rays generated at Argonne's Advanced Photon Source (APS) synchrotron. Forty instruments were studied, including twenty-three of the Eastern Astrolabes catalogued in the present volume. The remaining seventeen instruments studied included fourteen Western Astrolabes from the Adler collection, one astrolabe from a private collection, one fragment purchased specifically for metallurgical analysis, and for good measure a fake astrolabe purchased in Istanbul's Grand Bazaar.

1 See Brian Newbury, "A Non-Destructive Synchrotron X-ray Study of the Metallurgy and Manufacturing Processes of Eastern and Western Astrolabes in the Adler Planetarium Collection," Ph.D. dissertation, Lehigh University, December, 2004. This dissertation is available online at the web address given at the end of this appendix.

SELECTION OF INSTRUMENTS

Catalogue entries were written before the Argonne project was underway, so in a general sense they could have influenced the selection of instruments for X-ray analysis. In fact, that selection was largely governed by a desire to include reasonable-sized groups of astrolabes from important geographic regions (Isfahan, Lahore, Louvain) and well-known makers (Abd al-A'immah, Diyā' al-Dīn Muḥammad, Gualterus Arsenius). In retrospect there are a few instruments in our collection we wish we had analyzed but did not, notably the remaining few Lahore instruments. In this instance we did not realize the significance of our analysis until our experimental sessions were concluded.

TYPES OF ANALYSIS

Astrolabes in the study were disassembled, packed, and transported to Argonne National Laboratory's APS. Each component was sequentially mounted for three types of analysis using a high-energy (68 keV or more) X-ray beam.[1] We summarize them below; for a more detailed and authoritative discussion consult the Brian Newbury dissertation cited above.

Three analytic techniques were used:

X-ray fluorescence analysis (XRF) reveals the surface composition of the metal. This technique is well-known in historical metallurgy. Its chief disadvantage is the restriction to surface composition. Over time zinc leaches out of the surface of brass, so XRF analysis of instruments that are several hundred years old can significantly underestimate the proportion of zinc that was in the original brass. This can be problematic for studies of how the brass was made, as discussed below.

X-ray diffraction analysis (XRD) uses X-rays that penetrated the target entirely, but that are diffracted, or bent, as they pass through microscopic crystals within the metal. The resulting diffraction patterns provide a (highly abstract) picture of the interior of the metal, which is useful in two different ways. First, the pattern is sensitive to non-random orientation of the crystals, orientation which might have arisen, for example, under the extreme pressure of a rolling mill. Since rolled brass did not become available until the 18th century in most regions, one can rule out early dates for an astrolabe fabricated from rolled brass. Second, analysis of the diffraction pattern yields an estimate of the proportion zinc in the brass that is based upon the entire thickness of the metal, thus largely avoiding the problem of zinc leaching from the surface.

Traditional *radiography,* familiar from medical X-rays, measures the absorption of photons within the metal. Assuming that the metal is of uniform density, greater absorption implies greater thickness. The narrow beam of a synchrotron source is less than ideal for radiography, but by scanning across an astrolabe component one can plot its cross-section. Modern rolled brass immediately betrays itself by its uniform thickness.

BRASS-MAKING TECHNOLOGY

The fundamental difficulty in making brass is that zinc vaporizes at a lower temperature than copper liquifies. The oldest method of brass manufacture, used by the ancient Romans, was the cementation process, which relies upon the ability of vaporized zinc from ore to diffuse into small pieces of metallic copper within a small crucible. Thermodynamic considerations (basically, the zinc vapor diffuses back out of the brass) limit the proportion of zinc in the resulting brass to less than about 32% under normal circumstances. Brass with such low zinc content is relatively brittle and must be annealed (heated and cooled) repeatedly if it is to be extensively worked with a hammer without shattering—as was necessary when producing sheet brass in the years prior to the modern rolling mill.

It was thus surprising when we found several astrolabes from Lahore that exhibited zinc content of up to 40%. Such brass has far superior hot-working properties—and would therefore have been prized by the Lahore brass workers—but it can only be produced by melting together metallic zinc with copper, under suitable conditions. Although metallic zinc is known to have existed in the Islamic world, and was used to make jewelry and medicines, there was no evidence prior to this study that it was used in quantity for brass production. It is now clear that in the 17th century, brass technology in Lahore was far more advanced than brass technology in Europe.

EASTERN ASTROLABES STUDIED

The following table lists the Eastern Astrolabes in the present catalogue that were studied in the research project at Argonne's APS synchrotron.

Catalogue number (from catalogue entry)	*Accession number*	*Date*
4	A-396	mid 14th-15th century
5	M-37	1651 (?)
6	A-76	1720/1
7	A-87	1723/4
8	DPW-54	circa 1750
10	A-84	1130/1
12	A-91	1433/4
13	N-69	1601
14	A-79	1604/5
15	A-86	1637/8
16	A-70	1647/8
17	A-81	1660/1
18	A-78	1662/3
19	L-100	17th century
23	W-89	circa 1700
24	A-75	circa 1700
27	A-74	circa 1720
28	A-88	circa 1720
29	A-69	circa 1710
31	A-146	18th century
32	M-41	18th century
33	N-68	18th century
38	N-71	modern

Data from the project have been reduced, compiled, and placed on the web. See "Publications and data files," below.

MAJOR FINDINGS

Almost all the components of these instruments were brass, an alloy of copper and zinc with traces of lead, silver, tin, and other metals. Analysis of these trace elements, and comparison with known sources of ore, seems at first a promising line of research, but is defeated by the omnivorous early modern re-use of brass. Making brass was laborious and expensive, and it is evident that smelters happily availed themselves of any scraps of brass that were lying around.

Certain "impurities" were intentional. Astrolabe maters and rims were typically cast, rather than cut and hammered from plate as was the case with tympans and retes. Lead and tin are useful additives in metal that is intended to be cast, and indeed we encountered noticeably more lead and tin in cast components than in hammered components.

Zinc content of the sheet brass components (tympans and retes) was typically between 20% and 40%. Most of the components with zinc percentages over 33% were from Lahore astrolabes, as discussed above.

RELATION TO CATALOGUE ENTRIES

The Adler's Argonne research project was independent of the Adler's Catalogue project. This is unfortunate, but was inevitable given the scopes of the two projects, the different constraints on their schedules, and the different people involved.

When a catalogue entry differs from the Argonne results it is usually evident which should be preferred. Catalogue descriptions such as "brass" or "bronze" were based on visual examination, and are superseded by the more precise alloy compositions from Argonne. Dates in the catalogue, on the other hand, are based on inscriptions and expert knowledge, and should normally be preferred to any inferences from the Argonne data.

PUBLICATIONS AND DATA FILES

Dr. Brian Newbury's Ph.D. dissertation, giving a full description of the Argonne experiments and their analysis, is available as a set of PDFs for download from a page on the Adler website:

http://www.adlerplanetarium.org/history/research/newbury.shtml

Also available from that page are forty PDFs containing the reduced data for the instruments studied. All of these files contain images and are correspondingly large (up to 14 megabytes for dissertation chapters, and 1-3 megabytes each for the 40 data files). A broadband connection is recommended for downloading them.

Articles based on this project are in preparation for publication in scientific and scholarly journals. Preliminary results have already been published, with Brian Newbury as first author.

Concordance

Cat. No.	*Accession No.*	*Date*
1	M-35	1236/7
2	M-36	1240/1
3	A-71	13th century
4	A-396	14th or 15th century
5	M-37	1651(?)
6	A-76	1720/1
7	A-87	1723/4
8	DPW-54	*ca.*1750
9	A-77	1847/8
10	A-84	1130/1
11	N-70	1388
12	A-91	1433/4
13	N-69	June 1601
14	A-79	1604/5
15	A-86	1637/8
16	A-70	1647/8
17	A-81	1660/1
18	A-78	1662/3
19	L-100	17th century
20	M-39	17th century
21	A-40	1669/70 or 1688/9
22	A-82	1697/8 (Rete: 1830/1)
23	W-89	*ca.*1700
24	A-75	*ca.* 1700
25	A-37	1699/1700
26	A-90	1715
27	A-74	*ca.* 1720
28	A-88	*ca.* 1720
29	A-69	*ca.* 1710
30	W-151	18th century
31	A-146	18th century
32	M-41	18th century
33	N-68	18th century
34	DPW-52	*ca.* 1780
35	M-40	1792/3
36	A-83	Modern
37	A-73	Modern
38	N-71	Modern
39	M-38	Allegedly 1629/30
40	A-85	Modern
41	A-72	After 1856
42	G-36	Modern
43	A-153	Modern
44	A-80	Modern
45	A-89	Modern
46	A-212	Modern
47	A-204	Modern
48	A-68	18th century
49	W-102	19th century
A	W-94	
B	N-53	
C	W-103	18th century
D	W-93	19th century
E	M-466	1675/6
F	W-152	1800/1
G	A-27	

Cat. No.	Accession No.	Date
H	A-148	Late 18th century
I	W-101	
J	A-43	1864/5
K	W-209	
L	A-63	1783/4
M	A-67	18th or 19th century
N	A-66	22 February 1788
O	A-64	18th or 19th century
P	M-47	Modern
Q	M-14	17th century
R	A-140	17th century
S	A-115	17th century
T	A-114	1603/4
U	A-116	1822/3
V	A-39	18th century or later
W	A-38	18th century or later
X	A-112	Modern
Y	A-113	Modern
Z	A-142	Modern
AA	W-273	1799

Accession No.	Catalogue No.
A-27	G
A-37	25
A-38	W
A-39	V
A-40	21
A-43	J
A-63	L
A-64	O
A-66	N
A-67	M
A-68	48
A-69	29
A-70	16
A-71	3
A-72	41
A-73	37
A-74	27
A-75	24
A-76	6
A-77	9
A-78	18
A-79	14
A-80	44
A-81	17
A-82	22
A-83	36
A-84	10
A-85	40
A-86	15
A-87	7
A-88	28

Accession No.	Catalogue No.
A-89	45
A-90	26
A-91	12
A-112	X
A-113	Y
A-114	T
A-115	S
A-116	U
A-140	R
A-142	Z
A-146	31
A-148	H
A-153	43
A-204	47
A-212	46
A-396	4
DPW-52	34
DPW-54	8
G-36	42
L-100	19
M-14	Q
M-35	1
M-36	2
M-37	5
M-38	39
M-39	20
M-40	35
M-41	32
M-47	P
M-466	E
N-53	B
N-68	33
N-69	13
N-70	11
N-71	38
W-89	23
W-93	D
W-94	A
W-101	I
W-102	49
W-103	C
W-151	30
W-152	F
W-209	K
W-273	AA

Bibliography

Abbot (1937)	Abbott, Nadia. "Indian Astrolabe Makers." *Islamic Culture* 11 (Jan., 1937): pp. 144-46.
Adler Planetarium (1970)	*Adler Planetarium Guide Book.* Chicago, 1970.
Adler Planetarium (1973)	*A Guide to the Adler Planetarium.* Chicago, 1973.
Adler Planetarium (1977)	*The Adler Planetarium, 1976* (Annual Report, 1976). Chicago, 1977.
Adler Planetarium (1980)	*A Guide to the Adler Planetarium.* Chicago, 1980.
Architectural Digest (1973)	"Collecting Scientific Instruments." *Architectural Digest.* January-February, 1973.
Borgomale (1945)	Brogomeale, H.I. Rabino, Coins, Medals, and Seals of the Shahs of Iran 1500–1941. Hertford, England, 1945.
Brieux and Maddison (forthcoming)	Brieux, Alain. and Francis R. Maddison. *Répertoire des facteurs d'astrolabes et de leurs oeuvres: Première partie, Islam.* Paris: Centre National des Recherches Scientifiques, in press.
Chadenat (1956)	*Collection Ch. Chadenat: Instruments de Mathématiques Anciens…, 24 Mai 1956.* Paris, 1956.
Charette (2004)	Charette, Francois, *Mathematical Instrumentation in Fourteenth-Century Egypt and Syria: the Illustrated Treatise of Najm al-Din al-Masri.* Boston: Brill, 2003.
Chughtai (1994)	Chughtai, Arif Rahman. *Lutuf Ullah Muhandis: Chief Architect of Prince Dara Shikoh.* Jahangeer Book Club. Pakistan, 1994.
Combe, Sauvaget, and Wiet (1939)	Combe, Etienne; Jean Sauvaget; and Gaston Wiet. *Répertoire chronologique d'epigraphie arabe. X. Le Carire, 1931-54.*
Dershowitz and Reingold (1997)	Dershowitz, Nachum and Edward M. Reingold. *Calendrical Calculations.* Cambridge, England: Cambridge University Press, 1997.
Dor-Nev (1991)	Dor-Nev, Zvi, *Columbus and the Age of Discovery.* New York, 1991.

Downey (1961) — Downey, Glanville. *A History of Antioch in Syria*, Princeton: Princeton University Press, 1961.

Edwards and Signell (1982) — Edwards, Holly, and Karl Signell. Patterns and Precision: *The Arts and Sciences of Islam. Catalogue of an exhibition*. Washington, D.C., 1982.

Engelmann Catalogue (1924) — Engelmann, Max. *Sammlung Mensing. Altwissenschaftliche Instrument Katalog*, 2 vols. Amsterdam, 1924.

Evans (1999) — Evans, James. "The Material Culture of Greek Astronomy." *Journal for the History of Astronomy* 30 (1999): pp. 237-307.

Fox (1932) — Fox, Philip, "The Adler Planetarium and Astronomical Museum of Chicago." *Popular Astronomy* 40 (1932), pp. 125-55, 321-51, 532-49, 613-22.

Fox (1933) — Fox, Philip. *The Adler Planetarium and Astronomical Museum: An Account of the Optical Planetarium and a Brief Guide to the Museum*. Chicago, 1933. Reprinted 1942.

Franco (1945) — Franco, Garcia. *Catálogo critico de astrolabios existentes en España.* Madrid, 1945.

Gibbs and Saliba (1984) — Gibbs, Sharon and George Saliba. "Planispheric Astrolabes from the National Museum of American History" National Museum of American History. Washington D.C.: Smithsonian Institution Press, 1984.

Gingerich (1982) — Gingerich, Owen. "Astronomical Scrapbook: Fake Astrolabes." *Sky and Telescope* 63 (1982), pp. 465-468.

Gingerich *et al.* (1972) — Gingerich, Owen, David King, and George Saliba. "The 'Abd al-A'imma Astrolabe Forgeries." *Journal for the History of Astronomy* 3 (1972), pp. 188-198.

Goldstein (1976) — Goldstein, Bernard R., "The Hebrew Astrolabe in the Adler Planetarium" *Journal of Near Eastern Studies.* Vol. 35, No. 4 (Oct., 1976), pp. 251-260

Gunther (1932) — Gunther, Robert T. *The Astrolabes of the World.* 2 vols. Oxford, 1932. *Volume 1. The Eastern Astrolabes.*

Harrak (1999) — Harrak, Amir. *The Chronicle of Zuqnin*, parts III and IV. Toronto: Pontifical Institute of Medieval Studies, 1999.

Heilbronner Catalog (1922) — *Instruments de mathématiques.* Heilbronner sale, Hôtel Druot, Paris, March 8, 1922.

Higton (2001) — Higton, Hester. *Sundials: An Illustrated History of Portable Dials.* London, 2001.

ICA — Gibbs, Sharon L., Janice A. Henderson, and Derek J. de Solla Price. *A Computerized Checklist of Astrolabes.* New Haven, 1973. (Computer printout of ICA checklist.)

Khalili (1990) — Khalili, Nasser D. *A Selection of Islamic Art at the Brunei Museum*. Bandar Seri Begawan; Negara Brunei Darussalam: The Museum, 1990.

Kennedy and Kennedy (1987) — Kennedy, E.S. and M.H. *Geographic Coordinates of Localities from Islamic Sources*. Frankfort am Main, 1987.

King (1983) — King, David A. *al-Khwarizmi and New Trends in Mathematical Astronomy in the Ninth Century*. New York: Hagop Kevorkian Center for Near Eastern Studies, New York University, 1983.

King (1987)	King, David A. *Islamic Astronomical Instruments*, London: Variorum Reprints, 1987.
King (1991)	King, David A. "Strumentazione astronomica nel mondo medievale islamico," in Turner, G. L'E., ed., *Storia delle Scienze: Gli Strumenti*. Torino: Giulio Einaudi, 1991.
King (1994)	King, David A. "Astronomical Instruments between East aned West." *Kommunikation zwischen Orient und Okzident Alltag und Sachkultur.* Wien: Verlag der Österreichischen Akademie der Wissenschaften, 1994.
King (1995)	King, David A. "Making Instruments Talk: Some Medieval Astronomical Instruments and their Secrets." *Bulletin of the Scientific Instrument Society* 44 (1995): pp. 5-12.
King (1999)	King, David A. *World-Maps for Finding the Direction and Distance of Mecca: Innovation and Tradition in Islamic Science*. Boston: Brill, 1999.
King (2004-5)	King, David A. *In Synchrony with the Heavens: Studies in Astronomical Timekeeping and Instrumentation in Medieval Islamic Civilization*. Leiden and Boston: Brill, 2004-2005. 2 v.
Kunitzsch (1987)	Kunitzsch, Paul. "Al Khwarazmi as a Source for the Sententie astrolabii", in *From Deferent to Equant: A Volume of Studies in the History of Science in the Ancient and Medieval Newar East in Honor of E.S. Kennedy,* ed. David A. King and George Saliba. New York: New York Academy of Sciences, 1987.
Kurz (1975)	Kurz, O. *European Clocks and Watches in the Near East*. Brill, Leiden, 1975.
Libros del saber (1864)	*Libros del saber de astronomia del ray D. Alfonso X de Castilla*. Madrid, 1864.
Maddison (1976)	Maddison, Francis R. "14th Centennial of Islam Exhibition: April-August, 1976."
Maddison and Turner (1976)	Maddison, F.R. and Turner, A.J. *Science and Technology in Islam: An Exhibition at the Science Museum, London, 7 April–29 August 1976.* London: Crescent Moon Press, 1976.
Mayer (1956)	Mayer, L. A. *Islamic Astrolabists and Their Works*. Geneva, 1956.
McAllister (1970)	McAllister, Paul R. "Romance with the Stars," *Lake Forest Academy Antiques Show*, 1970, pp. 14-19.
Morley (1856)	Morley, William H. *Description of a Planispheric Astrolabe.* Reprinted in Gunther (1932) Vol. I pp.1–50.
Moskowitz (1984)	Moskowitz, Saul. *Historical Technology Catalog 127* Marblehead, Massachusetts: Fall, 1984.
al-Nadim (1988)	al-Nadim, Mohammad ben Eshaq (A.H. 377 – A.D.987). *Ketab al-Fihrist.* Ed. R. Tajuddad, 3rd edition, Teheran, 1988.
Nadvi (1937)	Nadvi, Sayyid Suleiman. "Indian Astrolabe-Makers." *Islamic Culture* 11: 4 (1937), pp. 537-539.

Nau (1899)
F. Nau, ed. *Le livre de l'ascension de l'esprit sur la forme du ciel et de la terre; cours d'astronomie rédigé en 1279 par Grégoire Aboulfarag, dit Bar Hebræus, publié pour la première fois d'après les manuscrits de Paris, d'Oxford et de Cambridge*. Paris: É. Bouillon, 1899.

Neugebauer (1975)
Neugebauer, Otto. *A History of Ancient Mathematical Astronomy*. 3 v. New York: Springer-Verlag, 1975.

Newbury (2004a)
Newbury, Brian Dale. "A non-destructive synchrotron X-ray study of the metallurgy and manufacturing processes of Eastern and Western astrolabes in the Adler Planetarium collection." Ph.D. dissertation, Lehigh University, Bethlehem, Pa., 2004. [Cited 4 October, 2007.] Available from http://www.adlerplanetarium.org/research/collections/websters/index.shtml.

Newbury (2004b)
Newbury, Brian Dale, *et al.* "Synchrotron Applications in Archaeometallurgy: Analysis of High Zinc Brass Astrolabes," *Powder Diffraction* 19 (2004), pp. 12-15.

Newbury (2006)
Newbury, Brian Dale., *et al.* "The Astrolabe Craftsmen of Lahore and Early Brass Metallurgy." *Annals of Science* 63 (April, 2006), pp. 201-213.

Paris Universal Exhibition Catalogue (1900)
L'exposition universelle internationale, Musée Retrospectif, classe 15. Paris, 1900.

Pingree (1967)
Pingree, David. "Review of Grahandana of Paramesvara", Edited and translated by K.V. Sarma. in Journal of American Oriental Society, Vol. 87, No. 4, 1964. pp. 337-339

Pingree (1970)
Pingree, David. "The Fragments of the Works of Al-Fazarī" *Journal of Near Eastern Studies*, Vol. 29, No. 2. (Apr., 1970), pp. 103-123.

Pingree (1975)
Pingree, D. "Al-Biruni's Knowledge of Sanskrit Astronomical Texts," *The Scholar and the Saint*, New York, 1975, pp. 67-81.

Pingree (1983)
Pingree, David. "Brahmagupta, Balabhadra, Pṛthūdaka and Al-Bīrūnī" *Journal of the American Oriental Society*. Vol. 103, No. 2 (Apr., 1983), pp. 353-360

Puig (1986)
R. Puig, *Al-Sakkaziyya. Ibn al-Naqqa, al-Zarqallub.* Barcelona, Universidad de Barcelona, 1986.

Raikva (1936)
"The Yantrarajagamavyakhya of Malayendu Suri." Edited by K. K. Raikva. Mumbayi, India, 1936.

Renaud (1942a)
Renaud, H. P. "Astronomie et Astrologie marocaine" *Hesperis XXIX*. 1942, pp. 41-63.

Renaud (1942b)
Renaud, H. P. "Quelques constructeurs d'astrolabes en occident musulman." *Isis* 34 (1942): pp. 20-23.

Rose (1968)
Rose, Paul Lawrence. "The Origins of the Proportional Compass." *Physis* 10 (1968): pp. 54-69.

Saliba (1977a)
Saliba, George. "Computational Techniques in a Set of Late Medieval Astronomical Tables", *Journal of the History of Arabic Science*, 1977, 1: pp. 24-32.

Saliba (1977b)
Saliba, George. "The Buffalo Astrolabe of Muhammad Khalil." *Quarterly Journal of the American University of Beirut* 26: pp. 11-18.

Samsó (1992) — Samsó, Julio. *Las cincias de los antiquos en al-Andalus*, Madrid, 1992.

Sarma (1994) — Sarma, S. R. "The Lahore Family of Astrolabists and their Ouvrage" *Studies in History of Medicine and Science*, 13, 1994, pp. 203-224.

Savage-Smith (1985) — Savage-Smith, Emilie. *Islamicate Celestial Globes: Their History, Construction, and Use*. Smithsonian Studies in History and Technology, no. 46. Washington, D. C., 1985.

Savage-Smith (1990) — Savage-Smith, Emilie. "The Classification of Islamic Celestial Globes in the Light of Recent Evidence." *Der Globusfreund* 38/39, pp. 23-35.

Savage-Smith (1992) — Savage-Smith, Emilie. "Celestial Mapping." In Harley, J. B., and David Woodward, *The History of Cartography. Volume Two, Book One. Cartography in the Traditional Islamic and South Asian Societies*. Chicago: University of Chicago Press, 1992. pp. 12-70.

Sayili (1960) — Sayili, Aydin. *The Observatory in Islam and its Place in the General History of the Observatory*. Ankara: Turk Tarih Kurumu Basimevi, 1960.

Schoy (1927) — Schoy, C. "'Ali ibn 'Isa, Das Astrolab und sein Gebrauch" *Isis* 9, No. 2 (June, 1927), pp. 239-254.

Sedillot (1844) — Sedillot, L. Am., "Mémoire sur les instruments astronomiques des Arabes." *Memoires présentés . . . à l'Académie Royale des Inscriptions et Belles-Lettres de l'Institut de France,* 1e ser. (sujets divers d'érudition), t. I, Paris, 1844.

Sezgin (1978) — Sezgin, Fuat. Tārīkh-i nigārish'hā-yi 'Arabī / ta'līf-i Fu'ād Sizgīn ; tarjumah, tadvīn va āmādah'sāzī, Mu'assasah-i Nashr-i Farhangistān; bih ihtimām-i Khānah-yi Kitāb. Tihrān, Sāzmān-i Chāp va Intishārāt, Vizārat-i Farhang va Irshād-i Islāmī, 1978.

Sotheby's (1968) — Sotheby's. *Important English and Continental Watches and Clocks and Scientific Instruments*. Catalogue of sale on Dec. 4, 1961. London, 1961.

Sotheby's (1985) — Sotheby's S.A. *Islamic Art.* Catalogue of sale on June 25, 1985. Geneva, 1985.

Stephenson, Bolt, and Friedman (2000) — Stephenson, Bruce; Marvin P. Bolt; and Anna F. Friedman. *The Universe Unveiled: Instruments and Images through History.* Chicago: Adler Planetarium and Cambridge: Cambridge University Press, 2000.

Suter (1900) — Suter, Heinrich. *Die Matematiker und Astronomen der Araber und Ihre Werke*. Leipzig, 1900.

Turner, A. J. (1985) — Turner, A. J. *Astrolabes, Astrolabe Related Instruments*. The Time Museum: Catalogue of the Collection, edited by Bruce Chandler. Vol. I: Time Measuring Instruments, part I. Rockford, IL, 1985.

Turner, G. L'E. (1991) — Turner, G. L'E. *Storia delle Scienze: Gli Strumenti*. Torino: Giulio Einaudi, 1991.

Turner, H. R. (1995) — Turner, Howard R. *Science in Medieval Islam: An Illustrated Introduction.* Austin, Texas: University of Texas Press, 1995.

Van Kleempoel (2005) — Van Kleempoel, Koenraad. *Astrolabes at Greenwich. A Catalogue of the Astrolabes in the National Maritime Museum, Greenwich*. Oxford and the National Maritime Museum, 2005.

Webster and Webster (1998) — Webster, Roderick S., and Webster, Marjorie K. *Western Astrolabes: Historic Scientific Instruments of the Adler Planetarium & Astronomy Museum*, vol. 1. Chicago: Adler Planetarium, 1998.

Wede (1960) — Wede, Karl F. *Ship Models, Marine Antiques and Rare Maritime Books.* Saugerties, N.Y. 1960.

Wede (n.d.) — Wede, Karl F. *The Age of Sail: Ship Models for the Connoisseur.* Saugerties, N.Y. n.d.

Wiedemann (1984) — Wiedemann, Eilhard. *Gesammelte Schriften zur arabisch-islamischen Wissenschaftsgeschichte*, Frankfurt-am-Main, Johann Wolfgang Goethe Universitat, 1984.

Wiet (1935) — Wiet, Gaston. *Exposition d'art persan.* Le Caire: Impr. pour la Société des amis de l'art par l'Impr. de l'Institut français d'archéologie orientale du Caire, 1935.

Yonge (1968) — Yonge, Ena L. *A Catalogue of Early Globes Made Prior to 1850 and Conserved in the United States.* American Geographical Society Library Series, no, 6. New York, 1968.

Index